CAPTURE PUMPING TECHNOLOGY

An Introduction

Pergamon Titles of Related Interest

ASHBY & JONES
Engineering Materials 1 & 2

HAYWOOD
Analysis of Engineering Cycles, 4th edition

HEARN
Mechanics of Materials, 2-Volume Set, 2nd edition

IIW
Joining/Welding 2000

KELLY
Concise Encyclopedia of Composite Materials

OSGOOD
Fatigue Design, 2nd edition

O'SHIMA & VAN RIJN
Production Control in the Process Industry

TAYA & ARSENAULT
Metal Matrix Composites

WU & CARLSSON
Weight Functions and Stress Intensity Factor Solutions

ZHANG
Theory and Technique of Precision Cutting

Pergamon Related Journals

(*free specimen copy gladly sent on request*)

Acta Mechanica Solida Sinica

Acta Metallurgica et Materialia

Engineering Fracture Mechanics

Fatigue and Fracture of Engineering Materials and Structures

International Journal of Engineering Science

International Journal of Impact Engineering

International Journal of Mechanical Sciences

International Journal of Non-linear Mechanics

International Journal of Solids and Structures

Journal of Applied Mathematics and Mechanics

Mechanism and Machine Theory

Scripta Metallurgica et Materialia

Vacuum

CAPTURE PUMPING TECHNOLOGY

An Introduction

KIMO M. WELCH
Brookhaven National Laboratory, Upton, NY, USA

PERGAMON PRESS
OXFORD · NEW YORK · SEOUL · TOKYO

UK	Pergamon Press plc, Headington Hill Hall, Oxford OX3 0BW, England
USA	Pergamon Press Inc., 395 Saw Mill River Road, Elmsford, New York 10523, USA
KOREA	Pergamon Press Korea, KPO Box 315, Seoul 110-603, Korea
JAPAN	Pergamon Press, 8th Floor, Matsuoka Central Building, 1-7-1 Nishi-Shinjuku, Shinjuku-ku, Tokyo 160, Japan

First edition 1991

Library of Congress Cataloging-in-Publication Data

Welch, Kimo M.
Capture pumping technology: an introduction/Kimo M. Welch.—1st ed.
p.   cm.
Includes bibliographical references and indexes.
1. Pumping machinery.   I. Title
TJ900.W45 1991   621.6′9—dc20   91-25026

British Library Cataloguing in Publication Data

Welch, Kimo M.
Capture pumping technology.
I. Title
621.6

ISBN 0-08-040197-X

Printed in Great Britain by BPCC Wheatons Ltd, Exeter

This book is dedicated to the memory of Arline H. Welch, my mother. In the early years of her life she was an ordained minister in the Salvation Army. In this work, and throughout her life, she dedicated herself to the caring, educating and loving of all children.

TABLE OF CONTENTS

TABLE OF CONTENTS

TABLE OF CONTENTS

TABLE OF CONTENTS

TABLE OF CONTENTS

PREFACE

This is a practical textbook written for use by engineers, scientists and technicians. It is not intended to be a rigorous scientific treatment of the subject material as this would fill several volumes. Rather, it introduces the reader to the fundamentals of the subject material, and provides sufficient references for an in-depth study of the subject by the interested technologist. The author has a lifetime teaching credential in the California Community College System. Also, he has taught technical courses with the American Vacuum Society for about 25 years. Students attending many of these classes have backgrounds varying from high-school graduates to Ph.D.s in technical disciplines. This is an extremely difficult class profile to teach. This book still endeavors to reach this same audience. Basic algebra is required to master most of the material. But, the calculus is used in derivation of some of the equations. The author risks use of the first person *I*, instead of *the author*, and *you* instead of *the reader*. Both are thought to be in poor taste when writing for publication in the scientific community. However, *I* am writing this book for *you* because the subject is exciting, and *I* enjoy teaching *you*, perhaps, something new. The book is written more in the vein of a *one-on-one* discussion with you, rather than the author lecturing to the reader. There are anecdotes, and examples of some failures and successes I have had over the last thirty-five years in vacuum-related activities. I'll try not to understate either.

Lastly, there are a few equations which if memorized will help you as a vacuum technician. There are less than a dozen equations and half that many *rules of thumb* to memorize, which will be drawn on time and again in designing, operating and trouble-shooting any vacuum system. These key elements are identified with the symbol "¤" in the text. The student will want to master the derivation of these concepts, where the daily user eventually accepts them as the *laws of physics*, and applies them.

ABOUT THE AUTHOR

Kimo Welch has worked in vacuum-related industries for the last 35 years. This background includes work in the microwave tube industry, at General Electric, Raytheon and Litton; in the vacuum components and equipment industry at Varian; and, work in the high energy physics *industry* at the Stanford Linear Accelerator Center, and the Alternating Gradient Synchrotron Department and Relativistic Heavy Ion Collider at Brookhaven National Laboratory. Also, he has a teaching credential in the community college system of the State of California, having taught courses in electrical engineering and mathematics. As a youngster, Kimo worked in the shipyards of Stockton, where he was a journeyman welder. He later spent time on the board, and also worked as a microwave tube technician. He eventually went on to obtain an undergraduate degree in physics from the University of the Pacific, and a Masters degree in Electrical Engineering at Stanford University. It is this practical and theoretical experience, in conjunction with his teaching experience, that makes this book of interest to the reader.

ACKNOWLEDGEMENTS

I am forever indebted to my wife, Junella, for her love, understanding and support in all things, including the writing of this book. Also, I express my sincere appreciation to Dr. J.V. Lebacqz, Dr. Robert Jepsen, Mr. James Lind, and Mr. Bert Ryland for their support and guidance in various stages of my career. I have indeed been privileged to work with such accomplished technologists and true gentlemen. The many vigorous and enthusiastic technical discussions which I have had with Henry J. Halama over the last three decades have shown *light* on the subject of this book. My sincere thanks to Marion V. Heimerle for typing and editing much of this work, and to Joseph E. Tuozzolo for editing the work and working many of the problems at the end of the chapters. Lastly, my thanks to the staff of the Brookhaven Libraries for their help, and to the Associated Universities, Inc., Brookhaven National Laboratory and the Department of Energy for the use of their libraries on weekends and evenings.

CHAPTER 1

BASIC THEORY

1.0 Introduction, Pump Classification

There are two types or classifications of UHV[1] (ultrahigh vacuum) pumps: 1) *momentum transfer* pumps; and 2) *capture* pumps. Examples of momentum transfer pumps include diffusion and turbomolecular pumps. Examples of capture pumps include all forms of cryopumps, sputter-ion and other getter pumps. This book will deal only with capture pumps. Momentum transfer pumps serve as a conduit for compressing gas, and conveying it along, to be exhausted at higher pressures, called *forepressures*. Because of this, momentum transfer pumps have no inherent capacity limitation; that is, no limitation in the amount of gas they can remove from the vacuum system.

1.1 Pump Capacity

When discussing vacuum pump capacity, there is sometimes confusion, as pump capacity has two meanings. One meaning relates to the total *amount* of gas that the pump can remove from the vacuum system prior to having to be serviced in some manner. The second meaning relates to the *rate* at which the pump can remove the gas from the vacuum system without exceeding its design limitation. The former is typically referred to, simply, as *pump capacity*. The latter is referred to as *throughput capacity*. Both of these concepts will be discussed at length.

All capture pumps share one common feature: they *store* gases which are pumped. Because of this, all capture pumps have a finite capacity for the amount of gas they can pump. This capacity differs depending on the species of gas being pumped, and the pressure at which it is being pumped. Once this capacity is exceeded, the pump must either be replaced, or refurbished by some process unique to the type of pump. The refurbishment process might require a complete rebuilding of the pump, as in the case of sputter-ion pumps, or merely executing some sort of pump warm-up cycle as in the case of cryopumps.

1.2 Understanding Pump Behavior

You have a significant advantage as a user of capture pumps if you have a quantitative *feel* for the capacity and throughput limitations of your pump. Without this insight, problems of misapplication occur, and there will often be misinterpretation of pump performance. Also, you are at an advantage if you are able to predict when the pump has exceeded its finite storage capacity, and you are able to recognize these symptoms. Better yet, if you can antici-

1

pate when this storage capacity will be exceeded, you will have control over your equipment rather than the equipment controlling you.

This system insight requires the ability to make estimates of the rate at which gas is being pumped. These are simple concepts, which I refer to as *counting molecules*. They require a basic understanding of the behavior of gases, and the models or devices of man used to describe the behavior of gas in a vacuum system (*e.g.*, pump speed, conductance, *etc.*) . If you have mastered these concepts, go on to Chapter 2. If not, take the time to study this material, as it will afford you a better understanding and control of your equipment.

1.3 The Ideal Gas Assumptions

The kinetic theory of gases predicts the macroscopic behavior of a gas by statistically averaging the behavior of individual particles or molecules of that gas. There is nothing complex or abstract about this theory, with the single exception that we are dealing with microscopic particles which we cannot see.

Not being a theoretician, I find comfort in the fact that the major advances in kinetic theory were almost totally empirical; that is, engineers and scientists *fussing around* in a laboratory or shop, over a period of several centuries, reached certain conclusions about the behavior of gas. For example, Newtonian mechanics - dealing with concepts of force, momentum, inertia, etc. - was, for the most part, founded on empiricism. Charles', Boyle's and Dalton's Laws were founded entirely on empirical observations. Of course, there were notable theoreticians who contributed to the kinetic theory. Maxwell's theoretical derivation of the distribution function was of singular importance. But, I am getting ahead of myself.

There are some assumptions, with regard to the behavior of gases, essential to the development of kinetic theory. These include:

A. Gas comprises a very large number of molecules or atoms.

B. The size of these particles or molecules is small compared to the space which they occupy.

C. Collisions of these molecules with their neighbors and with the walls of the container are elastic.

D. The behavior of these molecules may be described with Newtonian mechanics.

E. External forces on the molecules such as gravitational, electrostatic, etc., are negligible.

1.4 Definitions of Temperature and Pressure

For the most part, except at very low temperatures and very high pressures, the above assumptions are realistic. But, temperature and pressure are terms which we have not yet defined. There are several ways in which temperature may be defined. It is known that when molecules are contained in a vessel, the hotter the vessel, the greater the average velocity of the molecules. Therefore, perhaps we could use this average molecular velocity as a means of defining the temperature of the vessel (and gas). Steam and ice are common states in nature. Another approach in defining temperature might be to use the *ice point*, T_i, and *steam point*, T_s, of water as two *temperature states*, arbitrarily assigning $0°$ C to the ice point and $100°$ C to the steam point (*i.e.*, the scale is arbitrarily divided into 100 equal parts, as in the $°$ C scale). Defining pressure as *force per unit area* (*e.g.*, lbs./in^2, Nt/m^2, *etc.*), experiments conducted with different gases led to the experimental observation with regard to pressure, P, at steam and ice points:

$$\frac{T_s}{T_i} \;\triangle\; \lim_{P_i \to 0} \frac{P_s}{P_i} \approx 1.366.$$

The term "$\triangle$" is used by mathematicians to mean *is defined as*. Using this definition and requiring that $(T_s - T_i) = 100°$ K, leads to $T_i = 273.16°$ K and a value $T_s = 373.16°$ K. Again, the scale between the ice point and the steam point was conveniently and arbitrarily divided into 100 units.

The Ideal Gas Law

Constant pressure thermometry is essentially founded on Charles' Law from which it was determined that, within limits, the volume assumed by a fixed quantity of gas will vary linearly with temperature (Fig. 1.4.1).

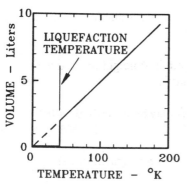

Figure 1.4.1. Charles' Law where the volume of a gas at constant pressure varies linearly with temperature.

Boyle determined that for a fixed quantity of gas at a fixed temperature, the volume occupied by that gas varied inversely with pressure (Fig. 1.4.2).

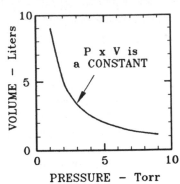

Figure 1.4.2. Boyle's Law where the product of the pressure and volume of an ideal gas is equal to a constant.

Avogadro, in experiments with various elements, discovered that they combined in chemical reactions in certain proportions. From this he was able to define what chemists now call a *mole* of something. One mole of anything is $\sim 6.023 \times 10^{23}$ of those things (*e.g.*, atoms, molecules, boxcars, *etc.*). According to Avogadro, the volume occupied by any gas, at a fixed temperature and pressure, is directly proportional to n, the number of moles of that gas. The symbol N_0 is frequently used in the literature to denote Avogadro's number. These three important findings are mathematically summarized as follows:

Charles' Law: $V \propto T$ (1.4.1a)

Boyle's Law: $V \propto 1/P$ (1.4.1b)

Avogadro's Law: $V \propto n$ (1.4.1c)

¤ 6.023×10^{23} particles $\approx$ 1 mole   (1.4.1d)

$$\frac{6.023 \times 10^{23} \text{ particles}}{1 \text{ mole}} \approx N_0.$$

Combining the empirical results of Charles, Boyle and Avogadro, we have:

$$V \propto n\,T/P. \qquad (1.4.2)$$

The algebraic manipulation of (1.4.2) and assigning of a proportionality constant "$\Re$", called the *Universal Gas Constant* in chemistry textbooks, yields the following:

¤ $PV = n\Re T.$ (1.4.3)

The value of the term "$\mathfrak{R}$" depends on the units selected for P, V and T (of course a mole is as defined by Avogadro). In vacuum-related work, the terms P, T and V may have different units. For centuries, chemists typically have used atmospheres (e.g., 0.5 Atm., 1.5 Atm., etc.) as the unit of pressure, P. Volume, V, was typically given in liters, and temperature in degrees Kelvin.

Vacuum technologists have made a *jumbled mess* of units in the short span of fifty years. I will use Torr as the unit of pressure (P), liter ($\mathcal{L}$) as the unit of volume (V), and degree Kelvin as the unit of temperature (T). To convert these terms to other units, you are referred to a basic vacuum technology handbook.[2,3] However, just this once: 1 Torr = 133.32 Pascals. The value of $\mathfrak{R}$, for the units I've selected, is then:

$$\mathfrak{R} \approx 62.36 \text{ Torr-}\mathcal{L}/\text{mole } °K. \tag{1.4.4}$$

1.5 Counting Molecules (or Atoms)

The simple expression given in (1.4.3), and variations of it, will be used throughout the text. Equations (1.4.1d), (1.4.3) and (1.4.4) are worth memorizing, as they are invaluable *tools* to the vacuum technologist in *counting gas molecules* pumped by a capture pump. Let's work a simple example using these three equations. Assume you have a sealed vessel where $V = 1.5 \,\mathcal{L}$, the pressure in the vessel is $P = 750$ Torr, and the temperature of the vessel (and gas) is $T = 300°$ K. How many gas molecules are contained in the vessel?

$$\frac{(P)(V)}{(\mathfrak{R})(T)} = n$$

Or,
$$\frac{(750 \text{ Torr})(1.5 \,\mathcal{L})}{(62.36 \text{ Torr-}\mathcal{L}/\text{mole } °K)(300° K)} \approx 0.06 \text{ moles.} \tag{1.5.1}$$

Using Avogadro's number (which you memorized), we know:

$$6.023 \times 10^{23} \text{ molecules} = 1.0 \text{ mole,}$$

or,
$$\frac{6.023 \times 10^{23} \text{ molecules}}{1.0 \text{ mole}} = \text{"1.0".} \tag{1.5.2}$$

We know that we can multiply any number by *one* and not change its value. If we multiply the result of (1.5.1) by (1.5.2) we conclude there are ~3.6×10^{22} molecules in the vessel.

1.6 Density, Pressure and Molecular Velocity

It is somewhat incredible that there were no restrictions placed on the type of gas, when stating the gas laws given in equations (1.4.1a) - (1.4.1d). Incredible, in that these laws apply to molecules and atoms of all sizes. The vessel in the example given in the last section could have contained small and relatively light helium atoms, with the atomic weight of ~4 amu (i.e., atomic mass units). Or, it could have contained very large and comparatively heavy xenon atoms, which have an atomic weight of ~130 amu. For the same P, V and T in the above example, there would be the same number of atoms in the vessel, regardless of the atomic weight of the gas.

The density of gas atoms, that is the number of atoms per cubic centimeter, would be the same whether He or Xe. What would happen to the number of molecules in the vessel if we increased the temperature of the vessel? Nothing, as it is a sealed vessel, and the number of molecules per unit volume would remain the same. However, according to (1.4.3), the pressure would increase with increasing temperature.

Pressure is defined in Section 1.4 as the force, F, exerted per unit area, A, on the walls of a vessel by a gas (*i.e.*, $P = F / A$). In order to determine the pressure, we need only use some sort of mechanical pressure gauge. This, however, tells us nothing about the density of gas in the vessel, unless we also know the temperature. Pressure has no meaning at the very center of the vessel, but density does. Therefore, in order to determine the density of gas molecules in a vessel, we must know P, V and T. This is the distinction between density and pressure.

We know that pressure on the walls of the vessel stems from collisions of the gas atoms on the walls. How is it that the much heavier Xe atoms produce the same force (*i.e.*, change in momentum per second) on the walls of the vessel as that imparted by the lighter He atoms? We can only conclude that the number of Xe atom wall collisions per second must be less than the wall collisions of the He atoms, under the same conditions. Knowing that there are the same number of atoms in the vessel, for the same P, V, and T, we can only conclude that the Xe atoms must be moving about the vessel at a much slower velocity than the He atom.

This proves to be the case; that is, there is a relationship between the velocity of the atoms (or molecules) of an ideal gas and the atomic (or molecular) weight of that gas. Also, we know that when we heat the gas in a container, the pressure increases on the walls of the container. Therefore, the velocity of the molecules must increase with temperature.

Molecular Velocities

In 1859 Clerk Maxwell theoretically derived an expression which made it possible to predict how molecular velocity varied with temperature and molecular weight. He predicted that in a very large collection of molecules, there would be a distribution of molecular velocities for one type of molecule at a given temperature, rather than one distinct velocity. His theoretical derivation was later empirically verified in a number of famous experiments. His findings, referred to as *Maxwell's Distribution Law*, are summarized in the following equation:

$$N_v = 4\pi N \left\{ \frac{m}{2\pi kT} \right\}^{3/2} v^2 \exp \left\{ - \frac{mv^2}{2kT} \right\} dv, \qquad (1.6.1)$$

where the terms are:

v = the velocity of the molecules or atoms in meters/sec,
N = the total number of molecules,
$N_v dv$ = the number of atoms found in the velocity interval v and $v + dv$,
k = Boltzmann's constant ($\sim 1.38 \times 10^{-23}$ joule/° K),
m = the mass of the molecule in kg.

For example, assume we have a collection of 10^6 molecules of H_2 in some container at a temperature of 273° C, and that we are able to measure the velocity of each of these molecules in the container. If we then made a plot of the number of molecules as a function of their velocity, we would be able to construct a plot similar to that shown in Fig. 1.6.1.

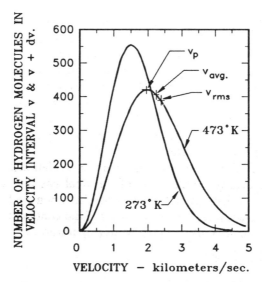

Figure 1.6.1. Maxwellian distribution of velocities of 10^6 hydrogen molecules at temperatures of 273 °K and 473 °K.

This figure shows two population vs. velocity curves. One is for when the container is at 273° K, the second is for when the container is at 473° K. We see that there is a *distribution* of molecular velocities for each temperature. These two curves are examples of Maxwell's Distribution Law.

There are three special velocities that are of interest. One is called the *root-mean-square* velocity, v_{rms}. The *average* velocity, v_{avg}, of the whole population is the second. The third is the *most probable* velocity, (*i.e.*, a small interval of velocities where we are likely to find the greatest number of molecules), v_p. Using (1.6.1), one can solve for these three velocities.[4,5] The results are as follows:

$$v_{rms} \approx 1.7 \{ \frac{kT}{m} \}^{1/2},$$ (1.6.2)

$$v_{avg} \approx 0.98 \ v_{rms},$$

and, $v_p \approx 0.82 \ v_{rms}.$

The relationship $(kT/m)^{1/2}$ appears in each of the velocities of interest. This relationship, or a variation thereof, is used extensively in making vacuum calculations. In (1.6.2), m is the mass of the gas molecule, in kilograms. In making vacuum calculations it is more convenient to use M, the *atomic* or *molecular weight* of a gas, rather than the mass in kilograms of the gas particle. This is another area of possible confusion to the student. Chemists insist on using units of grams, where physicists use units of kilograms. However, it can be shown that the mass of a gas, in kilograms, is simply:

$$m = 10^{-3} M/N_0 \text{ (kilograms)}.$$ (1.6.3)

We can now express v_{rms} in terms of the atomic or molecular weight of the gas species:

$$v_{rms} \propto (T/M)^{1/2}.$$ (1.6.4)

We will frequently use this relationship in making calculations in the following sections.

1.7 Vapor Pressure

The concept of vapor pressure is frequently misunderstood. It is discussed here as it is an extremely important concept as it relates to cryopumps, an important class of capture pumps.

We have thus far defined the meaning of pressure, density and temperature. Assume that we have a sealed box such as shown in Fig. 1.7.1, and that the box contains some element in a bulk (*i.e.*, solid or liquid) state. Assume that the walls of the box are at a uniform temperature T_{S1}, and that this temperature has been held constant for an extended period of time.

Above the bulk material in the box, some of the molecules or atoms of the bulk material will exist in a gaseous phase. These gaseous atoms will behave according to the ideal gas laws. The gaseous atoms will move about in the *empty* part of the box with a $v_{rms} \propto (T_{S1} / M)^{1/2}$. Both the bulk atoms and the gaseous atoms will be at the same temperature T_{S1}. The gaseous atoms will collide with the walls of the box, exerting pressure, and also return into the bulk material. At the same time, atoms within the bulk material with enough thermal energy will escape from the bulk material to exist for a while as gas above the bulk material.

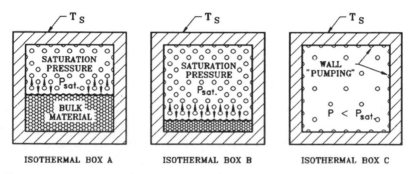

Figure 1.7.1. Three boxes at isothermal temperature T_S. Boxes A & B have sufficient bulk material to support a saturation vapor pressure above the material, where this is not the case in box C.

In time, an equilibrium state will be reached where the rate at which atoms leave the bulk material is the same as the rate at which other atoms impinge onto, and remain behind as part of the bulk material. The rate per cm^2 at which particles pass some plane - real or imaginary - is referred to as a *net flux* of particles. When the net flux of particles at the surface is zero, the bulk material is said to be in equilibrium with the gas. The gas above the bulk is said to be *saturated* with atoms, because, for every atom that escapes the bulk material to become part of the gas, another atom leaves the gas to become part of the bulk material. Each element has a specific saturation pressure, P_{S1}, for a given temperature, T_{S1}. Change the temperature of the box to T_{S2}, and the pressure above the bulk material will eventually change to a predictable saturation pressure, P_{S2}, uniquely characteristic of the element.

Figure 1.7.2 represents the situations where there is sufficient bulk material, in three cases, but the temperature is changed in each of the three instances. In this case, the density or pressure of gas varies, as a function of temperature, over the bulk material. If we were to plot the pressure of the given element as a function of temperature, we would find a smooth, monotonic curve unique to that element. The saturation pressure of that element is sometimes called the *saturation vapor pressure*, or the *equilibrium vapor pressure*, or just the *vapor pressure*.

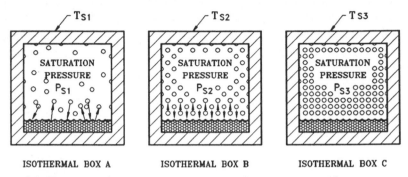

Figure 1.7.2. Three boxes at isothermal temperatures T_{S1}, T_{S2} and T_{S3}, where $T_{S1} < T_{S2} < T_{S3}$ and corresponding vapor pressures of the material are $P_{S1} < P_{S2} < P_{S3}$.

Examples of such vapor pressure curves for the common gases are given in Figure 1.7.3.[6] Some *smoothing* of the reference (6) data was done, where there were obvious errors in the tabulated data. There are similar curves for all of the elements. Note that the curves in these two figures don't seem to alter their *smooth* monotonic characteristics, when a material changes from a gas to liquid, or liquid to solid state. A summary of more current and precise vapor pressure data, reported by others for hydrogen, helium-3 and helium-4, is given by Haefer and shown in Fig. 1.7.4.[7] It is a common misconception that some mysterious process occurs at the point of phase change. This is not the case. We will revisit these two figures in considerable detail in some of the following chapters.

Surface Pumping

To achieve the true vapor pressure in a bounded environment, it is required that there be enough bulk material to supply sufficient atoms to support a saturated gas state. If there is not sufficient bulk material, the atoms in the box *sort of* behave according to the ideal gas law (*i.e.*, $PV = n\Re T$). This is not completely correct as, when the atoms impinge on the walls of the box, they tend to stay on these walls for a short period of time. If many billions of atoms are impinging on the walls, even though these atoms may dwell on the walls for say only a few μsec., the net effect is that the pressure will be less than would be predicted by the ideal gas law. The walls of the box have sort of a pumping effect, and at any instant in time, there will be a large population of molecules resident on the walls.

Pumping on a Liquid Cryogen

We noted in Fig. 1.6.1 that there was a distribution of velocities of H_2 molecules in the container, as defined by Maxwell's Distribution Law. The velocity of the molecule defines its kinetic energy. This implies that some of the molecules in the gas have much greater kinetic energies than others. Those *out on the tail* of the Maxwell distribution curve have much higher kinetic energy than the average population. There are other energy states possessed by the molecules besides their kinetic energy. These include their vibrational and rotational energy. There are similar energy distribution *states* for molecules and atoms resident in both solids and liquids.[8]

Molecules or atoms resident *out on the tail* of the distribution function have the greatest probability of escaping the bulk material and going into the gas phase. Departure of the higher energy molecules from the bulk, in essence has the effect of ever so slightly reducing the temperature of the bulk material.

For example, assume that by some process 10% of the molecules on the tail of the distribution function escaped the bulk material and became gas, and that this gas was removed from the container so that it could not return to the bulk. If we were to then construct a new distribution function for the balance of the bulk material, the average energy of the bulk material would be reduced because of the departure of the higher energy particles. A reduction of the average energy essentially implies a reduction in temperature of the bulk material.

Liquid cryogens (*i.e.*, very cold liquids) are commonly used in cryopumps. For example, liquid He is frequently used as a cryogen in special cryopumps.[9] It is also used widely in superconducting applications. Helium liquefies at a temperature of 4.2° K. A process commonly used to reduce the temperature of liquid He below this 4.2° K temperature is to pump on the gaseous He which escapes from the liquid with some sort of mechanical vacuum pump. Pumping away the higher energy He atoms which escape from the liquid He has the effect of reducing the temperature of the liquid He.[10]

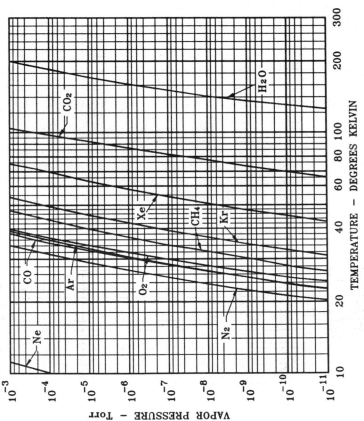

Figure 1.7.3. Vapor pressure of some of the common gases as reported by Honig and Hook.(6)

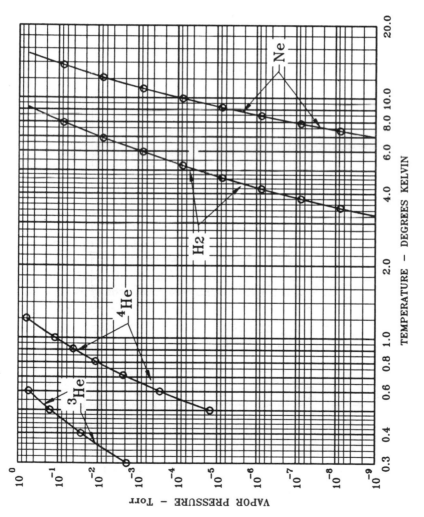

Figure 1.7.4. Vapor pressure of ^{3}He, ^{4}He, H$_2$ and Ne as given by Haefer.[7]

1.8 Mean Free Path

The mean free path of a particle in a medium is defined as:

The average distance a particle will travel in a medium before it suffers a collision with some other particle in that medium.

Refined calculations used to predict the behavior of gases in vacuum systems require that one be able to calculate the mean free path of gas molecules in the system. These calculations are used to determine the relative ease with which molecules will be conveyed or conducted from one place to another in a vacuum system. They are therefore called *conductance* calculations, and will be discussed in Section 1.12.

There are numerous other applications where the concept of a mean free path is of importance. For example, the solid state physicist is concerned about the mobility of electrons (and holes) in solids. The conductance (*i.e.*, the inverse of resistance) of an electron in a solid is directly related to the number of collisions the electron suffers as it traverses the material under the influence of an electric field. The electron microscopist is concerned about gas scattering of the electron beam in the microscope. In this case the particle is an electron and the medium is the gas in the microscope. Physicists working with plasmas and particle accelerators also are on the *long list* of technologists concerned with mean free path concepts.

Figure 1.8.1 depicts a chamber which is attached to a vacuum pump through an aperture. The chamber contains a large number of molecules which *bounce* off the walls, and which also collide with other gas molecules. If a molecule can bounce off the walls of the chamber many times prior to experiencing a collision with another gas molecule, then collisions with neighboring gas molecules will have little influence on whether or not the molecule eventually *finds* the aperture and escapes the chamber.

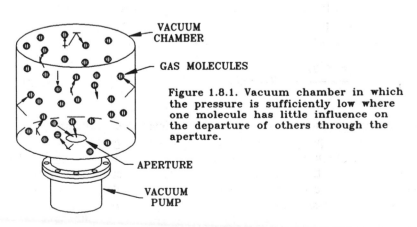

VACUUM
CHAMBER

GAS MOLECULES

Figure 1.8.1. Vacuum chamber in which the pressure is sufficiently low where one molecule has little influence on the departure of others through the aperture.

APERTURE

VACUUM
PUMP

With a very long mean free path, the probability of the molecules escaping the chamber will primarily be dependent on ν, the surface collision rate, and the area of the aperture. The velocity of a molecule is proportional to $(T/M)^{1/2}$. This means a molecule's *opportunity* to escape through the aperture diminishes with its molecular weight and increases with its temperature. If the mean free path is long compared to the dimensions of the chamber, the likelihood of a molecule escaping from the chamber is not influenced by its neighbors.

In vacuum terminology, when the pressure in a vacuum system is such that the behavior of one gas molecule - with regard to escaping through an aperture or moving along a conduit - is essentially independent of the behavior of neighboring gas molecules, this pressure range or interval is called the *molecular flow region*. This is analogous to being on an elevator in New York City. If the elevator is very crowded, you may have to disembark (momentarily) so that those behind you may get off at their selected floor. On the other hand, if there are few people on the elevator, all may disembark at will without bumping into other occupants.

It is reasonable to assume that the mean free path, λ, will be inversely proportional to the number of molecules per unit volume, N/V, and the cross-sectional area of these molecules, $\pi d^2 / 4$, where d is the diameter of the molecule. The diameter of gas molecules differ from species to species. Therefore, λ will differ with species and mixtures of different particles. The mean free path of some species in a medium of the same gas is given by the following equation:[3]

$$\lambda_a = kT / [2^{1/2} \pi P (d_a)^2], \tag{1.8.1}$$

where the subscripts "a" indicate the given species.

In our calculations we will frequently pretend that there is such a thing as an *air molecule*. This term actually means that on the average the mixture of all gases found in the atmosphere at sea level behaves collectively in some manner. For example, the average molecular weight of air is $M \approx 28.7$ amu. The mean free path of an *air molecule* in air, at $20°$ C, is:

$$\lambda_{air} = \frac{4.94 \times 10^{-3}}{P} \quad \text{Torr-cm.} \tag{1.8.2}$$

The actual constituents making up air, at sea level, are important to know when using capture pumps. For example, certain noble gases in air are more difficult to *sorption pump* - a term which will be explained in Chap. 5 - than other gases. Also, the presence of argon in air can cause problems with certain types of sputter-ion pumps. These phenomena will be discussed later.

Some of the air gases found at sea level, including average values for urban pollution, are given in Table 1.8.1.[11]

Table 1.8.1. Partial pressure of various gases at sea level.

GAS SPECIES	SYMBOL	PARTIAL PRESS. (Torr)	ATOMIC MASS UNITS (amu)*
Nitrogen	N2	$5.93 \times 10^{+2}$	28
Oxygen	O2	$1.59 \times 10^{+2}$	32
Argon	Ar	7.10×10^{0}	40
Carbon dioxide	CO2	5.04×10^{-1}	44
Carbon monoxide	CO	3.84×10^{-2}	28
Neon	Ne	1.38×10^{-2}	20
Helium	He	3.98×10^{-3}	4
Methane	CH4	1.22×10^{-3}	16
Krypton	Kr	8.66×10^{-4}	84
Sulfur dioxide	SO2	7.60×10^{-4}	64
Hydrogen	H2	3.80×10^{-4}	2
Nitrous oxide	N2O	2.28×10^{-4}	44
Xenon	Xe	6.61×10^{-5}	131
Water vapor	H2O	variable	18

*Approximate values of the most abundant isotopes.[11]

1.9 Thermal Conductivity of Gases

Mean free path considerations enter into an understanding of the thermal conductivity of gases. Also, an understanding of gas ionization processes, at reduced pressures, involve calculations of the mean free path of electrons in that gas. This will be discussed in the section on sputter-ion pumps. The thermal conductivity, K, of a gas or mixture of gases is given by:[12]

$$K = (1/3) \rho\ v_{avg}\ \lambda\ \epsilon\ c_v \qquad (1.9.1)$$

$$= \eta\ \epsilon\ c_v,$$

where ρ is the gas density, c_v the specific heat of the gas at constant volume. We have already defined λ and v_{avg}. The term η is defined as the *viscosity* of the gas. Excluding the ϵ term, (1.9.1) is derived from elementary kinetic theory. The term ϵ takes into account the *rotational* and *vibrational* energy states of various types of gas molecules, where

$$\epsilon = \frac{9\gamma - 5}{4}, \qquad (1.9.2)$$

and for monatomic gases $\gamma = 5/3$, where for polyatomic gases $\gamma \sim 1$.

The thermal conductivity of gases is particularly important in understanding the limitations of various forms of cryopumps. For example, most cryopumps have a finite refrigeration capacity. When pumping on a system with a cryopump, should the pressure exceed a certain value, it is possible to exceed this refrigeration capacity, and *swamp* the pump. Also, assuming that c_v is about the same for air and H_2, what would be the relative thermal conductivities of the two gases at the same pressure? This has important implications in the starting of cryopumps. But, that topic comes later.

At low pressures, the thermal conductivities of all gases are directly proportional to the pressure, over a very wide pressure range (*e.g.*, see Fig. 1.9.1). We note that at pressures less than $\approx 10^{-4}$ Torr, the thermal *losses* for the greater part are dominated by radiation between surfaces. For this reason, this represents the lower practical limit of thermocouple-type (TC) gauges, which depend on heat loss from an element as an inference of pressure. At pressures above a few Torr, the thermal conductivity becomes essentially independent of pressure. Note, however, that at all pressures the thermal conductivities of gases are proportional to v_{rms}.

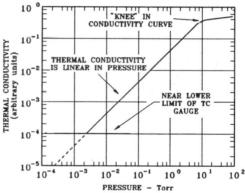

Figure 1.9.1. Thermal conductivity of air as a function of pressure.

Gas Flux Incident on a Surface

Using kinetic theory, it can be shown that the flux, v, on a surface, or through an imaginary plane in some vacuum vessel, is given by:[13]

$$v = N\left\{\frac{kT}{2\pi m}\right\}^{1/2}. \qquad (1.9.3)$$

The only new symbol here is N, the density of particles per cubic meter. Assume we wish to determine the rate at which molecules impinge on the plane defined by the aperture in Fig. 1.8.1, with a nitrogen pressure, P, in the chamber. The density of molecules in the chamber is simply the total number of molecules in the chamber, divided by the total volume. Using (1.4.1d) and (1.4.3), we can express the density of molecules in the chamber, in terms of molecules per liter, as:

$$N = \frac{PN_0}{\Re T} \tag{1.9.4}$$

Substituting this into (1.9.3), and noting there are $10^3\, \ell/\mathrm{m}^3$, we have:

$$\nu = 10^3\, \frac{PN_0}{\Re T}\, \{\frac{kT}{2\pi m}\}^{1/2}. \tag{1.9.5}$$

In (1.9.5) the flux density, ν, is given in particles per square meter. With a little algebraic manipulation, and using (1.6.3), this equation can be put in a form expressing ν, the flux density per cm^2, as a function of pressure and temperature:

$$\nu = 3.51 \times 10^{22}\, P/(MT)^{1/2}. \tag{1.9.6}$$

Assuming the chamber is filled with N_2, and the gas is at room temperature (i.e., $293°$ K), the rate at which N_2 molecules strike the wall of the box is simply:

$$\text{м} \quad \nu = 3.88 \times 10^{20}\, P \text{ molecules/cm}^2\text{-sec.} \tag{1.9.7}$$

Equation (1.9.7) is an extremely useful equation for developing a *feel* for the physics going on in your vacuum system. For example, assume that every molecule which strikes some special surface in your system sticks to that surface. In this case the molecule is said to have *unity sticking coefficient* on the given surface. Perhaps it is trapped by some chemical process occurring on the surface - this process is called *chemisorption*. Or, perhaps there is a *cold* surface in the system, so that when the molecule strikes the surface, sufficient energy is removed from the molecule so that weak attractive surface forces prevent it from escaping back into the vacuum system - this process is called *physisorption*.

If one molecular layer on the special surface comprises $\sim 10^{15}$ molecules/cm^2,[13] how long would it take, at a pressure of 10^{-6} Torr, for one molecular layer to build up on the surface by either chemisorption or physisorption? The answer is, less than three seconds. This can become a serious problem in some applications. For example, assume that you were subliming a special semiconductor material, such as As, onto a substrate for the purpose of *growing* some solid state device - a process called *molecular beam epitaxy* (MBE).[14] The material has to be sublimed at a fairly slow rate, say one Å/sec, so that it will *settle in* and become a well ordered structure. A molecule has dimensions of the order of 2-3 Å. If the material being sublimed is chemically active with the gas in the system, and the pressure is 10^{-6} Torr, this implies that for every molecule sublimed, a chemically active gas molecule arrives on the surface and has an opportunity to form an undesirable chemical compound. Because of this problem of possible gas contamination of a deposited film, MBE is typically accomplished at pressures ranging from 10^{-10} to 10^{-9} Torr.

1.10 Pumping Speed, a Convenient Abstraction

The concept of pumping speed is merely a mathematical abstraction, created by man as a convenient *tool* to describe the behavior of gas in a vacuum system.

Pump speed is defined as a volumetric displacement rate.

Note that this definition implies nothing about pressure, mass flow, gas density or anything that has to do with gas molecules. The symbol "S" is usually used in the literature to represent pump speed. To best understand the concept of speed, refer to Fig. 1.10.1. Assume the following:

1)   The piston shown in Fig. 1.10.1 forms a perfect gas seal with the cylinder within which it is contained.
2)   There is no outgassing from the components.
3)   The piston is attached to a crankshaft which can be rotated either manually or at some fixed rate.
4)   The volume above the piston is one liter when it is fully withdrawn from the cylinder, and zero liters when it is fully extended.
5)   There are two valves located in the piston head. One valve, V_1, is used to introduce gas into the piston head to some desired pressure. The second valve, V_2, automatically opens as the piston approaches the top of the cylinder - venting gas and closing the instant the piston reaches top dead center.

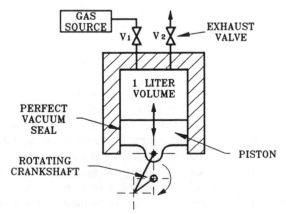

Figure 1.10.1. Mechanical model of a perfect vacuum pump where speed is merely $\Delta V / \Delta t$.

If we manually turn the crankshaft one full cycle, with all the above assumptions, there would be no gas contained above the cylinder head at the end of a cycle (*i.e.*, when the piston is again fully withdrawn). The amount of gas which was vented through V2 at the *top* of the cycle would simply be $P \times V$, where P is the pressure above the piston prior to the start of the cycle, and V is one liter, the volume of the cylinder head when the piston is fully withdrawn. The next time we *crank* the piston a complete cycle, the net amount of gas vented out of V2 will be zero, as none was contained in the one-liter volume at the start of the cycle. We may repeat the process, but, as long as V1 is closed, no gas is vented out of V2 in subsequent cycles.

Assume now that we turn on a motor which is attached to the crankshaft, and that the shaft rotates at one revolution each second. Clearly, the piston under this condition sweeps out the one liter volume of the cylinder every second. We say that the piston has a volumetric displacement rate of one liter per second. The rate the volume is *swept out*, S, is 1 $\mathcal{L}$/sec. This is the speed of the pump. I have described a perfect pump which has a speed of one liter per second. Every second ($\Delta t = 1$ sec.) a volume ($\Delta V = 1$ liter) is swept out of the cylinder head. Or,

$$S = \frac{\Delta V}{\Delta t} = \frac{\text{one liter}}{\text{one second}}. \qquad (1.10.1)$$

The above equation implies nothing about the flow of a gas. It is merely a statement of the rate at which the piston is sweeping out the volume; the volumetric displacement rate. If the rpm of the motor could be varied to say two revolutions per second, the volumetric displacement rate, or speed, would be:

$$\maltese \quad S = \frac{dV}{dt} = \frac{\text{two liters}}{\text{one second}}. \qquad (1.10.2)$$

This mechanical model, defining pump speed, is easy to visualize in applications involving mechanical pumps. These pumps either have pistons or vanes which sweep out volumes at a given rate. However, the model is totally applicable to all forms of vacuum pumps. We must stretch our imaginations a little when visualizing a *mechanical* model in the case of, for example, cryopumps or sputter-ion pumps. Capture pumps have no moving parts within the vacuum. How then does a *volumetric displacement rate* model apply? It applies only as a convenient *tool* for making calculations; it is an abstraction, rather than as a physical reality. However, I have found it convenient to *picture* capture pumps, *in my mind's eye* as consuming abstract or elemental volumes at some specified rate.

For example, assume the pump in Fig. 1.10.2 has a specified speed of

1000 $\mathcal{L}$/second. Picture this pump as consuming elemental volumes at a rate of 1000 $\mathcal{L}$/second, at the input flange of the pump. These imaginary volumes, as in the case of the mechanical pump model, are consumed by the pump at unspecified pressure. The pump may be a capture pump, turbomolecular pump, *etc*. This same abstraction applies to *all* forms of vacuum pumps.

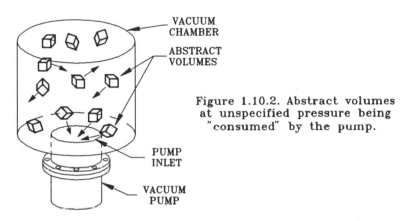

VACUUM
CHAMBER

ABSTRACT
VOLUMES

Figure 1.10.2. Abstract volumes at unspecified pressure being "consumed" by the pump.

PUMP
INLET

VACUUM
PUMP

We will return to the concept of pumping speed after first learning about the concept of throughput.

1.11 Throughput

Throughput is defined as a gas flow rate (d(PV)/dt).

For example, it might represent: 1) the rate at which gas is leaking into a vacuum system; 2) the rate at which molecules are outgassing from the walls of the system; or 3) the rate at which combinations of several sources of gas travel into a vacuum pump.

The symbol "Q" is frequently used in the literature to represent throughput. The units of throughput which we will use are Torr-$\mathcal{L}$/second (*i.e.*, $P \times V / t$). We note that the temperature, T, is not specified, and remember from (1.4.3) that knowing T is also required to be specific about the number of gas molecules involved. When data involving throughput are given in the literature, it is usually assumed that PV is specified at room temperature (*i.e.*, 293° K), unless otherwise noted.

Returning to the mechanical pump model of Fig. 1.10.1, assume now that the motor is disconnected, and the piston withdrawn so that the volume above the piston is one liter. Assume that this volume is pressurized to 10^{-3} Torr, by introducing gas through V_1. After a pressure of 10^{-3} Torr is achieved in the cylinder head, V_1 is closed. We assume in our model that the gas is neither cooled on expansion into the cylinder head, nor heated during the compression cycle (*i.e.*, it remains at 293° K). If, while invoking the five assumptions of the previous section, we manually "cranked" the pump through one complete cycle, 10^{-3} Torr-$\mathcal{L}$ of gas would be exhausted through V2 at the top of the cycle. This is merely $P \times V$ vented through V2 during some unspecified time interval.

If we now turn on the motor and repeat the process once a second, the rate at which gas is exhausted out of V2 (and introduced into the system through V1) would be $P \times \Delta V / \Delta t$. Or,

$$Q = P \times \frac{\Delta V}{\Delta t}$$

$$\text{¤} \quad Q = P \times S \qquad (1.11.1)$$

$$= 10^{-3} \quad \text{Torr-}\mathcal{L}/\text{sec}.$$

If we increase the pressure to 10^{-2} Torr between each cycle, the throughput, Q, would increase to 10^{-2} Torr-$\mathcal{L}$/sec., but the pump speed would remain the same, etc. The important thrust of this is that, for a constant pumping speed, throughput varies linearly with pressure. To be able to assume this was the prime motivator in creating the concept of pumping speed as a constant volumetric displacement or consumption rate.

1.12 Conductance, Another Convenient Abstraction

Conductance is another mathematical model or *tool*, devised to be able to describe the behavior of gas within the geometrical confines of a vacuum system. We were introduced to the concept of imaginary volumes entering the flange of a pump. The concept of conductance is similar. That is:

> *Conductance is a measure of the ease with which elemental,*
> *abstract volumes, at unspecified pressure, pass from one*
> *place to another place in a vacuum system, per unit of time.*

Remember, this is the definition of a model, rather than a physical reality. With this model, we pretend that little abstract volumes at zero pressure *bounce around* inside the constraints of a vacuum system. Mathematical equations comprising the model let us make calculations of the ease with which these volumes travel from place to place in the system. Then, we assume the little volumes are filled with gas at a specified pressure, and make a statement about the transport of the gas in the system. You will soon see that it is worth all this fuss.

For example, assume that the vacuum manifold in Fig. 1.12.1a is separated by a thin foil with an aperture in the center. Assume that d, the diameter of the aperture, is much less than D, the diameter of the manifolds. For reasons that will later be explained, assume that $D \gg \lambda_g$, the mean free path of gas species g in the vacuum system. For the moment, disregard the fact that there is gas in the system. A conductance model enables us to calculate the rate at which abstract volumes which pass from Manifold A, through the aperture of conductance C_a, and into Manifold B. It is reasonable to assume that the rate at which these elemental volumes accomplish this passage will be directly proportional to the area of the aperture. That is, in Fig. 1.12.1a,

$$C_a (A \rightarrow B) \propto \frac{\pi d^2}{4} \; \mathcal{L}/\text{sec.}$$

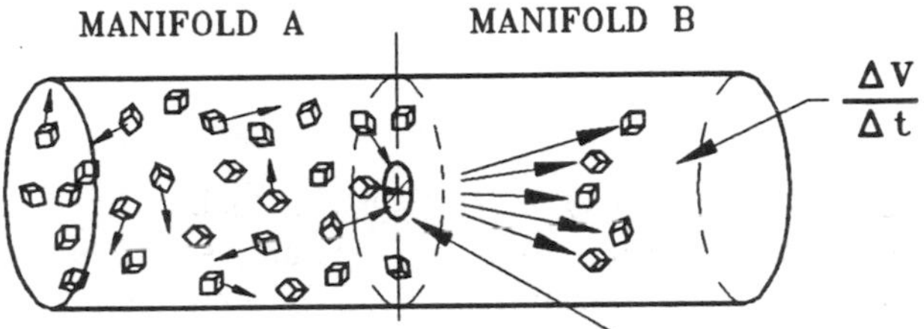

Figure 1.12.1a. The conductance of the foil aperture is the measure of the rate it will pass abstract volumes from Manifold A to Manifold B, at unspecified pressure.

Also, in Fig. 1.12.1b,

$$C_a (B \rightarrow A) \propto \frac{\pi d^2}{4} \; \mathcal{L}/\text{sec.}$$

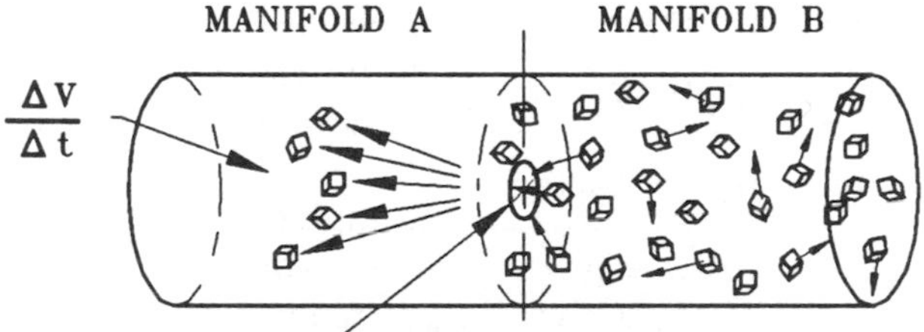

Figure 1.12.1b. The foil aperture has the identical conductance for the passage of abstract volumes, at unspecified pressure, from Manifold B to Manifold A.

Assume now that Manifold A is at a pressure P_A, and Manifold B is at a pressure P_B. As shown in Fig. 1.12.1c, we then pretend that the elemental volumes escaping from Manifold A to Manifold B retain the pressure of

Manifold A when arriving at Manifold B, and vice versa. Therefore, there is transport of gas in these elemental volumes in both directions.

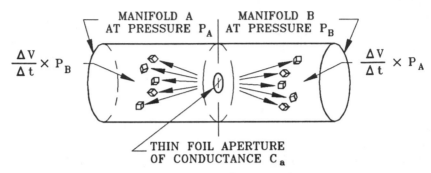

Figure 1.12.1c. A "model" for the net passage of gas through an aperture conductance separating manifolds at different pressures.

The net flow of gas in the system, Q_{net}, is:

$$Q_{net} \propto \quad [\ Q_A\ ] \quad - \quad [\ Q_B\ ],$$

$$\propto [\frac{\pi d^2}{4}P_A] \quad - \quad [\frac{\pi d^2}{4}P_B], \qquad (1.12.1)$$

$$Q_{net} \propto \quad [\frac{\pi d^2}{4}] \quad \{P_A - P_B\}.$$

1.12.1 Conductance Model Applications

Species and Temperature Dependence

We have used the above conductance model to predict the flow of gas through an aperture in a vacuum system. However, we have not specified the type of gas. We know from (1.6.4) that the v_{rms} of any gas species is proportional to $(T / M)^{1/2}$ of that species. We know, for example, that for a given T, H_2 molecules move about in a vacuum system much faster than do N_2 molecules. Common sense tells us that an H_2 molecule would more readily escape from Manifold A to Manifold B, than would an N_2 molecule. Therefore, to predict the exact amount of gas transported between the two manifolds, we must take into the account the gas species and temperature. This is a simple process, once the exact conductance equations (i.e., models) are known for one gas species.

Pressure Dependence of Conductance

In the example given in (1.12.1), we imposed the restriction that $\lambda_g \gg D \gg d$. The implication of this was discussed in Section 1.8. That is, the mean free path of one gas species g, in the gas of all other species present, was so great compared to the aperture diameter, d, that the presence of other gas species had little influence on the one of species g passing through the aperture. When the mean free path in a gas is large compared to the dimensions of the system, we call this the *molecular flow region*.

In the molecular flow region, a molecule travels from one place to another in the system in a totally random manner. It is by *accident* that the molecule *finds* a pump appending the system and is removed from the system. Providing a larger conductance between the source of the gas and the pump increases the probability that the molecule, in its random sojourn about the system, will *find* the pump.

There are equations similar to (1.12.1) for conditions where λ_g is comparable to or much less than the dimensions of the system. In these circumstances, the conductance expressions are no longer independent of pressure. As a consequence, the expression for the throughput becomes nonlinear in pressure. That is:

$$Q = C(P,d)\,[P_A - P_B], \tag{1.12.2}$$

where the expression $C(P,d)$ means C is a function of P and d. When the mean free path of gases in the system is comparable to or less than dimensions of the vacuum system, these nonlinear equations in P apply, and the calculations of system Q become *messy*. The pressure realms where these nonlinear equations apply are coined *viscous* and *turbulent flow regimes*.

Capture pumps are seldom operated at pressures where these special calculations are required. However, if one is constructing a large cryopumped space simulation chamber, where the initial pumpdown or *roughing* time is an important consideration, such calculations would be used. Also, for example, if one is manufacturing and selling large quantities of leak detectors, costs of vacuum manifolding might play a significant role in the competitiveness of your company. Formulas for making viscous and turbulent flow calculations may be found in references (2) and (3).

¤ A *rule of thumb* is that these nonlinear equations will always yield conductances (*i.e.*, $C(P,d)$) which are greater than will be derived using the molecular flow equations. Therefore, using the molecular flow calculations will yield conservative results in roughing times.

System Geometry Dependence for Molecular Flow

The conductance of an aperture in the molecular flow region, and for *air* at 293° K, is given by the following equation:

¤ $C_{a,air} = 11.6A$ $\mathcal{L}$/sec, (1.12.3)

where A = the aperture area in cm².

The conductance of a round manifold in the molecular flow region, and for *air* at 293° K, is given by the following equation:

¤ $C_{m,air} = 12.1 \dfrac{D^3}{L}$ $\mathcal{L}$/sec, (1.12.4)

where D = the diameter of the tube in cm,
and L = the length of the tube in cm.

This equation (*i.e.*, (1.12.4)) is sometimes called the *long tube formula* as it applies to tubes where $D / L \gtrsim 10$. At the risk of displeasing the more rigorous technologists, I offer you some advice to follow when making UHV vacuum calculations. Outgassing rates in vacuum systems may vary, in time, as much as five or six orders in magnitude. Because of this, though you may accurately know the pump speed, and have precisely calculated the various conductances of the system leading to the pump, you will do well to predict system pressures to within ×2. In such instances, calculating conductances to three decimal-place accuracy is a waste of time.

There are often approximations which afford sufficient accuracy to adequately predict the behavior of your system. For example, the equation for calculating the conductance of a long rectangular tube, for air at room temperature, is as follows:[2,3]

$$C_{\square} = 30.9 \frac{a^2 b^2}{L(a+b)} K \ \mathcal{L}/sec,$$ (1.12.5)

where a = rectangular tube width in cm,
              b = rectangular tube height in cm,
              L = length of rectangular tube in cm,
and, if a/b = 0.1   0.2   0.4   0.6   0.8   1.0
then K = 1.44 1.29 1.17 1.13 1.12 1.10, respectively.

Equation (1.12.5) is an exact expression for the conductance of a rectangular tube. However, in one of the problems at the end of this chapter, you will be asked to compare the conductance for air in a round tube (*i.e.*, using (1.12.4)), with diameter D and length L, with that of a rectangular tube of dimensions a, b and the same L, where:

$$a \times b = \frac{\pi \ D^2}{4}. \tag{1.12.6}$$

From this, you will find that solving for the equivalent D of (1.12.6) and using this D in (1.12.4), which you memorized, will usually provide more than sufficient accuracy in your calculations.

Molecular Conductance for Different Gases

In the previous sections it was emphasized that $C_{air} \propto (T / M_{air})^{\frac{1}{2}}$, and we may use (1.12.3) and (1.12.4), equations applicable to an *air* conductance, to derive the value of a conductance for other gases. Noting the constant *11.6* in (1.12.3), we must conclude that the expression $(T / M_{air})^{\frac{1}{2}}$ went into deriving the constant. That is:

$$11.6 = k \ (T / M_{air})^{1/2}, \tag{1.12.7}$$

where k is a constant. If we wish to derive the conductance of an aperture for some other gas with molecular weight M_g, we need only multiply *11.6* (*i.e.*, (1.12.3)) by the value $(M_{air})^{\frac{1}{2}}$ and divide the same expression by the value $(M_g)^{\frac{1}{2}}$. Refer to the atomic mass units of Table 1.8.1 for values of M_g for the different gases. Also, we remember from Section 1.8 that the value of M_{air} is ~28.7 amu.

Let's work out a couple of examples where we modify (1.12.3) to compensate for a different gas at a different temperature. Assume that we have calculated the conductance of an aperture for air, and we want to know the value of this conductance for helium. The atomic weight of helium, M_{He}, is 4 amu. Therefore,

$$C_{He} = \{C_{air}\} \times [\frac{M_{air}}{M_{He}}]^{1/2} \tag{1.12.8}$$

$$= \{11.6\,A\} \times [\frac{28.7 \text{ amu}}{4 \text{ amu}}]^{1/2}.$$

This conductance for He is the value at a temperature of 293° K. We may calculate the conductance of the aperture for He at, say, 400° K as follows:

$$C_{He,400° K} = \{C_{He,293° K}\} \times [\frac{400° K}{293° K}]^{1/2}. \tag{1.12.9}$$

1.13 Voltage, Current and Impedance Analogies

In the molecular flow region (*i.e.*, S and Q are *linear* in P), there is a *one-to-one and onto* correspondence - mathematicians call this an isomorphism - between vacuum conductance, pumping speed and throughput, and the electrical concepts (*i.e.*, linear circuit theory) of electrical conductance, voltage and current, respectively. This is not coincidental, for similar models were created much earlier to describe the flow of charge in electrical circuits. The correspondence is as follows, where the subscripts v are for vacuum values, and the subscripts e are electrical values:

$$C_v \leftrightarrow G_e = 1/R \quad \text{(electrical conductance of a component)},$$
$$S_v \leftrightarrow G_e$$
$$P_v \leftrightarrow V_e \quad \text{(voltage across a linear component)},$$
$$Q_v \leftrightarrow I_e \quad \text{(current, or charge flow rate)}.$$

This correspondence between *linear vacuum circuits* (*i.e.*, in the molecular flow region) and linear electrical circuits provides an aid in expanding use of the vacuum S, C and Q models using elementary circuit theory.[15] The definition of throughput, with constant T, was:

$$Q = \frac{d(PV)}{dt} = P\frac{dV}{dt} + V\frac{dP}{dt}. \tag{1.13.1}$$

Assuming steady state conditions (i.e., $dP / dt = 0$), yields:

$$Q = P \times S, \tag{1.13.2}$$

or, $$Q = \Re T\frac{dn}{dt}. \tag{1.13.3}$$

The electrical equivalence of (1.13.2) is merely:

$$I = V_e \times G = V_e \times 1/R, \tag{1.13.4}$$

$$Q = S P$$

$$I = G V_e$$

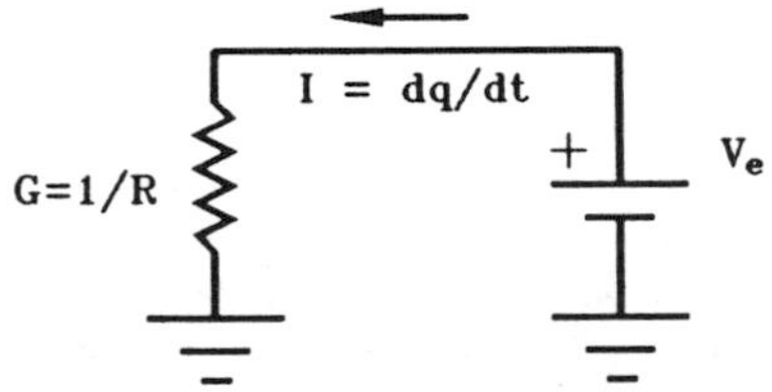

Figure 1.13.1. Electrical correspondence between V & S, G & C and I & Q.

where current, I, or charge flow (*i.e.*, dq/dt) is analogous to the flow of molecules, dn/dt, in a system. An electrical circuit which represents (1.13.4) is given in Fig. 1.13.1.

Were the circuit to have a second resistor, as in Fig. 1.13.2, we can combine the resistors into the reciprocal of one equivalent resistor with electrical conductance G_e:

$$R_e = R_1 + R_2,$$

$$\text{Or,} \quad \frac{1}{G_e} = \frac{1}{G_1} + \frac{1}{G_2}$$

$$\text{And,} \quad G_e = \frac{G_1 \, G_2}{G_1 + G_2}. \tag{1.13.5}$$

The current, I, in the circuit in Fig. 1.13.2 is:

$$I = V_e \, \frac{G_1 \, G_2}{G_1 + G_2}. \tag{1.13.6}$$

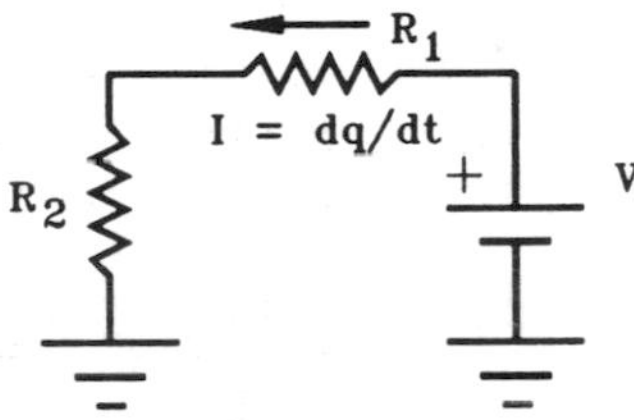

Figure 1.13.2.   The correspondence between electrical and vacuum conductances.

The corresponding vacuum *circuit* is shown in Fig. 1.13.3, where the vacuum conductance of a manifold leading to a pump, C_m, is combined with the pump speed, S_p, to yield an equivalent pump speed, S_c, at the chamber:

$$\frac{1}{S_c} = \frac{1}{S_p} + \frac{1}{C_m}$$

$$S_c = \frac{S_p \, C_m}{S_p + C_m}. \tag{1.13.7}$$

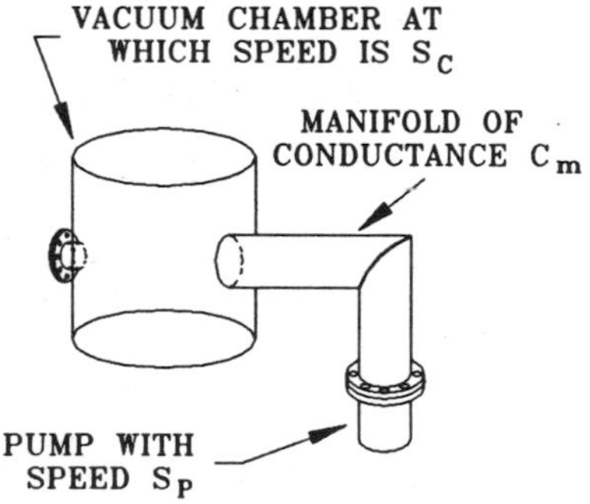

Figure 1.13.3. Speed produced at vacuum chamber for a given S_P and C_m.

Knowing the pressure in the chamber, P_c, we may calculate the throughput (*e.g.*, outgassing) from the chamber, Q_c, accordingly:

$$Q_c = P_c S_c = P_c \frac{S_p C_m}{S_p + C_m}. \tag{1.13.8}$$

Kirchhoff's first rule in electrical circuits originates from the assumption that charge is neither created nor destroyed in a circuit. There is an analogy in vacuum circuits, where we assume that mass (*i.e.*, *PV*) is preserved. For example, if we assume that the only source of gas in the system shown in Fig. 1.13.3 originates in the chamber, then the rate at which gas is leaving the chamber, Q_c, must be the same rate at which gas is entering the throat of the pump, Q_p. Therefore,

$$Q_c = Q_p,$$

$$\bowtie P_c S_c = P_p S_p. \tag{1.13.9}$$

We will later use (1.13.9) to diagnose possible problems in a vacuum system. Also, in the next section, you will learn how to take into account the added effect of an outgassing manifold between a chamber and pump. However, let's first pursue the electrical and vacuum analogies a little further.

The similarity between (1.13.6) and (1.13.8) is obvious. Also, we have used simple circuit theory to derive (1.13.8). Equation (1.13.8) is not totally correct as we have not taken into account conductance of the aperture leading into the pipe from the chamber. This aperture conductance, C_a, is combined with the conductance of the manifold, C_m, to yield a total conductance, C_t, of:

$$C_t^{-1} = C_a^{-1} + C_m^{-1}. \tag{1.13.9a}$$

In (1.13.9a), conductances in series are combined as reciprocals to yield a total conductance, C_t:

$$C_t = \frac{C_a C_m}{C_a + C_m}. \tag{1.13.9b}$$

Any number of these series vacuum conductances may be combined as reciprocals, as in (1.13.9a), to yield a total equivalent conductance, C_t, between the chamber and pump. Parallel conductances are combined by simple addition of the various component conductances.

The total conductance, C_t, for *air* may be calculated by using (1.12.3) and (1.12.4) to determine C_a and C_m, respectively. Then, the total speed delivered to the chamber, S_c, is given by:

$$\bowtie S_c = \frac{S_p C_t}{S_p + C_t}. \tag{1.13.10}$$

If conductances are required for other gases, or air at some other temperature, then equations equivalent to (1.12.8) and (1.12.9) may be used to modify C_t, the total conductance for air. It is not necessary to calculate each of the conductances for the new gas or temperature, and then combine the results.

1.14 Dalton's Law and Linear Superposition

Dalton's Law states that the partial pressures of gases in a mixture behave individually according to the ideal gas laws. For example, assume a vessel of volume V_1 contains partial pressures of argon and helium of pressures P_A and P_{He}, respectively. Assume a second vessel of volume V_2 is isolated from the first by means of a valve, and that it contains no gas (this is sometimes referred as being *under vacuum*). The total initial pressure in V_1 is simply $P_{i,A} + P_{i,He}$. Were we to open the valve between the two vessels, the amount of argon would remain the same (*i.e.*, $n\Re T$ before and after). The argon in V_1 would expand into V_2, and the final argon pressure, $P_{f,A}$, according to (1.4.1b) would be:

$$P_{f,A} = P_{i,A} \frac{V_1}{V_1 + V_2}.$$

Similarly, $\quad P_{f,He} = P_{i,He} \dfrac{V_1}{V_1 + V_2}.$

The final pressure is merely the sum of the individual partial pressures of helium and argon (*i.e.*, their *linear superposition*). Linear superposition theory applies to both electrical and vacuum *circuits*. For example, assume there is a time varying voltage source, $V_{e1}(t)$, in the circuit in Fig. 1.14.1a. The current in the circuit, $I_1(t)$, will vary as shown in Fig. 1.14.1b. Or,

$$I_1(t) = G \times V_{e1}(t). \tag{1.14.1}$$

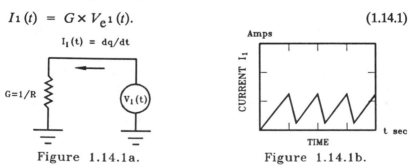

$$I_1(t) = dq/dt$$

$$G = 1/R \qquad v_1(t)$$

Figure 1.14.1a.   Figure 1.14.1b.

Similarly, we may solve for current, $I_2(t)$, as a function of a second voltage source, $V_{e2}(t)$, in the circuit of Fig. 1.14.2a (*i.e.*, the same conductance):

$$I_2(t) = G \times V_{e2}(t). \tag{1.14.2}$$

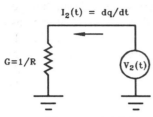

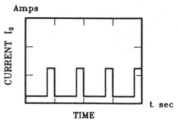

Figure 1.14.2a.   Figure 1.14.2b.

Were we to install both voltage sources in the same *linear* circuit, the principle of linear superposition implies that all we must do to find the total resultant current, $I_t(t)$, is add each of the individual current components given by (1.14.1) and (1.14.2). That is:

$$I_t(t) = I_1(t) + I_2(t) \tag{1.14.3}$$

$$= G \times V_{e1}(t) + G \times V_{e2}(t).$$

Or, $I_t(t) = G \times [V_{e1}(t) + V_{e2}(t)].$ \hfill (1.14.4)

This total resultant current as a function of time, $I_t(t)$, is depicted in Fig. 1.14.3b. Note that by the manipulation of (1.14.4), we may express the total voltage imposed on the circuits as the algebraic sum of V_{e1} and V_{e2}.

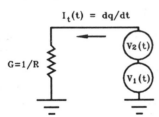

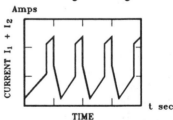

Figure 1.14.3a.   Figure 1.14.3b.

The correspondence in a linear vacuum *circuit* might relate to the following cases. Case #1: There are several sources of different gases, but each gas originates from one location in the vacuum system (*e.g.*, from within a chamber such as shown in Fig. 1.13.3). Case #2: There may be several sources of the same type of gas, but from different locations in the system. Case #3: Combinations of Cases #1 and #2. The first two cases will be discussed. Case #3 is part of the problem set.

CASE #1, Different Gases and Sources:

Assume that there is an air leak, Q_{air}, in a flange seal appending the vacuum chamber shown in Fig. 1.13.3. Also, assume that some source within the chamber has an H_2 outgassing rate, Q_{H_2}. What is the total chamber pressure P_c and P_p, the pump pressure? Assuming we know the pump speed for these two gases, what are the partial pressures of each of the gas species at both chamber and pump? Let's first solve the problem for the air leak. Then we will solve the problem for the H_2 outgassing. Using (1.12.3), (1.12.4) and (1.13.9b), we can calculate the total conductance for air, $C_{t,air}$, from chamber to pump. Knowing this conductance, we can then use (1.13.10) to calculate the equivalent speed for air, $S_{c,air}$, delivered to the chamber. We can then calculate the partial pressure of air in the chamber, $P_{c,air}$ accordingly:

$$Q_{air} = P_{c,air} S_{c,air}$$

$$Q_{air} = P_{c,air} \frac{S_{p,air} C_{t,air}}{S_{p,air} + C_{t,air}}. \tag{1.14.5}$$

The speed of all pumps varies with the type of gas being pumped. Therefore, (1.14.5) specifically refers to both the pump speed and total conductance for air. A similar equation may be developed for the hydrogen outgassing in the chamber:

$$Q_{c,H_2} = P_{c,H_2} S_{c,H_2}$$

$$Q_{c,H_2} = P_{c,H_2} \frac{S_{p,H_2} C_{t,H_2}}{S_{p,H_2} + C_{t,H_2}}. \tag{1.14.6}$$

Because the gas is now hydrogen rather than air, we compensated for both the conductance and pumping speed of H_2 in (1.12.6). We can solve for P_{total} by the algebraic manipulation of (1.14.5) and (1.14.6) so as to add the partial pressures of the two gases (i.e., P_{c,H_2} and $P_{c,air}$). We did the same to determine the sum of V_{e1} and V_{e2} in (1.14.4). In this case, the total pressure in the chamber (analogous to total voltage) is simply:

$$P_{c,total} = P_{c,air} + P_{c,H_2}$$

$$= \frac{Q_{c,air}}{S_{c,air}} + \frac{Q_{c,H_2}}{S_{c,H_2}}. \tag{1.14.7}$$

Using the conservation of mass law, it is a simple matter to calculate the partial pressures of air and hydrogen at the pump. That is:

$$Q_{c,air} = Q_{p,air}$$

$$P_{c,air} \times S_{c,air} = P_{p,air} \times S_{p,air}. \tag{1.14.8}$$

Also, $Q_{c,H_2} = Q_{p,H_2}$

$$= P_{p,H_2} \times S_{p,H_2}. \qquad (1.14.9)$$

The manipulation of (1.14.8) and (1.14.9) enables one to determine the total pressure at the pump.

$$P_{pump} = P_{p,air} + P_{p,H_2}$$

$$= \frac{Q_{air}}{S_{p,air}} + \frac{Q_{H_2}}{S_{p,H_2}}$$

CASE #2, Outgassing Manifold and Chamber:

It can be shown, using slightly more advanced linear circuit theory,[16] that the pressure as a function of length along a long, uniformly outgassing cylindrical manifold, such as shown in Fig. 1.14.4, is given by (1.14.10).

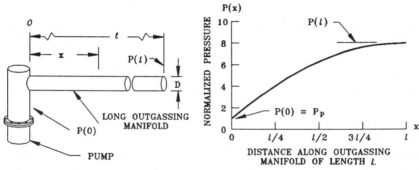

Figure 1.14.4. Pressure profile in a long, uniformly outgassing manifold.

$$P(x) = P_p + \frac{\pi q}{2kD^2} [2x\ell - x^2], \qquad (1.14.10)$$

where P_p = the pressure at the pump,
   x = the distance from the pump in cm,
   D = the diameter of the manifold in cm,
   ℓ = the length of the manifold in cm,
   q = the outgassing rate Torr-$\mathcal{L}$/sec-cm^2,
and k = 12.1 (i.e., $f((T/M)^{1/2})$ in (1.12.4)).

Returning to Fig. 1.13.3, assume that the total chamber outgassing is Q_c and that the outgassing rate of the manifold between chamber and pump is q. Assume that the gas, in both cases, is air. The total outgassing rate of the manifold, Q_m, is merely the inner area of the manifold, A, multiplied by q, or:

$$Q_m = [A] q$$

$$Q_m = [\pi \times D \times \ell] q.$$

Therefore, the pump pressure is:

$$P_p = [Q_c + Q_m]/S_p.$$

The pressure in the chamber due to outgassing of the manifold, $P_m(\ell)$, is found by calculating the pressure at $x = \ell$ in (1.14.10).

$$P_m(\ell) = P_p + \frac{\pi q}{2kD^2} [\ell^2] \qquad (1.14.11)$$

The pressure in the chamber due to outgassing in the chamber, Q_c, is:

$$P_c = Q_c \frac{S_p C_t}{S_p + C_t}. \qquad (1.14.12)$$

The total pressure in the chamber due to both chamber and manifold outgassing, P_t, is just the linear superposition of (1.14.11) and (1.14.12). Or,

$$P_t = P_m(\ell) + P_c.$$

1.15 Selective and Variable Pumping Speed

It is noted in (1.14.5) and (1.14.6) that pump speeds vary depending on the type of gas. This is true for all pumps. Much more will later be said about this as it relates to specific types of capture pumps. At high pressures, the speed of all high vacuum pumps (*i.e.*, capture pumps, momentum transfer pumps, *etc.*) deteriorates for reasons unique to the type of pump. For example, the oil in the jets of a diffusion pump *stack* may start to randomize; the rotor on a turbomolecular pump may start to slow down; the thermal capacity of a cryopump may be exceeded, *etc.*

The *mechanical model* of pump speed stressed that the speed of a pump was treated as being independent of the pump pressure. However, we know that all pumps have both high and low applicable pressure limits. For example, even the best mechanical pump has a *base* or *blank-off* pressure of $\sim 10^{-4}$ Torr. The *base* or *blank-off* pressure of a pump is the minimum pressure which can be achieved when the pump is valved off at the input port.

All pumps have a minimum blank-off pressure (irrespective of many of the published data sheets). Also, this blank-off pressure may vary with the amount and type of gas which has been pumped. Because of this, we must modify our mechanical model of pump speed to compensate for these effects. The high, operating pressure limitations of different capture pumps will be treated in their respective sections. The low pressure limitation of any pump is a bit simpler to model.

Assume that you have measured the blank-off pressure of some pump. The implication of a blank-off pressure, $P_\ominus$, is that the usable speed at this pressure is *zero*. Speed of the pump as a function of P, the operating pressure, is given by the following equation:

$$\text{и} \quad S = S_{\max} \left\{ 1 - \frac{P_\ominus}{P} \right\}, \qquad\qquad (1.15.1)$$

where $S_{\max}$ is the maximum inherent speed of the pump. This maximum speed limitation primarily stems from geometrical or mechanical limitations of the pump. However, it will also depend on the condition of the pump. With capture pumps, $P_\ominus$ is critically dependent on the amount and type of gas which has been pumped. However, (1.15.1), an example of which is given in Fig. 1.15.1, is valid for all types of vacuum pumps.

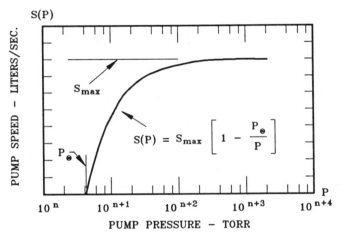

Figure 1.15.1. Pump speed as a function of base pressure.

1.16 Measuring Pump Speed

Perhaps a half dozen methods exist for determining the speed of a pump. Four of these methods are now common practice in industry, and are reported in the literature. They include 1) Rate of Pumpdown; 2) Single Dome; 3) Three Gauge Dome; and, 4) Fischer-Mommsen Dome methods. Each method has certain advantages and limitations. All of these methods of measuring pump speed require the use of calibrated vacuum gauges for the applicable test gas.

1.16.1 Rate of Pumpdown

By monitoring the rate a pump evacuates a vessel (sometimes called the *pumpdown rate*), we attempt to deduce the speed of the pump in question. Assume a vessel of known volume V is attached to a pump with a valve and manifolding having a total conductance C_t. Assume that there is a gauge attached to the vessel, and the indicated pressure of this gauge is $P(t)$. The equivalent speed delivered to the vessel is S_{eq}. Figure 1.16.1 represents such a system.

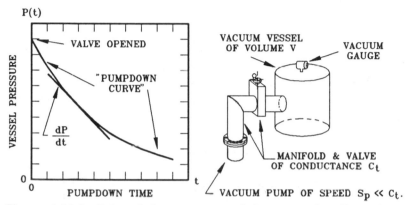

Figure 1.16.1. Determining pump speed by measuring the rate at which pressure decreases in a vacuum vessel of known volume.

This method of making speed measurements is usually used to measure the speed of roughing pumps, rather than UHV pumps. Because of this, calculating C_t, as the pressure "passes" through the various pressure regions, becomes very difficult. That is, C is a function of geometry, pressure and time. It is difficult to compensate for $C(P,x,t)$. Therefore, the test apparatus is usually designed so that $C_t \gg S$. Then, from (1.13.10), we may make the approximation $S \sim S_{eq}$. During pumpdown, the rate at which gas (*i.e.*, $P(t) \times V$) is removed from the vessel, Q_v, is:

$$Q_v = \frac{d(VP)}{dt} = P\frac{dV}{dt} + V\frac{dP(t)}{dt}. \tag{1.16.1}$$

The volume of the vessel is a constant (*i.e.*, $dV/dt = 0$). Therefore,

$$Q_v = V\frac{dP(t)}{dt}. \tag{1.16.2}$$

If we neglect outgassing from the manifold, the gas which is being evacuated from the vessel, Q_v, is the same gas which is entering the pump (*i.e.*, $S \times P_p$).

With $C_t \gg S$, then $P \sim P_p$, and:

$$Q_v \sim S P(t). \tag{1.16.3}$$

Setting (1.16.2) equal to (1.16.3) and integrating from the time the valve is opened to some time, t, in the test yields:

$$\int_0^t (S/V)\, dt = \int_{Po}^{P(t)} dP(t) / P(t),$$

¤ or, $P(t) = Po\ e^{-(V/S)t}, \tag{1.16.4}$

where Po is the pressure in the vessel at $t = 0$. Equation (1.16.4) is called the system *pumpdown equation*. Taking the natural log of both sides, and solving for S, we find:

$$\text{¤}\quad S(t) = \frac{V}{t}\ \ln \frac{Po}{P(t)}. \tag{1.16.5}$$

A plot of pressure, $P(t)$, as a function of time is called a *pumpdown curve*. We may select any Po and $P(t)$ along this curve to determine $S(P)$, pump speed, as a function of pressure.

One assumption made in solving for (1.16.4) was that pump speed was independent of time (*i.e.*, pump pressure). This is not the case, for it was noted in (1.15.1) of the previous section that speed is a function of pressure, particularly when operating the pump near its blank-off pressure. If speed is a function of pressure, and pressure is a function of time, then speed must be a function of time. Therefore, we may not treat S as a constant, as was done in development of (1.16.4). Using (1.15.1), where $P_\odot$ is the blank-off pressure of the pump, the equivalent of combining (1.16.2) and (1.16.3) becomes:

$$S_{max}\left\{ 1 - \frac{P_\odot}{P(t)} \right\} P(t) = V \frac{dP(t)}{dt}.$$

When rearranging this equation, integrating, and solving for $P(t)$, we find:

$$P(t) = P_\odot + Po\ e^{-(S/V)t}. \tag{1.16.6}$$

Another, far more serious, limitation of this pumpdown-rate method of speed measurement is that it neglects the effects of outgassing in the vessel, o-rings and manifolding. Outgassing rates in the vessel, $Q_0(P,t)$, may stem from several sources which vary both as a function of time and pressure. If we could predict these, we could modify (1.16.2) and (1.16.3) accordingly:

$$S_{max} \left\{ 1 - \frac{P_\odot}{P(t)} \right\} P(t) = V \frac{dP(t)}{dt} + \sum_{i=1}^{k} Q_i(P,t), \qquad (1.16.7)$$

where $\Sigma Q_i(P,t)$ is the sum of "k" outgassing sources. Outgassing of the chamber will result in very large errors at reduced pressures. Solution of (1.16.7), for some specific sources, is given as a problem at the end of this chapter. The outgassing problem is somewhat remedied by baking the system and using pure N_2 for the test gas.

We may also use a simple graphical method for determining pump speed. Assume the chamber shown in Fig. 1.16.1 has a 100 $\mathcal{L}$ volume, has negligible outgassing and has been filled with N_2 to a pressure $P_0 = 100$ Torr. Assume that we wish to test the speed of a mechanical pump with a blank-off pressure $P_\odot$ of 10^{-4} Torr. Assume we measure an average vessel pumpdown curve, for the pump under test, similar to that shown in Fig. 1.16.2.

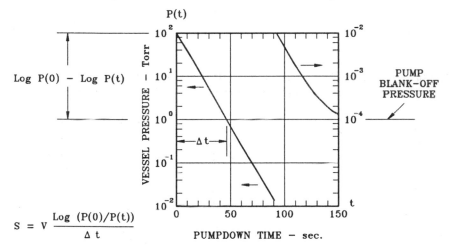

Figure 1.16.2. Graphical determination of pump speed using the rate at which pressure decreases in a known volume.

Setting (1.16.2) equal to (1.16.3) we obtain:

$$S(P,t) = \frac{V}{P(t)} \times \frac{dP(t)}{dt}. \qquad (1.16.8)$$

Of course, $dP(t)/dt$ is merely the slope of the pumpdown equation at some chosen $P(t)$. By graphically differentiating this curve at say $P = 1.0$ Torr, we find that $\Delta P(t)/\Delta t \sim 0.1$ Torr/sec. Using these values of P, $\Delta P(t)/\Delta t$ and V in (1.16.8) we determine the pump speed is approximately 10 $\mathcal{L}$/sec.

Note that we used the same gauge to determine both ΔP and P in (1.16.8). If there is a calibration error in the gauge reading, a gauge correction constant for both ΔP and P would cancel in (1.16.8), assuming gauge linearity.

Therefore, though we know the absolute speed of the pump at the indicated pressure, we do not know the absolute pressure at that reading. In a plot similar to Fig. 1.15.1, there would be some uncertainty in the placement of $S(P)$ along the P-axis.

The response times of the gauge and gauge electronics may cause errors in speed measurements. Also, problems of gauge outgassing will obscure the data if tubular ionization gauges are used to determine vessel pressure during a high vacuum pumpdown measurement. An example of this will be given in the chapter on cryopumping.

1.16.2 Single Gauge Dome Method

This method, requiring an apparatus similar to that illustrated in Fig. 1.16.3, has been used for many decades to measure the pumping speed of diffusion pumps.[17] Gas is introduced into the dome, at a known rate, through the variable leak, V_L. On entry into the dome, the gas is directed by a small internal pipe so as to *bounce* off the top of the dome. This is done to minimize gas beaming effects within the dome.

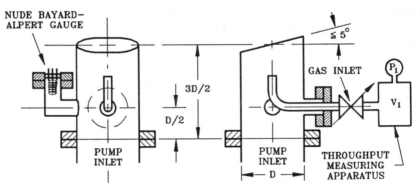

Figure 1.16.3. Single gauge dome, speed measuring apparatus.

The top of the dome is sloped so that oil which backstreams from the diffusion pump, when condensing on this surface, will run down the sides of the dome rather than drop down onto the pumping stack and cause erratic pressure bursts.

The rate at which gas is introduced into the dome, Q_d, must be measured by some means. This is often done by measuring $\Delta P / \Delta t$ of a known volume, V_1, filled with gas at a known pressure, $P_1 (t)$. The throughput into the dome is given by:

$$Q_d = V_1 \frac{dP_1}{dt}. \tag{1.16.9}$$

It is assumed that the pressure in the dome, P_d, is the pump pressure, P_p. Therefore, the speed of the pump is:

$$S(P) = \frac{V_1}{P_d} \times \frac{dP_1}{dt}. \tag{1.16.10}$$

Note that both of the gauges must be calibrated for the test gas, and volume V_1 must be known, in order to accurately determine the pump speed. A stopwatch is usually used to determine $\Delta P_1/\Delta t$. Use of a strip chart recorder to plot $P_1 (t)$ will make possible the graphical differentiation of this value with respect to time.

1.16.3 Three Gauge Method

The three gauge method of making speed measurements requires the use of an apparatus similar to that shown in Fig. 1.16.4. This technique came into prominence in the early 1960s, at which time it was used to make speed measurements on sputter-ion pumps.[18,19] Three vacuum gauges are required.

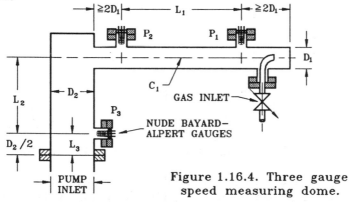

Figure 1.16.4. Three gauge speed measuring dome.

Pump throughput, Q_p, is determined by measuring the pressure difference along tube L_1, of calculated conductance C_1. This manifold is a *long tube*. Therefore, (1.12.4) is used to determine C_1. Assuming air as the test gas, throughput is then:

$$Q_p = C_1 [P_1 - P_2] \tag{1.16.11}$$

$$= 12.1 \frac{(D_1)^3}{L_1} [P_1 - P_2].$$

The speed of the pump is found by setting $Q_p = S_p P_p$. We can't directly measure the pump pressure. But, we can calculate this value. Note that manifold L_3 is also a portion of a *long tube*. This means that the portion of its conductance from the gauge to the pump flange is given by:

$$C_3 = 12.1 \frac{(D_2)^3}{L_3}. \qquad (1.16.12)$$

The throughput into the pump is also given by the following:

$$Q_p = C_3 [P_3 - P_p]. \qquad (1.16.13)$$

Setting (1.16.13) equal to (1.16.11) and solving for P_p, we find:

$$P_p = P_3 + \frac{C_1}{C_3} [P_1 - P_2]. \qquad (1.16.14)$$

Of course, $S_p P_p = C_1 [P_1 - P_2].$ \qquad (1.16.15)

Solving (1.16.15) for S_p, and substituting (1.16.14) for P_p, yields the following expression for S_p:

$$S_p = \frac{C_3 C_1 [P_1 - P_2]}{C_3 P_3 - C_1 [P_1 - P_2]}. \qquad (1.16.16)$$

There are numerous problems associated with using this method of speed measurement. It is advisable to bake out the system. This is true for most speed measuring apparatus. However, the long tubes in this configuration add significantly to the error in speed measurement (see problem at end of chapter).

Secondly, the three vacuum gauges must be calibrated to precisely plot S_p as a function of P_p. If all the gauges were "normalized" with respect to each other, the magnitude of S_p could be precisely determined, though the S_p curve would be translated by some error along the P–axis. Gauge normalization is required as usually the indicated pressure, P_{ni}, of any of the n gauges will not agree, though all are at the same absolute pressure, P_{abs}. That is:

$$P_{abs} \neq P_{1i} \neq P_{2i} \neq P_{3i}. \qquad (1.16.17)$$

Gauge normalization is accomplished by shutting off the pump and back-filling the test apparatus with the test gas. Normalization constants, k_2 and

k_3 for example, are then found for two of the three gauges, by requiring that at constant pressure:

$$P_1 i = k_2 P_2 i = k_3 P_3 i.$$

Remember, these are the indicated gas pressures. Most commercially available Bayard-Alpert Gauges (BAG) will be accurate to within $\pm$ 20 % for N_2 or air. However, note that there are significant differences in ionization gauge sensitivities for the different gases.[21,22]

A problem at the end of this chapter asks the student to calculate the errors in a hydrogen pump speed measurement which would stem from a known trace of argon in the test gas. Speed measurement errors will be exacerbated by the presence of trace contaminant gases, as apparatus conductance calculations differ for various gases, as well as gauge calibration.

1.16.4 Fischer-Mommsen Dome

The Fischer-Mommsen Dome method of speed measurement requires the use of an apparatus similar to that shown in Fig. 1.16.5. It is sometimes called the two gauge method. It was developed at CERN[22] for making speed measurements on sputter-ion pumps. For this reason, it is often called the *CERN* method. The apparatus has been slightly modified in subsequent AVS and ISO standards.[23,24] In the late 1970s it also became a popular apparatus for making speed measurements on cryopumps. Data reported in reference (25) were taken using a modified CERN Dome.

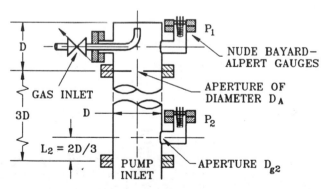

Figure 1.16.5. Fischer–Mommsen speed measuring dome.[22]

The geometry of the dome was determined through *Monte Carlo* calculations. This amounts to using a computer to theoretically predict the random sojourn of numerous molecules launched into the system at the input port. These calculations predict the proper location and diameter (i.e., L_2 and D_{g2}) of the aperture leading to the vacuum gauge indicating pressure P_2. Theoretically, the indicated value of the pressure reading P_2 is equivalent to the pressure P_p at the pump flange. That is, we need not take into account, in speed calculations, that the gauge is a distance equivalent to length L_2 in Fig. 1.16.5. There is some error in this assumption depending on the *capture coefficient* of the pump.[22] The pump capture coefficient is merely the probability that a molecule on entering the pump inlet will be captured (*i.e.*, will be pumped).

The diameter of the aperture separating the upper and lower chambers of the dome, D_a, is selected to maintain the pressure difference between the two chambers to prescribed values. Therefore, some pretest estimate of the probable pump speed is needed. If the size of this aperture is too small, the pressure difference across the aperture will result in P_1 being much greater than P_2. The throughput introduced into the dome is simply the product of the pressure difference between the two chambers and the conductance C_a. Or,

$$Q_p = C_a [P_1 - P_2]. \qquad (1.16.18)$$

In that P_p corresponds P_2, the pump speed is simply:

$$S_p = C_a \frac{[P_1 - P_2]}{P_2}. \qquad (1.16.19)$$

Note, as in the case of the three gauge method, the two gauges used in this apparatus must be normalized with respect to each other. This becomes a nuisance when making speed measurements over pressures of several orders in magnitude, and with different gases.

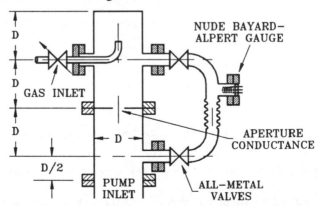

Figure 1.16.6. Modified CERN speed measuring dome.

At Varian, in 1976, I modified the conventional CERN configuration shown in Fig. 1.16.5, and installed two all-metal valves at the locations of the gauge ports (*i.e.*, see Fig. 1.16.6). These two valves were connected to a single gauge which was used, by manipulation of the valves, to determine both of the indicated pressures P_1 and P_2.

Assuming gauge linearity with pressure, use of an apparatus similar to that shown in Fig. 1.16.6 eliminates the need of determining gauge normalization constants. However, the gauge must still be calibrated for the test gases to accurately locate the *speed curve* on the *S-P* coordinate system.

Three Gauge versus Fischer-Mommsen Results

Over the years, I've found that the three gauge method will yield indicated pump speeds which are 10% - 15% higher than results of the Fischer-Mommsen apparatus, for the same pump. Similar disparities are reported elsewhere.[26] This is true for all of the gases.

Also, for reasons which are discussed in the chapter on cryopumps, the theoretical speed of a cryopump for H_2O should be within a few percent of actual measured speeds. However, cryopump speed tests for H_2O, with the Fischer-Mommsen method, will yield results which are approximately 20% lower than the theoretically predicted speed values. I quickly emphasize that, because of apparatus wall pumping effects, making H_2O speed measurements is very difficult and subject to errors.

In the final analysis, as long as the method of speed measurement is defined when reporting results, the departure of this measuring equipment from absolute speed is of little consequence. The Modified CERN Dome method of speed measurement is far more convenient to use than any other method of which I am aware. Test results also prove to be very repeatable.

1.17 System Diagnostics with Any Pump

There is a formal method in the use of logic whereby if: 1) we claim an assumption to be true; 2) we thereafter test the assumption by making measurements or calculations which are known to be exact or true; and then, 3) we reach a conclusion which is a contradiction to the original assumption, then the original assumption is false.[27]

For example, assume that we make a claim that the average mass of each of ten apples in a bowl is greater than 0.1 kg (the assumption). Thereafter, we precisely weigh each of the ten apples (the measurement), and find when we sum their masses and divide by ten (the calculation), that the average mass is 0.08 kg. We have reached a contradiction (*i.e.*, 1.0 kg $\neq$ 0.8 kg), and our original assumption was disproven.

Similar use of logic may be used to *troubleshoot* a pump attached to a vacuum system. Assume that a problem exists with a production sputtering system similar to that depicted in Fig. 1.17.1. Say that the pressure in the vacuum chamber is a factor of ten (10) higher today than it was yesterday. We do not keep log books on each piece of equipment in the factory. But, we should! The operators observed that the chamber base pressure today is 2×10^{-6} Torr, where yesterday, they recall it was 2×10^{-7} Torr.

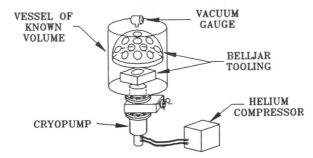

Figure 1.17.1. Coating system with problems in base pressure.

Assume that last night the production engineers put new tooling in the chamber, and as part of a scheduled maintenance program, the maintenance personnel installed a refurbished compressor on the cryopump. The old compressor was due for an adsorber change. Today, because of the pressure problem, the maintenance people leak checked the system and found no leaks! Also, the temperature of the second stage of the cryopump was ~10° K (*i.e.*, it was "OK"). The maintenance people claim: "The tooling is dirty or has a virtual leak!" The production people claim: "The tooling is OK, but the maintenance people screwed up the pump!"

Rather than become embroiled in these sometimes emotional confrontations, we will resort to a formal logical process in an attempt to locate the problem. We will start with the assumption that: 1) there is nothing wrong with the pump; 2) we will then make measurements and calculations based on this assumption; and, 3) if we reach a contradiction, the original assumption is incorrect. Remember that in UHV-type applications, we should expect errors in measurement accuracy to be typically as much as ± 20% of results predicted by theoretical calculations.

The manufacturer's data sheet indicates a pump speed, S_p, of 1000 $\mathcal{L}$/sec for air. We may make calculations for any gas in testing the claim. But, for now, we will assume that the gas is air. Assume we calculate the total conductance leading from chamber to pump to be $C_t = 1000\ \mathcal{L}$/sec. Using (1.13.10), we calculate the equivalent speed delivered to the chamber to be:

$$ S_c \quad = \quad \frac{S_p\,C_t}{S_p + C_t} \quad = \quad 500\ \mathcal{L}/\text{sec}. $$

Then, the total assumed throughput into the chamber, $Q_{t,A}$, and the conclusion drawn by the original assumption of the value of S_p, is given by:

$$Q_t = [500 \, \mathcal{L}/\text{sec}] \times [2 \times 10^{-6} \, \text{Torr}],$$

$$= 10^{-3} \, \text{Torr-}\mathcal{L}/\text{sec}.$$

Measuring Throughput by Rate-of-Pressure-Rise

We may use another method for directly measuring the throughput, $Q_{c,m}$, into a chamber of volume V_c. Assume, after pumping for an extended period of time, we close the valve between the pump and chamber. We can conclude something about the rate at which gas is coming into the chamber - either as the result of outgassing or because of leaks - by monitoring the pressure in the chamber, P_c, as a function of time. That is, using (1.16.2),

$$Q_{c,m} = V_c \frac{dP_c}{dt}. \qquad (1.17.1)$$

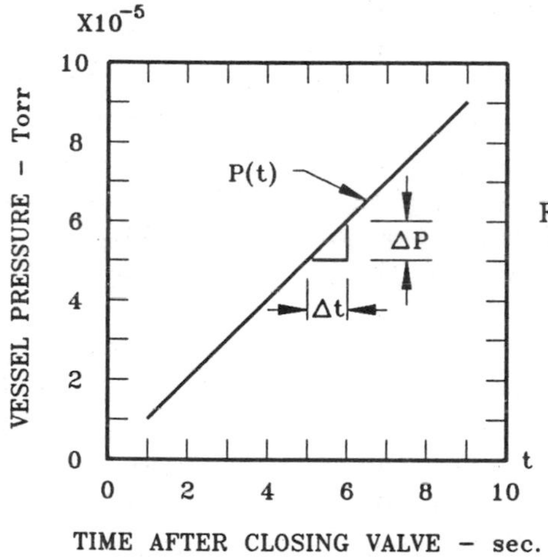

Figure 1.17.2. Determining the outgassing or leak rate into a vessel by measuring the rate−of−pressure−rise.

This method is called the *rate-of-pressure-rise* method of measuring throughput. Assume that $V_c = 1000 \, \mathcal{L}$, and if we plot $P_c(t)$ we obtain a curve such as shown in Fig. 1.17.2. Graphically differentiating the curve, we determine $\Delta P_c / \Delta t \sim 10^{-5}$ Torr/100 sec; or, 10^{-7} Torr/sec. Then,

$$\frac{Q_{c,A}}{Q_{c,m}} \approx \frac{P_c \dfrac{S_{c,A} C_c}{S_c + C_c}}{V_c \dfrac{\Delta P_c}{\Delta t}}, \qquad (1.17.2)$$

$$\approx \frac{10^{-3} \, \text{Torr-}\mathcal{L}/\text{sec}}{10^{-4} \, \text{Torr-}\mathcal{L}/\text{sec}} \neq 1.0 \pm 20\%$$

Or, $Q_{c,A} \neq Q_{c,m}.$ →←   □

Note that the same gauge was used to measure both P_c and $\Delta P_c/\Delta t$. There-fore, an error in gauge calibration in (1.17.2), for some unknown gas, may be excluded. We have made the assumption of a value for the pump speed, have tested the assumption, and therein reached a contradiction. Therefore, the original assumption was incorrect, and the pump is for some reason not delivering the anticipated speeds to the chamber. From this we conclude that either there is some obstruction between the pump and chamber, or the pump is defective.

Had we found agreement, to say within ±20% between $Q_{c,A}$ and $Q_{c,m}$, we could safely assume that the problem did not reside in the pump. If a leak could not be found in the system, the disparity between $Q_{c,A}$ and $Q_{c,m}$ probably is caused by some internal outgassing source. I've found that even the manufacturers of vacuum systems fail to use this simple technique in logic to troubleshoot vacuum systems *on the floor*. Also, it is best to keep formal log books of the history and performance of complicated vacuum systems.

1.18 Electrical Discharges in Gases

It is reasonable for you to ask about the possible relevance of gas ionization processes in a book on capture pumps. Of course, you know that sputter-ion pumps require electrical discharges to function, and some of the material in this section will be drawn on in the chapter dealing with these pumps. However, it is even helpful to know a little about electrical discharge pro-cesses when designing and using certain types of cryopumps. For example, closed-loop gaseous helium cryopumps have sealed compressors which provide high pressure, helium gas to the cryopump. This high pressure gas is conveyed through hoses to an *expander*, where refrigeration is produced. There are electrical motors in both the sealed compressor and in the expander. If there is a leak in the high pressure He system, the pressure may drop below some value at which electrical breakdown will occur. This electrical arcing could cause irreparable damage to the sealed compressor, or damage the expander motor.

Another problem of electrical breakdown has been encountered by many of us as we tried to leak check in proximity to a sputter-ion pump high voltage feedthrough while the pump was energized. The feedthrough assembly sputtered and arced when we introduced the He, but seemed to function properly in the presence of air. Why is this?

Electrical discharges may constitute the momentary sparking or *burn-off* of metallic *whiskers* in the presence of an electric field - in this case the metal atoms are ionized and form part of the gaseous discharge. With a robust power supply, large currents may be drawn through a sustained discharge.

Electrical discharges may be extremely weak and imperceptible to the eye, as in the case of vacuum ionization gauges operating at a low pressures. Or, an intensely glowing discharge - this is the origin of the expression *glow discharge* - may be visible to the eye, and occupy a large volume, as in the case of certain types of thin film sputtering operations, or in the starting of certain types of sputter-ion pumps. Our visual perception of the discharge stems from light which is given off as a consequence of electrons recombining with ions in the discharge, rather than the process of ionization.

Under certain circumstances when starting diode sputter-ion pumps, the intensely glowing discharge may fill the entire volume of the vacuum system. This can cause serious damage to sensitive electronic equipment contained within the system, or even promote the polymerization of hydrocarbons on insulators or once optically bright surfaces.

Such glow discharges can also cause safety problems. For example, several decades ago I constructed a small sputter-ion pump system to process microwave tubes. This was in the *early days* of ion pumps, and before I knew anything about the concepts of pump speed, throughput, *etc.* The tubes being processed had oxide cathodes which gave off copious quantities of CO and CO_2 during what was called *cathode conversion*. This excess gas caused an ion pump throughput problem (*i.e.*, the pump *swamped*). We installed an LN_2 (liquid nitrogen) trap in the system, with the hope that this would help with CO and CO_2 pumping. (The student will immediately refer to the vapor pressure curves for CO and CO_2 at LN_2 temperatures; not much help for CO, but beneficial for CO_2.) The LN_2 trap was vertically suspended in a *well* forming part of the vacuum chamber, and its flange rested on a slightly oversized o-ring. On roughing the vacuum system and attempting to start the ion pump, the volume was initially filled with an electrical discharge. Positive ions collected on the reservoir, with a charge build-up equivalent to the starting potential. Had I placed the probes of a voltmeter between the reservoir and ground, I would have measured the full potential used in starting the ion pump (*i.e.*, several hundred volts). The LN_2 reservoir, insulated by the oversized o-ring, developed the full potential. I inadvertently discovered the problem when refilling the trap. A shocking experience.

Lastly, we must use some form of gauge to measure pressure in our vacuum systems. Measuring pressures $\leq 10^{-6}$ Torr requires the use of ionization gauges. Such gauges are used to ionize gas. The ion current which is produced by the gauge is, over a very large pressure range, directly proportional to the pressure within the system. We will learn, in Chap. 2, that sputter-ion pumps are very similar to a specific type of ionization gauge.

It soon becomes evident that when designing or using many forms of capture pumps, having a better understanding of the physical processes involved in the ionization of gases will be very useful.

Ionizing Gases

When the outer electrons of atoms or molecules are *ripped* off the particle by some process, the particle is said to be *ionized*. If one electron is ripped off the particle, we refer to the ion as being *singly ionized*; if two electrons are removed, it is said to be *doubly ionized, etc.* The most common process of ionization in vacuum systems occurs when energetic electrons collide with gaseous molecules or atoms. If there are numerous such collisions of electrons within the gas, and there is some form of sustaining electric field, the volume in the vacuum system may become highly populated with both ions and electrons. Once an electron *parts company* with the molecule, it is said to be a *free* electron. A gas which is rich in free electrons and ions, in the presence of an electric field, conducts current and is called an electrical discharge. It is sometimes called a *plasma*. However, the word plasma is usually used to describe a volume containing the same number of electrons and ions of opposite charge. In this instance, on the average, it is electrically neutral.

High Pressure Electrical Discharges

In electrical discharges there is a transient state, where the discharge is initiated by a single electron. An avalanche process occurs, such as illustrated in Fig 1.18.1, where an initial electron, by ionizing collisions, creates additional free electrons, etc.

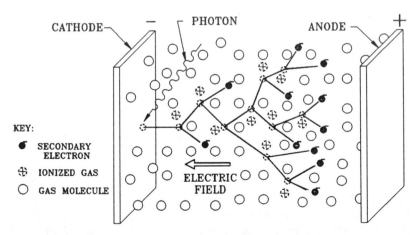

Figure 1.18.1. Townsend discharge initiated by a photoelectron in the presence of an electric field.

The avalanche process is called a *Townsend Discharge*. The first electron may have been launched into the vacuum as a consequence of field emission from some very sharp, metallic, whisker-like point, with extremely high electric fields.[28] The electron may have been thermally emitted from some type of hot filament or cathode;[29] or, it may have been dislodged from a surface by a photon or some cosmic particle which penetrates the walls of the vacuum envelope. Also, technologists have used radioactive sources, which are *beta* or *alpha emitters* (*i.e.*, by radioactive decay, they emit either electrons, positrons or alpha particles), to initiate discharges.[30,31]

If, for example, a photon dislodges an electron from the plate to which the negative potential was applied, the electron would be accelerated across the gap by the electric field, and impinge on the opposite plate. An electron dislodged from a surface by a photon is called a *photoelectron*. If, as it traverses the distance between the plates, the electron comes very close to a gas molecule, and if it has accumulated sufficient energy in the acceleration process, it might ionize the gas molecule. The gas molecule - now a positive ion - would be accelerated toward the negative plate (called the cathode). But, now there are two free electrons being accelerated toward the positive plate (called the anode), the initial photoelectron, and the electron stemming from the ionization of the gas molecule. Each probably have different energies, but both are picking up velocity as they progress toward the anode. The distance the initial electron traveled prior to striking the gas molecule is simply λ_{eg}, where the subscripts denote the mean free path of an electron, "e", in a particular gas, "g". Because of ionizing collisions with gas molecules, the electron current, j_e, progressively increases closer to the anode.

This current may be calculated in the following manner.[32] Assume that there are n electrons entering the back of the imaginary 1 cm^2 surface shown in Fig. 1.18.2.

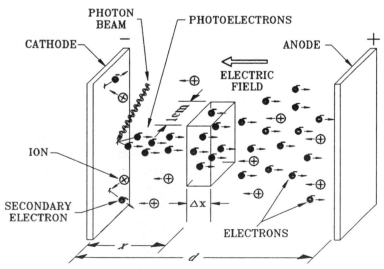

Figure 1.18.2. Electron and ion currents in an electrical discharge initiated by secondary and photoelectrons.

Assume that this square is Δx cm thick. Define α as the number of ionizing collisions which occur as n electrons travel the distance Δx. We know that α will be inversely proportional to the mean free path of the electron in the gas, and that it will vary as some function of the electric field, $\mathbf{E}$, between the plates. That is:

$$\alpha = \frac{\varphi(\mathbf{E})}{\lambda_{eg}}. \tag{1.18.1}$$

Then, the number of free electrons produced, Δn (dn), for a given distance, Δx (dx), is simply:

$$dn = \alpha\, n\, dx.$$

Or, $n = c\, e^{\alpha x}$.

The constant of integration c is found by setting $x = 0$; that is, it is the number of photoelectrons initially emitted from the cathode, n_0, each cm^2. If q_e is defined as the charge of an electron, then the product $n_0 \times q_e$ is simply j_0, the photoelectron current density drawn from the cathode. That is, the current density anywhere between the two plates is:

$$j = j_0\, e^{\alpha x}. \tag{1.18.2}$$

Equation (1.18.2) assumes that the electric field is uniform between the two plates (i.e., there are no regions of field depression due to space charge). For every free electron created in the space between the two plates, a positively charged ion is created (i.e., we assume gas is singly ionized). These ions are accelerated toward, and bombard the cathode. When charged particles bombard a surface, they will dislodge and create free electrons.[33,34] The electrons which are knocked loose from a surface, as a consequence of the bombardment of *primary* ions or electrons, are called *secondary electrons*. The ratio of the number of secondary electrons produced for a given number of primary particles bombarding a surface is called the *secondary emission coefficient*, and is usually given the symbol γ in the literature. Define the following:

n_s = the number of secondary electrons created / sec-cm^2 of cathode,
n_0 = the number of photoelectrons emitted / sec-cm^2 of the cathode,
n_c = the total number of electrons emitted / sec-cm^2,
n = the number of electrons reaching the anode / sec-cm^2.

By the above definitions, clearly,

$$n_c = n_0 + n_s. \tag{1.18.3}$$

The number of free electrons produced between the two plates is simply (n − n_c). Also, for every free electron produced in the space between the two plates, there is an ion produced (assuming only singly ionized gas). Therefore, the number of secondary electrons produced by ion bombardment of the cathode is,

$$n_s \quad = \quad \gamma(n - n_c). \tag{1.18.4}$$

Substituting (1.18.4) into (1.18.3) yields,

$$n_c \quad = \quad n_o + \gamma(n - n_c). \tag{1.18.5}$$

Substituting (1.18.2) into (1.18.5), and assuming $e^{\alpha d} \gg 1$, the electron current reaching the anode is found to be:

$$j_e \quad = \quad j_o \frac{e^{\alpha d}}{1 - \gamma e^{\alpha d}}. \tag{1.18.6}$$

We see that the electron current becomes very large as the denominator of (1.18.6) approaches zero. In fact, electrical breakdown occurs as the quantity $(1 - \gamma e^{\alpha d})$ approaches zero, or,

$$e^{\alpha d} \quad = \quad 1/\gamma. \tag{1.18.7}$$

We have not yet clearly defined the mathematical expression for α. We merely indicated that it must be inversely proportional to λ_{eg} and be related to some function of the electric field, $\mathbf{E}$. With considerable insight into experimental evidence, Cobine made the following assumption,[3 5]

$$\alpha \quad = \quad A P e^{-BP/E}, \tag{1.18.8}$$

where P is the pressure, $\mathbf{E}$ the electric field, and A and B are some constants which we can later assign values to based on experimental evidence. If we assume the field between the two plates is uniform, we may express the field in terms of the voltage applied to the plates, V_s, divided by the distance between them (i.e., $\mathbf{E} = V_s / d$). Here the subscript "s" denotes the *sparking potential*, or the voltage at which breakdown occurs. Taking the natural log of (1.18.7), substituting (1.18.8) into this expression, and solving for V_s, yields:

$$V_s \quad = \quad \frac{B P d}{\ln [\{A P d\}/\{\ln (1/\gamma)\}]}. \tag{1.18.9}$$

Equation (1.18.9) is known as *Paschen's Law*. We noted at the beginning of this section that electrical breakdown is observed near sputter-ion pump high voltage feedthroughs in the presence of He, but not in the presence of air.

This leads us to believe that (1.18.9) may be different for different gases. Of course, this is the case. That is, the constants "A" and "B" vary depending on the gas species. Setting $\partial V_s / \partial (Pd) = 0$, to solve for the minimum breakdown voltage, we find,

$$(Pd)_{min} = \frac{e}{A} \ln \frac{1}{\gamma} \qquad (1.18.10a)$$

$$V_{s,min.} = \frac{e}{A} B \ln \frac{1}{\gamma}, \qquad (1.18.10b)$$

where $e \sim 2.72$. Experimental data for several gases are given in Table 1.18.1.[36] Using these data, we may solve for the values of "A" and "B" for different gases. These values may then be substituted into (1.18.9), and the breakdown voltage determined for different gases as a function of pressure and electrode spacing. This is left as an exercise at the end of this chapter. (Hint! The term "$\ln(1/\gamma)$" cancels when substituting general expressions for "A" and "B" in (1.18.9).)

Table 1.18.1. Minimum sparking potentials and pressure-distance product for different gas species. [36]

Gas	$V_{s,min}$ (volts)	$(Pd)_{min}$ (Torr-cm)
Air	327	0.567
A	137	0.9
H_2	273	1.15
He	156	4.0
CO_2	420	0.51
N_2	251	0.67
N_2O	418	0.5
O_2	450	0.7

The sparking potential will vary significantly depending on the shape of the electrodes. However, assume for the moment that high voltage feedthrough of a sputter-ion pump may be represented by parallel electrodes. Assume that when we were leak checking around the feedthrough of the sputter-ion pump we introduced approximately one atmosphere of He near the high voltage feedthrough. With these assumptions, and using the values of "A" and "B" for He and air in the equivalent of (1.18.9), we can solve for the ratio of the breakdown voltage for He vs. air. We find that,

$$\frac{V_{He}}{V_{air}} \approx 0.09. \hspace{3cm} (1.18.11)$$

Equation (1.18.9) is found to be valid for pressures greater than ten atmospheres. Almost all of the high pressure He compressors used in gaseous helium cryopumps, mentioned earlier in this section were originally produced for use in domestic air conditioners. They were intended for use with a type of Freon®, which proves to have very good electrical breakdown properties - four or five times better than air. It is now obvious why we must take great care not to operate these compressors at reduced He pressures. Also, we see in (1.18.11) that the high voltage feedthroughs of an ion pump will be able to stand off only about 10% of the voltage in the presence of He *vs.* air.

Low Pressure Ionization Processes

In (1.18.1) we noted that the number of ionizing collisions which occurred as a function of the distance an electron traveled in a gas was inversely proportional to the mean free path of the electron in the gas. If we assume that the gas molecules have a certain diameter, say a few Å, and are stationary compared with the electrons, and if we further assume that the electrons occupy only a point in space (*i.e.*, their diameter is *zero*), simple calculations indicate that,

$$\lambda_{eg} \approx 5.7 \lambda_g. \hspace{3cm} (1.18.12)$$

Equation (1.18.12) is an approximation for, on the average, the distance an electron will travel in a gas between each collision with a gas molecule. We cannot assume that when an electron collides with a gas molecule it will necessarily result in the molecule being ionized. The probability of an electron ionizing a gas molecule is inversely proportional to λ_{eg}. But, it also depends upon such variables as the energy of the ionizing electron, and the energy required to *tear loose* the outer electron from the gas molecule.

The probability of a gas molecule being ionized by an electron is termed the molecule's *ionization cross section*. The symbol σ is used in the literature to denote this property. A very comprehensive study of gas ionization cross sections was done by Rapp and Englander-Golden.[38] An energetic electron beam is launched into a chamber containing the test gas. It is probable that some gas molecules will lose more than one electron in collisions with primary electrons. Because of this, the gas is said to have a *total ionization cross section*, σ_T. This total ionization cross section is given by:

$$\sigma_T = \frac{I^+}{I^- N \ell}, \hspace{3cm} (1.18.13)$$

I^+ = the ion current produced by gas ionization,
I^- = the ionizing electron beam current,
N = the density of the gas,
ℓ = the distance the electron beam travels in the gas.

A test apparatus similar to that shown in Fig. 1.18.3 is used to make measurements of the total ionization cross section of different gases.[37] The test gas is introduced into the ionizing chamber to a pressure of $\approx 5 \times 10^{-5}$ Torr.

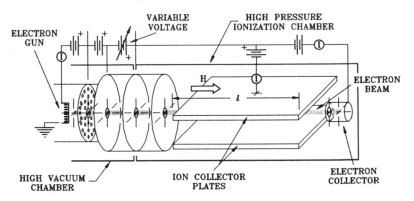

Figure 1.18.3. Test apparatus used to determine the total ionization cross sections of various gases as reported by Tate and Smith.[37]

A collimated electron beam of varying energies is launched into one end of the apparatus. The electron beam is strongly focussed by a magnetic field. The electrons travel a distance ℓ within a high pressure ionization chamber to the electron collector. Ions created by the electron beam as it travels the length of the chamber are collected on the negatively biased ion collector plates.

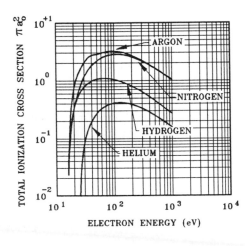

Figure 1.18.4. Total ionization cross sections as measured by Rapp and Englander–Golden.[38]

Figure 1.18.4 is a plot of typical results of such measurements for some of the more common gases.[38] In this figure, total ionization cross sections are plotted as a function of the electron beam voltage. It is a common convention to express σ_T in units of πa_0^2, where a_0 is the *first Bohr orbit radius* of the hydrogen atom (i.e., 5.29×10^{-9} cm).

Ionization Gauge Sensitivities

There are two forms of ionization gauges which are widely used. The first is the familiar Bayard-Alpert gauge (BAG); the second is the Penning ionization gauge (PIG; sometimes called a cold cathode ionization gauge). An outstanding reference on the physics of these and other types of gauges is given by Redhead, Hobson and Kornelsen.[39] As mentioned, these gauges are used to ionize gas, and deduce total pressure by the amount of ion current produced, as a function of pressure. If we rearrange (1.18.13) it will take a more familiar form:

$$I^+ = I^- (\sigma_T \ \ell) N. \tag{1.18.14}$$

Substituting (1.9.4) into (1.18.14) yields:

$$I^+ = I^- \{ (\sigma_T \ \ell) \frac{N_0}{\Re T} \} P \tag{1.18.15}$$

$$I^+ = I^- \{ s \} P. \tag{1.18.16}$$

The term inside the brackets of (1.18.15) is defined as the *sensitivity, s,* of the Bayard-Alpert gauge, and has the units of $Torr^{-1}$. The term I^- is called the *gauge emission*, and is the adjustable electron current emitted from the filament. Years ago, gauge sensitivity was defined as the product $I^- \times s$. However, this led to manufacturers of gauges proclaiming very high gauge sensitivities, when in fact all they did was increase the gauge emission for the same gauge geometry as their competitor.

Electrons are launched into the gauge structure of a BAG, where they orbit about the electrodes in such a manner that the equivalent length ℓ is very large. Therefore, the probability of ionizing gas is greatly increased, for a given pressure. The value of ℓ will vary depending on the geometry of the Bayard-Alpert gauge. The energy of the electrons accelerated from the cathode of a Bayard-Alpert gauge is usually fixed, and of the order of 200 volts. These gauges are usually calibrated to indicate the pressure, in Torr, for N_2. If we normalize Rapp and Englander-Golden's data to N_2, the relative sensitivity of a gauge for the various gases will be given by values in the following table. Data shown in the two far right columns are summaries of relative gauge sensitivities.[40,41] Work by Rapp and Englander-Golden are more current.

Table 1.18.2. Ionization cross sections of various gases normalized to the cross section of N2 and compared with relative vacuum gauge sensitivities.

GAS	100 eV	200 eV	300 eV	BAG	PIG
N2	1.00	1.00	1.00	1.00	1.00
H2	0.37	0.32	0.30	0.33	0.49
He	0.15	0.15	0.16	0.15	0.25
Ne	0.25	0.34	0.38	0.32	−
Ar	1.13	1.05	1.03	1.09	−
O2	1.06	1.12	1.14	1.10	0.95
CO	1.05	1.04	1.04	1.05	−
CO2	1.39	1.43	1.46	1.36	−
CH4	1.45	1.33	1.30	−	−
C2H4	2.29	−	−	−	−

Data taken from references 38, 40 and 41.

Penning ionization gauges are called *cold cathode gauges* because they do not have hot cathodes or filaments which emit ionizing electrons. There is a self-sustaining discharge in the gauge which is replenished by secondary electrons which are emitted as the result of ion bombardment of the cathode. In this case the expression for sensitivity, s, has the units Amps per Torr, and the ion current, I^+, is given by the following:

$$I^+ = s P. \tag{1.18.17}$$

We will discuss (1.18.17) at considerable length in the next chapter.

Problem Set

1) Construct a table of the room temperature root-mean-square velocities of H_2, He, Ne, H_2O, N_2, CO and CO_2.

2) What is the vapor pressure of He, H_2 and Ne at $10°$ K?

3) The vapor pressures of all other gases are less than or equal to what value at $20°$ K?

4) Assume that the inner surface of a box, with a height, width and length of 10 cm, is covered with one monolayer of gas, and that the pressure within the box is 10^{-8} Torr. What would be the pressure if the temperature of the box was $300°$ K and the surface gas was liberated into the volume of the box?

5) Prove the last sentence of Section 1.13. (Hint! A legitimate proof might be to first claim that the sentence is incorrect, and then reach a valid contradiction.)

6) What is the ratio of the thermal conductivity of H_2 vs. air at room temperature, at a pressure of 10^{-2} Torr, assuming equivalent values for γ and λ for the two gases?

7) Assume that the room temperature partial pressure of H_2O in your vacuum system is 10^{-6} Torr. Assume that there is a unity sticking coefficient for this gas on some surface within the system. How long would it take to build up one monolayer of H_2O on the surface? Under the same conditions, how long would it take if the gas was H_2? Xe?

8) Assume that each time the piston in Fig. 1.10.1 is fully withdrawn, the volume above the piston is back-filled to a pressure P. The piston sweeps out the volume at the rate of $\Delta V/\Delta t$. Construct a table of throughput vs. pressure for $P = 0$, 10^{-4}, 10^{-3}, 10^{-2} Torr, and values of $\Delta V/\Delta t = 1, 2, 100$ and 10^3 liters per second.

9) Construct a table of the minimum conductances of an aperture, with a diameter of 1.0 cm, for the gases H_2, He, H_2O, *air*, N_2, CO and CO_2 at room temperature.

10) Construct tables of the molecular conductances of a round manifold with diameter D, and length L, where $D = 5$, 10 and 20 cm and $L = 242$ cm, for the gases in the previous problem, but at $400°$ K.

11) Calculate the molecular flow conductance of a rectangular tube for room temperature air, where $a = 10$ cm, $b = 25$ cm and $L = 250$ cm. What is the conductance of a round tube under these same conditions, where the diameter of the tube is D and $\pi D^2/4 = ab$?

12) Assume the diameter of the manifold in Fig 1.13.3 is 15 cm and the length of the manifold between the chamber and pump flange is equivalent to 200 cm. Assume that H_2, H_2O and CO are outgassing from within the chamber because of some process, but not from other surfaces with the vacuum system. If the speed of the pump for H_2, H_2O and CO_2 is 2000, 4000, and 1200 $\mathcal{L}$/sec, respectively, what are the partial pressures of these gases at the pump flange and in the chamber, if the outgassing rate of each is 10^{-6} Torr-$\mathcal{L}$/sec?

13) The same conditions exist as in the previous problem. However, the manifold between the pump and chamber has an H_2O outgassing rate of $q = 10^{-9}$ Torr-$\mathcal{L}$/sec/cm^2. What would the chamber and pump pressures be for the three gases?

14) A long manifold of diameter D and length L has a roughing pump attached to one end and a sputter-ion pump attached to the other end. Most of the bulk gas has been evacuated from the manifold by the roughing pump. We want to eventually turn on the sputter-ion pump, but we must wait until the pressure at sputter-ion pump reaches a value of $P(L,t)$. The manifold has an outgassing rate of $q(t)$/cm^2. The roughing pump has a speed of S_r. Express the length of the manifold between the two pumps as a function of S_r, $q(t)$, $P(L,t)$ and D.

15) Assume that $\sum Q_i = Q_0 t^{-1} + Q_1 t^{-0.5}$ in (1.16.7), where Q_0 and Q_1 are constants. Solve (1.16.7) for the general expression of $P(t)$.

16) You are conducting H_2 speed measurements of a sputter-ion pump using a Fischer-Mommsen Dome. Assume that the speed of the pump for H_2 is 150 $\mathcal{L}$/sec and for Ar it is 10 $\mathcal{L}$/sec. The diameter of the dome aperture is 1.5 cm. Unknown to you, your bottle of H_2 has been contaminated with 1.0% Ar. What is the percent error in the speed measurements as a consequence of the presence of Ar? Remember to take into account the relative gauge sensitivities for H_2 and Ar.

17) You are calibrating a quadrupole gas analyzer. You have a bottle containing a high-pressure mixture of several different gases, each of known partial pressure. A variable leak valve is connected between the bottle and the upper chamber of the modified CERN Dome shown in Fig. 1.16.6. You have replaced the BAG, normally used with the dome, with your gas analyzer. Assume that $C_A \ll S_p$ for all the gases, and that the rate at which the gases pass through the variable leak is proportional to $(M_x)^{-1/2}$, where M_x is the molecular weight of each gas species. Prove that the ratio of the partial pressures of each of the gases introduced into the dome will be the same as their partial pressures in the bottle.

18) Solve for the values of "A" and "B" for each of the gases listed in Table 1.18.1.

References

1. Kaminsky, M. S., Lafferty, J. M., Dictionary of Terms for Vacuum Science and Technology, Surface Science, Thin Film Technology, Vacuum Metallurgy, Electronic Materials (American Institute of Physics, New York, 1980).
2. Guthrie, A., Wakerling, R. K., Vacuum Equipment and Techniques (McGraw-Hill Book Company, Inc., New York, 1949).
3. O'Hanlon, J. F. , A User's Guide to Vacuum Technology (John Wiley & Sons, Inc., New York, 1980).
4. Halliday, D., Resnick, R., Physics For Students of Science and Engineering (John Wiley & Sons, Inc., 1963), p. 516 ff.
5. Sears, F. W., Thermodynamics (Addison-Wesley Publishing Company, Inc., London), p. 223 ff.
6. Honig, R. E., Hook, H. O., "Vapor Pressure Data for Some of the More Common Gases", RCA Review, 21, 360 (1960).
7. Haefer, R. A., Cryopumping, Translated by J. Shipwright, and R.G. Scurlock, (Clarendon Press, Oxford, 1989).
8. Simon, F. E., Kurti, N., Low Temperature Physics (Pergamon Press, Ltd., London, 1952), p. 32 ff.
9. Benvenuti, C., "Characteristics, Advantages and Possible Applications of Condensation Cryopumping," J. Vac. Sci. Technol. 11, 591 (1974).
10. Barron, R., Cryogenic Systems, (McGraw-Hill Book Company, New York, 1966), p. 65 ff.
11. Butcher, S. S., Charlson, R. J., An Introduction to Air Chemistry (Academic Press, New York, 1972), p. 4.
12. Dushman, S., Scientific Foundations of Vacuum Technique (John Wiley & Sons, Inc., New York, 1949), p. 44 ff.
13. Dushman, S., op. cit., p. 17 ff.
14. Pamplin, B. R., Bachrach, R. Z., Editors, Aspects of Epitaxy, Vol. 12 (Pergamon Press, London, 1986), p. 45 ff.
15. Leach, D. P., Basic Electric Circuits (John Wiley & Sons, Inc., New York, 1969).
16. Welch, K. M., "The Pressure Profile in a Long Outgassing Vacuum Tube", Vacuum, 8, 271, 1973.
17. Hablanian, M. H., "Recommended Procedure for Measuring Pumping Speeds", J.Vac. Sci. Technol. A5(4), Jul/Aug, 2552 (1987).
18. Rutherford, S. L., "Pumping Speed Measurements on Sputter-ion Pumps", Vacuum, 16, 643 (1966).
19. Tom, T., Munro, D. F., "Speed Measurements of Ion Getter Pumps by the 'Three-Gauge' Method", Trans. of 3rd Int. Vac. Cong., 1965, (Pergamon Press, London, 1966), p. 377.

References

20. Brombacher, W. G., A Survey of Ionization Vacuum Gauges and Their Characteristics, NBS Technical Note No. 298, 1967.
21. Redhead, P. A., Hobson, J. P., Kornelsen, E. V., The Physical Basis of Ultrahigh Vacuum, (Chapman and Hall, Ltd., London, 1968), p. 336 ff.
22. Fischer, E., Mommsen, H., "Monte Carlo Computations on Molecular Flow in Pumping Speed Test Domes", Vacuum, 17, 309 (1967).
23. International Organization for Standardization, ISO/TC 112/SC 3(Sec-20)44, April, 1970.
24. American Vacuum Society Standards (tentative), AVS 4.1-4.8, 4.10, J. Vac. Sci. Technol., 8, 664 (1971).
25. Welch, K. M., Flegal, C., "Elements of Cryopumping", Industrial Research/Development, March 1978.
26. Denison, D. R., McKee, E. S., "A Comparison of Pumping Speed Measurement Methods", J. Vac. Sci. Technol., 11(1), 337(1974).
27. Ball, R. W., Principles of Abstract Algebra (Holt, Rinehart, and Winston, New York, 1963), p. 31.
28. Wells, O. C., et al., Scanning Electron Microscopy (McGraw-Hill Book Company, New York, 1974), p. 94 ff.
29. Spangenberg, K. R., Vacuum Tubes (McGraw-Hill Book Company, Inc., New York, 1948), p. 24 ff.
30. Komiya, S., Sato, H., Hayashi, C., "Triggering Sputter Ion Pump in Extreme High Vacuum", Proc. 13th Nat. AVS Symp., 1966 (Herbick and Held Printing Company, Pittsburgh, Pennsylvania, 1967), p.19.
31. Downing, J. R., Mellen, G., Rev. Sci. Inst. 17, 218 (1946).
32. Cobine, J. D., Gaseous Conductors (Dover Publications, Inc., New York, 1958), p. 144 cf.
33. Bruining, H., Physics and Applications of Secondary Electron Emission (Pergamon Press, London, 1954).
34. Colligon, J. S., "Ion Bombardment of Metal Surfaces", Vacuum, 11, 272 (1961).
35. Cobine, J. D. op. cit., p. 162.
36. Thomson, J. J., Thomson, G. P., Conduction of Electricity Through Gases Vol. II (Dover Publications, Inc., New York, 1969), p. 487.
37. Tate, J. T., Smith, P. T., "The Efficiencies of Ionization and Ionization Potentials of Various Gases Under Electron Impact", Phys. Rev. 39, 270 (1932).
38. Rapp, D., Englander-Golden, P., "Total Cross Sections for Ionization and Attachment in Gases by Electron Impact. I. Positive Ionization", J. Chem. Phys. 43(5), 1464 (1965).

References

39. Redhead, P. A., Hobson, J. P., Kornelsen, E. V., op. cit, pp. 253 - 366.
40. Hultzman, W.W., "Characteristics and Performance of Several Mass Spectrometer Residual Gas Analyzers", National Aeronautics and Space Administration, Lewis Research Center, Cleveland, Ohio, Report No. NASA TN D-7554, 1974.
41. Conn, G. K. T., Daglish, H. N., "Cold Cathode Ionization Gauges", Vacuum 3(1), 24 (1953).

CHAPTER 2

SPUTTER–ION PUMPING

2.0 Introduction

Two types of sputter-ion, or *electronic*, pumps came into commercial use in the last few decades. The first of these, introduced on a commercial scale by the Granville-Phillips Company, was called the *Electro-Ion* pump, and was a simplified version of two varieties of pumps reported on earlier by Herb, *et al.*, and referred to as an *Evapor-ion* or *Orbitron* pump.[1,2] A version of Herb's Evapor-ion pump was offered for sale for a brief time by Consolidated Vacuum Corporation. Leybold also offered an electrostatic pump for sale for a brief period.[3] These types of pumps were called *electrostatic pumps* because of the exclusive use of electrostatic fields in their operation rather than both electric and magnetic fields. They are referred to in the past tense, as they were soon overshadowed by a second type of pump of much simpler construction. For completeness, a bibliography of some of the salient publications dealing with electrostatic pumps is given in Appendix 2A. No further discussion is given on these pumps in this chapter.

The main emphasis of this chapter will be on the second variety of sputter-ion pumps, first developed on a commercial basis by Varian Associates, and called by that company the VacIon® pump.[4] It is a sputter-ion pump which makes use of both electric and magnetic fields to achieve the pumping action of a low pressure, electrical discharge. These electrical discharges are referred to as *Penning Discharges*. The origin of all of these *electronic* pumps came from studies of various forms of vacuum ionization gauges and the phenomenon of sputtering of metals by gas ions; that is, observations of the somewhat annoying pumping effects of vacuum ionization gauges and gas pumping effects during sputtering *started it all*.

Hundreds of technical papers have been published on low pressure Penning discharges in the last half-century. About 10% of these have attempted to theoretically model the low pressure characteristics of Penning discharges. Redhead noted that none have had complete success in this effort,[5] though some of these papers have combined empirical and theoretical results and met with reasonable success. Great care must be used in interpreting data of any one or two publications, toward the formulation of a *unified* low pressure Penning discharge model. This is because slight changes in cell geometry and symmetry, magnetic field and its uniformity, electric field, pressure and gas species can significantly alter the properties of the discharge. I will first qualitatively describe a Penning discharge, discuss its properties as they relate to varieties of ion pumps, and then eventually discuss some of the theoretical considerations of these discharges.

2.1 The Penning Cell

In 1934 Gaede published a review paper on various types of vacuum gauges with which he had been working.[6] The emphasis of his work to that point was on the refinement of what came to be known as the Knudsen Gauge.[7] For completeness, in that paper he reported on experiments with gauges including an ionization manometer, configured as shown in Fig. 2.1.1. He recognized, as reported much earlier by Philips,[8] that if one applied a magnetic field axial to the electrodes to which a high voltage was applied, at low pressures (*i.e.*, ≤ 10^{-3} Torr) a stable electrical discharge would be initiated. He noted that the magnetic field tended to trap the electrons in the void between the electrodes, stating "... the emitted electrons (from the cathode) follow along the lines of the magnetic field (within the anode) and thus oscillate until they strike a gas molecule." Presumably the current "drawn" by the electrical discharge was noted to be linear with pressure. However, Gaede rejected this type of gauge for general applications. He noted that when the gauge was used, "... the gas pressure sank (*i.e.*, decreased) rapidly about 30 parts as though a pump were attached."

Gaede's gauge, though showing striking similarity to cold cathode gauges later described in the literature, was actually a hot filament gauge. That is, the filaments forming the cathodes of the gauge were thermal emitters. The two leads connected to the anode facilitated outgassing of this electrode by resistance heating.

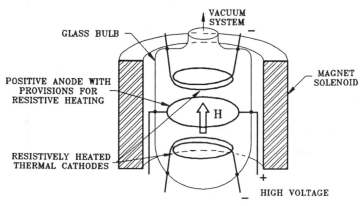

Figure 2.1.1. Gaede's magnetically confined discharge gauge.

Such hot filament gauges had been reported on decades earlier.[9] Gaede's contribution was in using a magnetic field to intensify and confine the discharge. Had Gaede turned off the filaments, with sufficient magnetic and electric field, the discharge would have been sustained within the electrodes over a large pressure interval.

Penning was also studying such discharges - that is, low pressure electrical discharges sustained by the use of electric and magnetic fields. I believe that Penning's initial work, rather than being related to gauging, was directed at using the sputtering action of these discharges to produce thin films. He submitted an application for a United States patent on such sputtering apparatus in 1936,[10] the same month he published a paper on the subject in Physica,[11] and he was awarded a German patent on such apparatus in 1935. In a review paper, Guentherschulze noted that this ion sputtering phenomenon had been reported on as early as 1858.[12] However, Penning identified a practical application of this phenomenon.

Two of the configurations claimed in Penning's patent are shown in Fig. 2.1.2. (The words *claim* and *claimed* are used in a legal sense in this book). In one, a hollow cylindrical tube served as the anode. Cathode plates of the desired sputtering materials - perhaps two different materials - were placed at each end of the cylindrical anode. The substrate, to be coated with the sputtered material, was placed as close to the anode cylinder as possible. At low pressures, ions created by the swirling cloud of electrons trapped within the anode volume bombard and sputter the cathodes. The selected cathode material is spewed onto the substrate located within the anode.

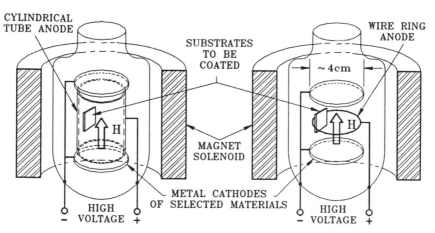

Figure 2.1.2. Examples of two of the sputter-coating cells noted in Penning's 1936 patent application.

In his patent application, Penning pointed out that as positive ions bombard the cathode surfaces, sputtering of these surfaces occurred. He referred to this sputtering as cathode *disintegration*. Though it was not included as a claim in the above patent, he noted therein that "Cathode disintegration may also be used for the purpose of reducing the pressure of a gas in a closed chamber. The cathode particles disintegrated combine with gas molecules and thus bring about a reduction in pressure."

Penning experimented with numerous other configurations, including what later came to be known as *magnetron* and *inverted magnetron* sputtering devices. Later variations of the Penning cell, some radical departures in scale and configuration, were found to produce very satisfactory sputtered films (*e.g.*, see 1 3 ,1 4 ,1 5). The original work by Penning proved to be the advent of the sputtering technology which is presently used to produce functional, decorative and semiconductor related thin films having applications in a wide variety of major industries.

Penning did not report on using a hot cathode as a source of electrons in these crossed field discharges. However, he verified Gaede's findings regarding current vs. pressure in magnetically confined discharges. Because of the absence of a *hot* filament, these electrical discharges became known as *cold cathode* discharges. With the advent of the magnetron rf oscillator, they were later sometimes described as *crossed field* discharges. This is because, similar to the magnetron, the electric and magnetic fields applied to the discharge apparatus, or *cell*, are somewhat orthogonal to each other. The analogy between the Penning cell and a magnetron proved appropriate, as you will later see. Qualitatively understanding the properties of a Penning electrical discharge was the starting point of understanding the workings of ion pumps. In early 1937 Penning published an article characterizing several gauge configurations, one of which was a minor variation of the original Gaede ionization manometer.[1 6] Of course it did not use a *hot* cathode filament. In this article he made mention that the pumping of gas occurs because of the electrical discharge.

As a practical gauge and pump, these devices were first reported on by Penning in 1949.[1 7] He commented on the need to thoroughly degas the gauge configuration in order to get *consistent* (not accurate) results, as he noted that the gauge *burnt away* gas. The property of the gauge to burn away gas was used in a leak detecting scheme he reported on in this paper. He reported that "..., owing to the discharge, a certain amount of the gas disappears. In the state of equilibrium the amount burnt away per second by the discharge is obviously equal to the sum of that leaking in ...". The property of these electrical discharges to consume gas had been reported on decades earlier.[1 8]

Four years later, in April 1953, Gurewitsch and Westendorp applied for patent on various forms of single-cell pumps, and the use of different cathode materials.[1 9] Rather than *curse the darkness* - that is the nuisance of the pumping effects of these *gauges* - they took the best that both Penning and

Gaude had to offer and patented it as a pump! Their patent included various single-cell configurations, some including the use of a hot filament. They described the physics of the devices as " ... compared to that of a cylindrical anode magnetron operating near cut-off voltage." They claimed use of various types of cathode materials (*i.e.*, Ti, C, Mg, Al and stn. stl.) and the use of a heated filament in some of their diode pumps, to aid in the build-up (*i.e.*, ignition) of the space charge in the cell. They subsequently reported results on use of both the Ti and C cathodes, indicating that the use of Ti cathodes aided considerably in the pumping of hydrogen.[20] They reported speeds of ~0.03 $\mathcal{L}$/sec for H_2.

Incidentally, most magnetron oscillators are effective pumps. When working as an engineering associate at Raytheon, in the late 1950's, I conducted an experiment where I deliberately introduced a continuous flow of H_2 into a high power magnetron during high voltage testing. The magnetron continued to operate stably, at an H_2 pressure of ~10^{-4} Torr for several hours. Arcing and missing pulses only occurred after removal of the H_2 leak.

In 1955, Gale reported on speed measurements on a single cell gauge/pump used to append a high voltage generator.[21] He reported measured pump speeds ranging from 0.01 to 0.03 $\mathcal{L}$/sec for various gases. But, I'm *getting ahead of myself.*

2.2 I^+/P Characteristics in Penning Cells

The linear relationship between pressure and ion current in Penning cells, at pressures $\lesssim 10^{-3}$ Torr, resulted in the first widespread interest in what then came to be known as a Penning gauge. For several decades it was also known as a Philips gauge, perhaps in honor of C.E.S. Philips,[8] or because Penning worked for the company N.V. Philips, located in the Netherlands.

Over the years, technologists attempted to establish parameters, including the geometry of the cell, magnetic field and applied voltage, which would yield well behaved (*i.e.*, constant and *smooth*) I^+/P or sensitivity characteristics as defined by (1.18.17) of Chap. 1. Conn and Daglish published two classical empirical studies of Penning discharges of that era.[22,23] In the second paper, published in 1954, they systematically determined the gauge sensitivity, s, as a function of pressure, P, magnetic field, B, and voltage, V, of a number of Penning cell configurations, including those first described by Penning. Serious discharge instabilities were reported for various configurations (*i.e.*, the current *jumped about* in certain pressure regions). Penning first reported seeing such instabilities at pressures $\leq 10^{-4}$ Torr. He termed these as "small jumps" in indicated pressures, and reported smoothing them out in his published data.[17] The Penning cell characterization studies of Conn and Daglish were limited to pressures in the range of 10^{-5} to 10^{-2} Torr. Their description of the visual and electrical characteristics of the discharges at various pressures is a tribute to their scientific technique. Their

1953 paper showed a remarkable insight into the mechanisms and characteristics of Penning discharges.[22] They noted at that time the similarity
between these gauges and magnetrons, and even made measurements of the
rf signals produced by the discharges.

One very important finding of Conn and Daglish was the discovery of the
cell configuration that later became known as the *triode*. This triode configuration (potentials reversed) is shown in Fig. 2.2.1. It did not evidence the
instabilities characteristic of the remaining configurations, termed diode
configurations. Of course, no one at that time, including Conn and Daglish,
recognized the significance of their triode discovery. Though showing figures
and test results of several diode configurations in that paper, they omitted a
figure of the triode, for which they gave only the test results.

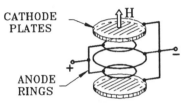

CATHODE
PLATES

ANODE
RINGS

Figure 2.2.1. The first "triode"
Penning cell as reported
by Conn and Daglish.

Jepsen was the first to point out the relationship of sputter-ion pump speed,
S, to the "I^+/P" characteristics of the cells of the pump.[24] This was
dramatically demonstrated in later work published by Rutherford.[25,26]
He quantitatively showed the direct correspondence between S and I^+/P of
sputter-ion pumps of various cell sizes. Rutherford and others at Varian
showed how I^+/P tended to significantly decrease at lower pressures for a
given cell geometry.[27,28] But, for the moment let's explore how a diode
Penning cell works.

Figure 2.2.2 represents a single-cell diode pump. The cathode plates of
this pump are typically made of a chemically active metal such as Ti, Ta, or
combinations thereof.[20,29] The anode cylinder and pump housing (not
shown in the figure) are usually made of a nonmagnetic stainless steel. The
anode is operated at a positive, dc voltage with respect to the cathode plates;
it is usually of the order of 3 to 7 kV. The magnetic field, applied axially to
the anode, is usually 0.1 - 0.2 Tesla (*i.e.*, 1000 to 2000 gauss). Single cell
pumps of this type have speeds of the order of 0.1 - 2.0 $\mathcal{L}$/sec.

If a free electron is somehow launched into the volume within the anode, it
becomes trapped by the *Lorentz force* of the magnetic field. That is, with the
appropriate magnetic field, it is physically impossible for this electron to
reach the positive anode. This *first* electron may have been dislodged from
some surface by a cosmic particle, or it may have originated from field
emission from the cathodes. In any event, this initial electron, trapped by the
magnetic field, swirls about within the anode volume.

The mean free path of the first electron for ionizing collisions may be very large, and it may travel great distances in its curved orbits within the anode cylinder (*e.g.*, several hundred kilometers at a pressure of 10^{-9} Torr). However, it will eventually experience an ionizing collision with a gas molecule. This collision will result in at least one additional free electron orbiting within the anode volume, *etc.* A Townsend avalanche process ensues (the *striking* of the discharge) until, at pressures ≲ 10^{-4} Torr, a dense and somewhat uniform, steady-state cloud of negative space charge (*i.e.*, electrons) is *stored* within the anode volume. These stored electrons cause the electrical potential within the volume to become depressed to a value approaching that of the cathodes.[30,31] The potential depression causes there to be a component of electric field in the anode volume which is radial to the axis of the anode cell, similar to that suggested by Fig. 2.2.2.

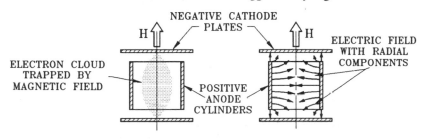

Figure 2.2.2. The negative space–charge cloud of electrons, trapped by the magnetic field of the Penning cell, causes there to be a large component of radial electric field.

During the transient buildup of the negative space charge, positive ions bombard the cathodes and dislodge secondary electrons. These secondary electrons aid in the buildup of negative space-charge in the anode volume.

At steady-state, the negative space-charge cloud remains constant within the volume, and consistent with the applied voltage.[32] Additional secondary electrons are prevented by the space-charge cloud from entering the anode volume. In fact, a mechanism must exist for depletion of the trapped electron space charge cloud as additional free electrons are created with creation of each positive ion. That is, at steady-state, for every positive ion which bombards the cathode, at least one electron must reach the anode.

The mechanism permitting electrons to eventually reach the anode, while crossing magnetic lines of force, has been *coined* by plasma physicists as <u>cross field mobility</u> or *migration*.[33] It means that electrons are given the mobility to cross magnetic lines of force as the result of collisions with gas molecules. They are deflected out of their quasi-circular orbits due to collisions with gas molecules, and some eventually reach the anode. This collision process, and the existence of spurious rf *noise* generated by the discharge,[34,35,36,37,5] makes it possible for electrons, with potential energies nearer that of the anode, to also reach the cathode.

Define the total amount of electron space charge stored in an anode cell as Σq. At a constant pressure, a greater cell sensitivity, s, means that for the same pressure there is a greater rate of gas ionization and, therefore, more current generated by the discharge. It is reasonable to assume that the rate at which gas is ionized within the cell will be directly proportional to the amount of space charge stored in the cell. If we were able to double Σq in a given cell configuration, we would be able to double I^+, (*i.e.*, the ion current created by the discharge). That is,

$$I^+ = k_1 \Sigma q P, \tag{2.2.1}$$

where k_1 is some constant of proportionality. We recognize $k_1 \Sigma q$ as being the gauge sensitivity, s. It is also reasonable to assume that there is some linear relationship between the rate of gas ionization and the rate at which gas is pumped (*i.e.*, the throughput, Q) in a Penning cell.[3 8] This is to say that for a given gas species,

$$Q = k_2 I^+, \tag{2.2.2}$$

where k_2 is some constant of proportionality. We have defined throughput as $Q = SP$. Using this relationship, (2.2.1) and (2.2.2) we find:

$$S = k_1 k_2 \Sigma q \tag{2.2.3}$$

That is, the speed of the pump, S, is directly proportional to the total amount of electron space charge, Σq, stored in the Penning cell (cell hereafter). Speed dependence on pressure is not indicated in (2.2.3). This is known not to be the case; that is, the speed (and amount of stored space charge) is pressure dependent. If we plot the slope of the $\log_{10}$ of current *vs.* $\log_{10}$ of pressure for most Penning discharges we empirically observe:

$$\frac{\log I_2 - \log I_1}{\log P_2 - \log P_1} = \frac{\log I_2/I_1}{\log P_2/P_1} = n. \tag{2.2.4a}$$

Substituting I^+ for I_2 and P for P_2, and solving for I^+ as a function of P, we find:

$$I^+ = k_1 P^n = k_1 \Sigma q(P) P \tag{2.2.4b}$$

where $\Sigma q(P)$ means that the stored charge in the cell varies as a function of pressure. The value of n varies with all of the parameters listed in Table 2.2.1. Depending on these parameters, there will be distinct and abrupt changes in n with pressure. The slope of (2.2.4a) will vary in value, over the pressure range of 10^{-12} to $\sim 5 \times 10^{-4}$, from $1.5 \geq n \geq 1.0$, respectively.

Solving (2.2.4b) for $\Sigma q(P) P$, we find that the speed of a pump as a function of pressure is:

$$S \quad = \quad k_1 \, k_2 \, P^{n-1} \hspace{4cm} (2.2.5)$$

The constant k_2 (see (2.2.2)) has a special significance as pointed out by Jepsen.[24] It is the efficiency with which gas is pumped by a sputter-ion pump. That is:

$$Q \quad = \quad \eta I^+ \; = \; SP \Rightarrow \eta \; = \; SP / I^+ \hspace{3cm} (2.2.6)$$

The term η or the ratio SP / I^+ appears frequently in the literature. Equation (2.2.6) clarifies its meaning. That is, it is merely the number of gas molecules pumped for each ion generated by the discharge.[39] The dependence of I^+ on n (i.e., (2.2.4b)) was most apparent in Penning's original data,[17] and has been the subject of over 50 subsequent publications - some relating to sputter-ion pumps (e.g., 26,28) and others dealing with various forms of vacuum gauges, (e.g., 40) as discussed in a review paper by Redhead.[5]

We conclude that anything we can do to increase the amount of electron space charge stored in the cell will go to improve the pumping speed of the cell. Recognition of this space charge dependence has led to some rather unique approaches in designs of sputter-ion pumps.[41,42] It has also prompted numerous studies of the space charge characteristics of these cells. This will be discussed at length in Section 2.13. However, for the time being, the parameters which significantly alter Σq and cell pumping speed are given in Table 2.2.1. Typical values used in commercial sputter-ion pumps are also given in this table.

Table 2.2.1. Parameters effecting Penning cell sensitivity
and sputter-ion pump speed and typical values.

Anode Voltage	V	3.0 - 7.0	kV
Magnetic Field	B	0.1 - 0.2	T
Cell Diameter	d	1.0 - 3.0	cm
Cell Length	ℓ	1.0 - 3.2	cm
Anode/Cathode Gap	a	0.6 - 1.0	cm
Pressure (P^n)	P	$1.0 \lesssim n \le 1.5$	Torr.

Qualitative Pump I/P Characteristics

I first will qualitatively describe how the sensitivity of pump cells varies with the parameters listed in Table 2.2.1, and then discuss construction of various forms of sputter-ion pumps. Figure 2.2.3a shows how I/P varies as a function of magnetic field, B, at a given pressure (Welch, SLAC 1969). The dimensions of the pump cell are given in the figure. We see that there is a value of B below which there is no discharge in the cell. This is called the *cutoff* field.

On discharge ignition, the cell I/P, and therefore I^+, increases linearly with increasing **B** field until it reaches a maximum value. This maximum value, dependent on anode voltage, is called the *critical* magnetic field. A further increase in **B** beyond the critical magnetic field will result either in a constant value of I/P, or possibly a gradual fall-off in its value. In this case, the words *cutoff* and *critical magnetic* field do not have the same meaning as with magnetrons. The *sharpness* or kurtosis of the I/P peak (*i.e.*, *knee*) at the critical magnetic field, increases with pressure. Because of the possible drop-off in I/P beyond the critical field, further increases in **B** can actually result in less pumping speed for the same voltage and cell configuration.

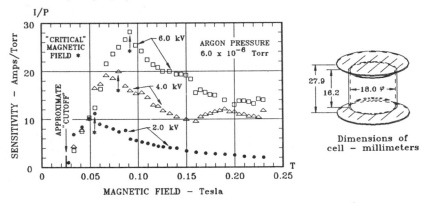

Figure 2.2.3a. Penning cell sensitivity at 6.0 x 10⁻⁶ Torr argon as a function of magnetic field and anode potential.

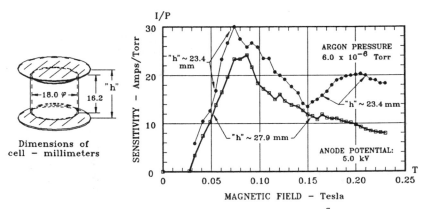

Figure 2.2.3b. Penning cell sensitivity at 6.0 x 10⁻⁶ Torr argon as a function of cathode to anode spacing and magnetic field.

The spacing between the cathodes and the anode also results in variations in I/P. This effect is shown in Fig 2.2.3b, where changes in I/P, are given for anode-to-cathode spacings of 3.1 and 5.9 mm, with the same magnetic field and anode potential. In Section 2.10 it will be shown that if the gaps are too narrow between the cathode plates and a multi-celled anode array it will restrict the passage of gas into the cells. This results in an apparent reduction in cell sensitivity.

The increase in I/P with V is linear up to the critical magnetic field. Then, I/P becomes nonlinear in V. This effect is shown in Fig. 2.2.4 for the same cell configuration given in Fig. 2.2.3a. Within limits, an increase in pump voltage usually results in an increase in pump speed.

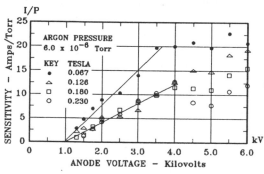

Figure 2.2.4. Penning cell sensitivity as a function of anode voltage, and for various magnetic fields.

The correspondence of I/P with speed, S, is graphically demonstrated in Figs. 2.2.5 and 2.2.6 where S and I/P, respectively, for H_2 and N_2, are given for a multi-cell element pump as a function of magnetic field.[43]

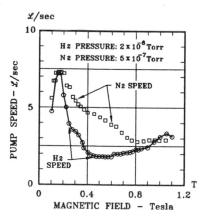

Figure 2.2.5. Pump speed for H_2 and N_2 as a function of magnetic field.

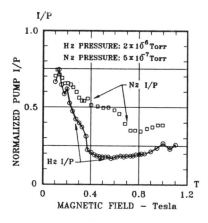

Figure 2.2.6. Normalized pump sensitivity for H_2 and N_2 as a function of magnetic field.

For reasons given in Section 2.13, the magnitude of I/P also varies with cell length, ℓ. As noted earlier, some of the first Penning cells were constructed with wire anodes. That is, the length of these anode cells was approximately

zero. For example, the *triode* of Fig. 2.2.1 has an I/P for air, at 10^{-5} Torr, of ~1.6 A/Torr. This would correspond to *end effects* of a cylindrical anode of finite length. Therefore, I/P is finite for an anode cylinder of near zero length, and increases linearly with anode length thereafter. There are three caveats regarding the length of the anode cell. 1) Should the cell be too long, then the conductance leading to the volume within the cell (*i.e.*, where the ionizing electrons are stored) becomes restricted. Therefore, though the amount of space charge stored in the cell may increase with cell length, the cell *sensitivity model* is no longer applicable, as the space charge is less accessible to the gas. 2) For a longer cell, a greater amount of magnetic material will be required to achieve a magnetic field for the optimum I/P.
3) The third consideration has to do with optimizing the surface area within the anode - where most of the chemically active gases, excluding H_2, are pumped. This is discussed below.

The value of I/P also varies as a function of the cross section of the gas species. That is, the sensitivity of the cell varies with the type of gas. This does not necessarily mean that because the sensitivity for a given gas is low, that the speed for that gas is correspondingly low. This is often to the contrary. For example, the approximate relative cell sensitivities of $Ar:N_2:H_2$ are 1.2:1.0:0.49 respectively (see Table 1.18.2). However, a conventional diode pump, with Ti elements, will pump these gases with speeds in the ratio approximately 0.02:1.0:1.4.[44,45] The implication here is that different gases are pumped more effectively than others for the same ion current. For example, for the same N_2 and H_2 pump currents (note: this corresponds to different pressures for the two gases), the H_2 throughput will be almost x3 greater. You will have to *think about this* effect for a moment. The key is in the ratio of the number of gas molecules pumped to the number of ions created in the discharge. This is called the *pumping efficiency*,[44] (*i.e.*, the term "η" in (2.2.6)).

Lastly, I/P varies as a function of pressure for different cell diameters. This has been discussed at length in the literature.[26,46,28] This was dramatically demonstrated in a paper published by Rutherford in 1963.[26] In this paper he reported measurements of I/P of pump modules having cells with diameters of 1.27, 2.54 and 5.08 cm. In each case, the *cell aspect ratio*, defined as the cell length, ℓ, to diameter, d, was held constant at $\ell/d = 1.5$. Rutherford's results are shown in Fig. 2.2.7. As a generalization, for a fixed magnetic field, the larger the cell diameter, the greater the I/P at lower pressures.

¤ As a very rough *rule of thumb*, for N_2: I/P ~10 A/Torr for a typical sputter-ion pump cell at 10^{-6} Torr. Therefore, a properly designed pump with ten cells would have an I/P of ~100 A/Torr, at this pressure.

On a lighter note, while at Varian, I developed a sputter-ion pump specifically for operation at very low pressures. The diameters of the cells and magnetic field were chosen for this purpose. We submitted this pump, for evaluation purposes, to a potential customer, with the condition that the cell geometries of the design must not be divulged. Keeping technical developments a secret is similar to keeping a secret in a church choir; it is not possible. Several months later, I was asked to referee a paper on sputter-ion pumps. It contained dimensions of the cells and speed data on this very pump and other competitor's pumps. The results proved most favorable to Varian. The paper was published. The authors never knew that I was one of the referees.

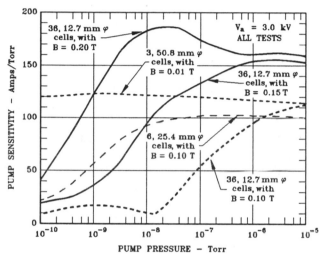

Figure 2.2.7. Rutherford's I/P measurements with a cell l/d of 1.5, vs. pressure, and for various cell diameters and magnetic fields.[26]

2.3 Pumping Speed Abstraction

The definition of the speed of a sputter-ion pump is the same as that described in Chap. 1 for a mechanical pump. We must use our imaginations a bit to envision how volumetric displacement rate applies to a Penning cell. Envision elemental abstract volumes, at unspecified pressure, finding access to the discharge volume within the cell. The rate at which these volumes enter the cell is influenced by the conductance of the gap between the cathode plates and anode cylinder. In fact, this is the definition of conductance. The conductance accessing the anode cell and the amount of space charge therein is crucial to achieving a high pumping speed.

The speed of a sputter-ion pump is analogous to a farmer of old cutting grain with a scythe. If the farmer walks along and sweeps his scythe at a constant rate, the rate at which the grain falls before the scythe is dependent only on the density of grain stalks in his path. In this illustration the swirling scythe, which is being swept at a constant rate, is analogous to the constant and swirling electron space charge cloud. The stalks of grain, perhaps of varying density, are analogous to gas, perhaps at varying pressure.

In Section 1.14, I discussed the implications of Dalton's law and how it related to the application of linear superposition in calculating system partial pressures for a given pump speed (*e.g.*, problem 12, Chap. 1). The assumption of linear superposition proves correct for turbomolecular pumps, diffusion pumps, cryopumps and most other types of pumps. It is also true, to the first order, for stably operating sputter-ion pumps. However, because of the complexity of sputter-ion pumping mechanisms, both the sequential and simultaneous pumping of different gases impacts on the speed of each of the species (*e.g.*, see 4 7 - 5 0). Also, as will be seen, it is possible for the speed of sputter-ion pumps to be selectively negative for some species (*e.g.*, noble gases) while, at the same time, positive for others.

2.4 The Making of Sputter-Ion Pumps

2.4.1 Pumping Mechanisms and Materials

English journal publications reporting on *electronic* pumping mechanisms, alone, number in the hundreds. Excellent review papers have been published on pumping effects in hot-filament ion gauges (*i.e.*, forms of electrostatic capture pumps).[5 1] Approximately 100 additional papers exist which report on the pumping mechanisms of sputter-ion pumping. Were I to reference all of these publications and treat this subject in depth, this alone would require a complete volume. Rather, I will describe these *electronic* pumping effects with simple models. There are two primary mechanisms responsible for the electronic pumping of gas. One is *physisorption* the other *chemisorption*. One need not have command of quantum physics to understand these simple mechanisms (Richard Feynman is quoted as saying "I think I can safely say no one understands quantum mechanics", anyhow). Both mechanisms to one degree or another are enhanced by *sputtering*. Sputtering, physisorption and chemisorption will be qualitatively described.

Sputtering and Sputter-Yield

When an ion bombards a solid (*e.g.*, metal) surface, it can be very disruptive to that surface, causing a great deal of damage on an atomic scale. For example, an argon ion with a few hundred volts energy will cause considerable damage to a metal lattice to a depth of ~100 Å.[5 2] This damage may be in the formation of small dislocations in the lattice structure. These dislocations can serve as conduits for the diffusion of H_2 and the inert gases. Damage may also take form in the *splashing* off of some of the *ground zero* metal atoms onto adjacent surfaces. This *splashing* process is called *sputtering*, and is the primary mechanism leading to sputter-ion pumping.[3 8]

The cathode material being bombarded is often called the *target material*. The effectiveness of the primary particle in sputtering or *splashing off* the target material is called the *sputter yield*. It is the ratio of the average number of target material atoms sputtered per bombarding ion. Sputter yields of ≳ 1.0 are common. Yields depend on the momentum, charge and mass of the ion, the angle at which it strikes the surface - this is called the *angle of incidence* - and properties of the target material. The angle of incidence is the angle formed by the trajectory of the ion with an imaginary line perpendicular to the target surface. Sputter yields have been published for many gas-metal combinations.[5 3,5 4]

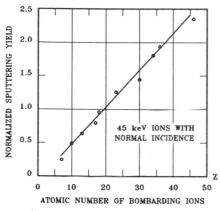

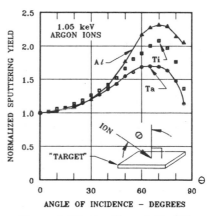

Figure 2.4.1. Sputter—yield of Si, normalized to Si sputter—yield for Ar, vs. Z of bombarding ions.[55]

Figure 2.4.2. Sputter—yield with incident angle of ion.[53]

The following generalizations regarding sputtering may be made as they relate to sputter-ion pumping: 1) The greater the ion momentum (*i.e.*, pump voltage), the greater the sputter yield. 2) The lighter the gas ion, the less the sputter yield for the same energy and target material. For example, He and H_2 ions have very little sputter yield compared to N_2 and Ar ions accelerated through the same voltage. Figure 2.4.1, illustrates the dependence of sputter yield on ion mass, as reported by Andersen and Bay.[5 5] 3) The sputter yield, for the same gas ion and energy, does not necessarily differ markedly for different target materials. Oechsner outlines the theoretical models reported by others which explain this characteristic.[5 3] This effect relates to the surface binding energy of the target material and the amount of energy per unit volume which can be deposited into the lattice of the target material by the bombarding particle. 4) The sputter-yield gradually increases with the angle of incidence of the bombarding particle to about 70°, then decreases *quickly* with increasing angle (*e.g.*, see Fig. 2.4.2).[5 3] Brubaker was the first to point out the significance of this effect as it related to sputter-ion pumping.[56]

Variation in sputter-yield with angle of incidence is analogous to the effect of a flat stone skipping over water. There is a modest *plop* when we throw the stone on edge directly into the water (*i.e.*, zero angle of incidence). However, if we throw stones of similar shape in the water, at progressively increasing angles of incidence, they will make progressively larger splashes as they enter the water. Finally, there is an angle of incidence which is less than 90° , but at which the stone does not break the surface of the water. It makes a very slight splash and skims along the surface. An even greater angle of incidence results in even less perturbation to the water.

The flux or intensity of material *splashed* off the surface of the cathodes, by the bombarding gas ions, is said to follow the *cosine law* for near zero angles of incidence. This merely means that the flux intensity is greatest perpendicular to the cathode surface, and that it gradually diminishes in proportion to the cosine of the angle ϕ, shown in Fig. 2.4.3. The flux intensity will also vary as r^{-2}, the length defining a surface at a distance r from the sputter source. These two relationships are given in (2.4.3a). For reasons which will become clear, a large proportion of the chemically active and noble gases are pumped on the anodes. Therefore, the size (*i.e.*, surface area) and shape of the anodes is important.[47] The cosine-law effect has important implications as it relates to the cell aspect ratios; that is, the ratio of the cell length to diameter (*i.e.*, ℓ/d). It is known that with magnetic fields ≤ 0.15 T, and at low pressures, the density of ion current is greatest at the cathodes at points subtending the axis of the anode structure (*e.g.*, see [57,58,59,60]). Assume, therefore, that most of the material sputtered from the cathodes originates at these points. One can calculate the flux intensity of sputtered material through an imaginary surface (see Fig. 2.4.3) by the following:

$$\nu \;\; = \;\; k_4 \, r^{-2} \, \cos\phi \tag{2.4.1a}$$

$$\nu \, dA \;\; = \;\; (k_4 \, r^{-2} \, \cos\phi)\,(r^2 \, \sin\phi \; d\phi \; d\beta) \tag{2.4.1b}$$

The rate at which sputtered material is deposited on the anodes, R, is then:

$$R \;\; = \;\; (N/\pi)\,\sin^2\phi. \tag{2.4.1c}$$

I have defined the quantity, N, as the total rate at which material is sputtered from the cathode. Such preferential cathode sputtering (centered on the anodes) will be evident to anyone who takes apart and services an extensively used pump. Using (2.4.1), the proportion of material sputtered on the anodes *vs.* the total amount sputtered is calculated to be, for ℓ/d, as shown in Fig. 2.4.4.[61]

Some sputtering of the cathode plates occurs off-center of the anodes. Ions are known to strike the cathodes beyond the projected anode radius.[57] However, at low pressures, the ion current density under the *shadow* of the anode cells is negligible compared to that on-center with the anodes. Therefore, there is a preferential erosion of cathode material toward the centers of the anodes, and a net build-up of sputtered material in the regions nearest the anodes.[59]

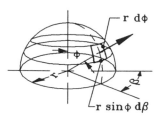

Figure 2.4.3. Elemental area through which flux passes.

Figure 2.4.4. Sputter–coating of anode cell vs. cell l/d.

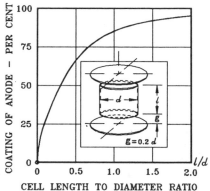

Physisorption and "Binding Energies"

Physisorption is the *catch-all* term used to describe all gas capture pumping mechanisms, excluding chemisorption. For example, if a gas molecule or atom is buried and trapped under a coating of sputtered film, this is called physisorption. If a high energy, neutral atom is buried deep within the metal lattice, this too is called physisorption. Lastly, if a gas molecule has no chemical affinity for a metal, but *finds* defects within the metal lattice into which it diffuses (*e.g.*, Ar), or is small enough to diffuse into interstitials of the lattice (*e.g.*, He) - in both cases because of a concentration gradient created by the bombardment of noble gas ions - this too is called physisorption. If that gas is being adsorbed into the bulk of the metal, it could also be called absorption. Noble or inert gases are pumped exclusively by combinations of these three physisorption mechanisms. The physisorption of noble gases, and the formation of mobile bubbles of these gases in metals, became of interest to those studying fission reactor materials.[62,63] Of course, the pumping effect due to van der Waals forces is another form of physisorption. These mechanisms will be discussed as they relate to specific types of pumps.

Physicists and chemists many times will model the flow or dynamics of physical processes in terms of *energy states* and changes in these states. This only becomes an intuitive process after years of thinking in these terms. We lay people tend to think in terms of concepts made tangible by our day-to-day experiences. For example, time, mass, velocity, and force are concepts most of us have a *feel for*. Therefore, when told that two bodies will have a mutual attraction for each other, related in some way to the distance at which they are separated, we can accept this as a condition of nature, and carry on. Such is the case with two atoms or molecules. When they come in close proximity to each other, there are competing forces of attraction and repulsion between the two atoms. Though we have never seen *force fields* (*e.g.*, gravitational or electrostatic fields), we daily experience their effects. And, we can accept that there must be some sort of force field emanating from each atom into free space. The force of the attractive field between two atoms varies as $1/r^7$, where the mutually repelling force varies as $1/r^{13}$, where r in each case is the separation between the atoms. That is, the force, $\mathbf{F}(r)$, between the two atoms varies as:

$$\mathbf{F}(r) = \frac{f}{r^{13}} - \frac{g}{r^7}, \qquad (2.4.2)$$

where f and g are positive constants which vary in value depending on the atomic or molecular species. In this equation, the repulsive force is given a positive value (*i.e.*, it requires the input of work to overcome) and the attractive component is given a negative value. The work which must be done, ΔE, to bring the two atoms close together by a distance Δr, from a great distance, is simply:

$$\Delta E = - \mathbf{F}(r) \times \Delta r$$

The negative sign implies that the particles are initially attracted to each other and therefore do work on the system (*i.e.*, give up heat), rather than the system having to do work on the pair of atoms. The work required to bring the atoms together from an infinite distance is simply:

$$E(r) = - \int_{\infty}^{r} \mathbf{F}(r)\, dr = \int_{r}^{\infty} \mathbf{F}(r)\, dr$$

$$E(r) = \frac{f}{12\, r^{12}} - \frac{g}{6\, r^6}, \qquad (2.4.3)$$

Equation (2.4.3) is said to be the energy state of the system (i.e., two atoms or molecules). Both (2.4.2) and (2.4.3) are plotted in Fig. 2.4.5, which shows both the force and *energy well* for two hydrogen molecules.[64]

Similar, comparatively weak attractive forces exist between all molecules and atoms, the constants f and g varying depending on the gas species. The depth of the energy well, varying from species to species, is called the *heat of adsorption*. It is merely a measure of the tenacity or strength with which the two particles are bonded. For this reason, it is often called the binding energy of the two atoms. Equation (2.4.2) is the more familiar *van der Waals* force equation. It expresses the force responsible for the cryosorption, condensation and liquefaction of all gases.

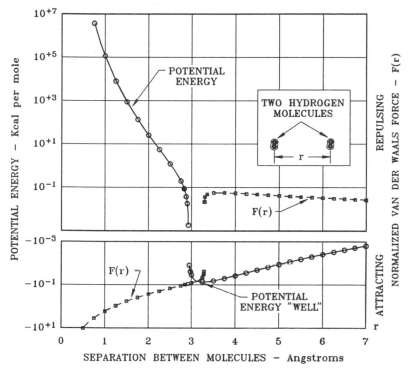

Figure 2.4.5. Potential energy and force as a function of distance of separation of two hydrogen molecules.[64]

Equation (2.4.3) is the van der Waals energy equation. The fact that the energy equation is negative in sign means that energy or heat is given off in the process of the combining of atoms or molecules. In fact, all physisorption processes are exothermic. When we supply refrigeration for the pumping of gases, the refrigerant or refrigeration system must be able to absorb calories stemming from the heat of sorption; but, this is covered in another chapter. It is sufficient to note that cryocondensation and cryosorption (*e.g.*, at temp. ≤ 293° K) are included in the pumping mechanism termed physisorption. And, the dominant mechanism responsible for this form of physisorption is van der Waals forces between the molecules.

Van der Waals forces also exist between liquid and solid surfaces and gases. An atom or molecule coming near to the surface will be attracted to that surface, *fall* into the potential well and stick to that surface for a finite period of time. To calculate these forces with reasonable accuracy, one need only assume a certain density of atoms per unit volume of the bulk material, and then integrate to find both the force of attraction and energy terms for a gas molecule some slight distance away from the surface. These calculations were made by Redhead, *et al.* (*i.e.*, (6 4)): This book is out of print. But, the student of UHV technology should seek it out in some good technical library.) Energy units most frequently used to describe the strength of the bond between a surface and some gas molecule are electron volts/atom or kcal/mole of gas. One electron volt per atom corresponds to ~23 kcal/mole. As a *rule of thumb*, the magnitude of heats of physisorption are usually ⩽ 10 kcal/mole, where heats of chemisorption are ≥ 10 kcal/mole.

Chemisorption

The above calculations were made for H_2 molecules, rather than atoms of hydrogen. Had we brought two atoms of hydrogen together, the potential well would have been much *deeper*. It would have represented a *chemical bond* between the two atoms. For example, N_2 is a chemical bond of one nitrogen atom to another nitrogen atom; CO a chemical bond of a carbon atom to an oxygen atom, *etc.* In fact, all atoms of gases, except the inert gases, when found in their natural states, are chemically bound to either similar atoms of the same gas (*e.g.*, N_2, H_2, O_2), or other atoms (*e.g.*, CO, CO_2, H_2O, CH_4). Such chemical bonds stem from the sharing of valence electrons in the outer *orbital shells* of the atoms. These gases also form stable chemical compounds with many of the active metals.

Chemisorption is the removal of gas from the system (*i.e.*, pumping) through the formation of chemical bonds between the gas and some chemically active metal (*e.g.*, Ta, or Ti cathode materials). Therefore, the cathode materials must have a *chemical affinity* for the gas. There are two chemisorption mechanisms which occur in ion pumps. The first involves dissociative pumping of the gas through the formation of stable chemical compounds; the second involves the pumping of a gas by dissociation and diffusion into the cathode plates and, to a lesser degree, the anodes.

The first, chemisorption model, depicted in Fig. 2.4.6, occurs as follows: 1) ions bombard the cathode surface and sputter cathode material onto some nearby surface (*e.g.*, the anode); 2) the chemically *active* film of sputtered atoms, residing on these surfaces, *awaits the arrival* of neutral gas molecules; 3) gas molecules which impinge onto these *active sites* are physisorbed for a brief time; 4) gas molecules dissociate on the surface either spontaneously - the fundamental mechanism - or by achieving some slight activation energy; so, 5) the dissociated gas forms a stable chemical compound with the film (*e.g.*, $N_2 + 2Ti \rightarrow 2TiN$). Of course, the gas molecule may already be present on the surface, residing there because of physisorption, at the time the sputtered cathode material arrives. The gas molecule may require an energy *nudge* before it is dissociated and combines with the sputtered cathode material. In a *nutshell*, this is pumping through sputtering and chemisorption by a *dissociative* process.(6 5)

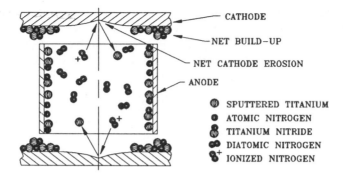

Figure 2.4.6. Chemisorption of nitrogen by surface dissociation and combination with chemically active, sputtered titanium films.

With the exception of CO, all diatomic gases are dissociatively adsorbed on metals.[64] The process is sometimes represented by an energy diagram, taken from Bond,[66] and shown in Fig. 2.4.7. Curve 1 represents the energy diagram for the physisorption of a diatomic molecule on a clean metal surface. Curve 2 represents the energy diagram for the chemisorption of each atom of a diatomic molecule. Energy E^* in this diagram represents the energy required to dissociate the molecule into atomic species. We see that the closer the molecule of Curve 2 comes to the surface, the lower the molecular dissociation energy becomes. This phenomenon is the process which facilitates chemisorption on metal surfaces.

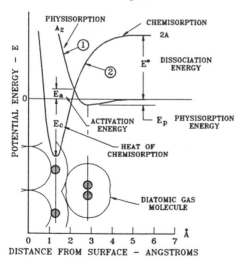

Figure 2.4.7. Potential energy diagram for the dissociative chemisorption of a gas molecule.[66]

The large positive value of E^* does not represent, in this case, a mutual repulsive force between the metal surface and the molecule. Rather, it represents a measure of the work required to dissociate the molecule as a function of its distance from the surface of the metal. In Curve 1, E_p represents the physisorption energy, where E_c represents the depth of the chemisorption potential well. If the physisorbed molecule could gain just enough energy to escape the shallow E_p well and reach the point of intersection of the two energy curves, it could dissociate and *fall into* the chemisorption potential well. The energy *barrier*, E_a, is very small in value compared to E^*. Energy E_a is referred to as the *activation energy* for the dissociative chemisorption of the gas by the metal surface. The gas molecule might have been energetically disturbed (*i.e.*, become *metastable*) due to a prior collision with an energetic ion or electron.[6 7] In this encounter, the molecule is given just enough energy by the collision that when it later lands on a surface, on which a chemically active atom resides, it is readily dissociated and chemisorbed. Depending on the surface material and gas molecule, E_a could have a value less than zero. In such an event, gas molecules would directly combine with the sputtered films without first requiring the intermediate dissociation process by some energetic particle. For a clean surface, unpopulated by gas, E_a will have the smallest value, and it will increase in magnitude with increasing gas coverage of the surface.

Some gases have sufficient chemical affinity for materials to spontaneously dissociate when encountering an atom of a metal. For example, O_2 spontaneously dissociates and combines with a number of metals (*e.g.*, In, Ti, Aℓ, Cr, *etc.*). However, once a 50 - 100 Å oxide layer develops on the surface of the metal, the oxide becomes a diffusion barrier to the formation of additional oxide. Sputtering away these oxides will replenish active metal sites.

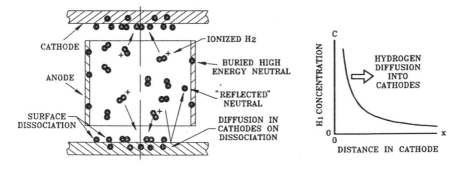

Figure 2.4.8. Chemisorption of hydrogen in the pump cathodes by dissociation and diffusion, and physisorption in anode by high–energy neutral burial.

The second chemisorption process, depicted in Fig. 2.4.8, also requires a chemical affinity between the cathode material and the gas. In this mechanism, on impact with the metal surface, the gas ions dissociate into an atomic state (e.g., $H_2^+ + e^- \rightarrow H + H$). A concentration gradient develops in the bulk material as a consequence of the dissociated surface gases. The gas then diffuses, in an atomic state, into the cathode material. The cathode plates serve as a bulk reservoir for the pumping of gas. Hydrogen is very prevalent in vacuum systems. Therefore, it is important that we understand the limitations of sputter-ion pumps in pumping this gas. I will return to this subject later in the chapter.

The relative activity of metals for chemisorption of different gases, as reported by Bond,[66] in an extension of Trapnell's work,[68] is given in Table 2.4.1. We see that Ti and Ta would serve as comparatively good sputter-ion pump cathode materials, where it is noted that aluminum would not seem to be a very good cathode material.[69,70,71] However, this was recently investigated, and there is some evidence, contrary to the indication in Table 2.4.1, that both CO and N_2 are chemisorbed on Al.[72,73] Some success has been reported in the use of pumps constructed with one Al and one Zr cathode,[74] and pumps constructed with cathodes of Ti-Zr-Al, though the alloy composition was not noted.[75] Also, Lu has recently reported on work with Al-Y and Al-Ce composite cathodes.[76]

Table 2.4.1. Metal and semi-metals adsorption
properties for various gases.[66,68]

Group	Metals	O_2	C_2H_2	C_2H_4	CO	H_2	CO_2	N_2
A	Ca, Sr, Ba, Ti, Zr, Hf, V, Nb, Ta, Cr, Mo, W, Fe,[1] Re[2]	($\Rightarrow$ decreasing heats of adsorption $\Rightarrow$) A: Gas Adsorbed NA: Gas Not Adsorbed						
		A	A	A	A	A	A	A
B_1	Ni, Co[2]	A	A	A	A	A	A	NA
B_2	Rh, Pd, Pt, Ir[2]	A	A	A	A	A	NA	NA
C	Al, Mn, Cu, Au[3]	A	A	A	A	NA	NA	NA
D	K	A	A	NA	NA	NA	NA	NA
E	Mg, Ag,[1] Zn, Cd, In, Si, Ge, Sn, Pb, As, Sb, Bi	A	NA	NA	NA	NA	NA	NA
F	Se, Te	NA	NA	NA	NA	NA	NA	NA

Notes: 1) Adsorption of N_2 on Fe and O_2 on Ag is activated at 0° C.
2) Metal probably belongs to this group, but film behavior unknown.
3) Au does not adsorb O_2.

As noted, all sorption processes are exothermic. That is, heat is given off during the process. The relative absolute values for the heats of chemisorption for the various gases decrease as shown in Table 2.4.1. The heats of chemisorption for various metals, used in sputter-ion pumps at one time or another, have absolute magnitudes accordingly:[64]

Ti,Ta > Nb > W,Cr > Mo > Fe > Mn > Ni,Co > Rh > Pt,Pd > Cu,Au.

An unending variety of materials is discussed in the literature as possible candidates for sputter-ion pump cathodes (*e.g.*, see Holland (7 7)). Suggestions for the *possible* use of many materials have far outweighed actual experimental evidence of the use of these materials. I will not make an attempt to summarize much of this speculation for you.

2.4.2 Diode Pumps

The Need for "Clean" Pumping

Varian Associates played a major role in the commercial development of sputter-ion pumps. The principal product of Varian Associates in the 1950's was microwave tubes. Many of these tubes had thermal emitting, oxide cathodes. These cathodes were fragile devices in that if they were contaminated by diffusion pump oils, cathode emission characteristics would be significantly degraded. Because of this, technologists sought alternate methods of pumping these vacuum tubes during their final stages of high-temperature bakeout.

Some klystron amplifiers were >1 m in length, operated at very high beam voltages (*e.g.*, 100-200 kV) and generated peak rf power in the MW range. After *aging* the tubes on diffusion pumped vacuum systems, they had to be *pinched off* the systems and shipped somewhere to a radar transmitter. The need of sustaining high vacuum in these devices during subsequent operation was further motivation for the creation of *clean*, portable pumps which permanently appended the microwave tubes.

Where the magnetron oscillator pumped very effectively on itself, linear-beam microwave tubes were less effective self-pumps. The presence of gas in these linear beam devices caused spurious oscillations and noise in the rf output. This required that better high temperature bakeout techniques be developed and appendage pumps be permanently attached to some tubes. Numerous scientists and engineers at Varian worked on these requirements.

The assumption leading to the development of the larger sputter-ion pumps was elegantly simple: If one pump cell had a measurable pumping speed of S, would not an integer, n, of these cells have $n \times S$ pumping speed? Reikhrudel and others tested this assumption and reported on the first multi-cell sputter-ion pump in 1956.[6 9] Their sputter-ion pump had cells arrayed in series, analogous to the pages of a book, where the odd numbered *pages* comprised the cell cathodes and the even numbered *pages*, anode rings. The cathodes of this multi-cell pump were operated at the same negative potential with respect to the anode cylinders, sandwiched between the cathodes. The entire assembly, fitted in a long glass cylinder, was inserted into a solenoid which provided the magnetic field. They conducted tests with

Ta, Mo, Ni and Aℓ cathodes, finding only the latter to be unsatisfactory in the pumping of "air", He and Ne.

Varian engineers constructed their first multi-cell pump with a flat array of parallel anodes, sandwiched between two cathode plates. Hall, Helmer and Jepsen submitted an application for patent of such a multi-cell pump on July 24, 1957.[4] It was reported on by Hall a year later.[78] A diagram of the first commercial pump of this nature is shown in Fig. 2.4.9. The anode, cantilevered on the center rod of the high voltage feedthrough, comprised the plainer array of 36, 1.3 cm square anode cells, 1.9 cm in length. The cathodes were made of Ti and held in place by spacers. The pump housing was made of 304 stn. stl. (*i.e.*, stainless steel) and TIG (*i.e.*, Tungsten Inert Gas) welded.

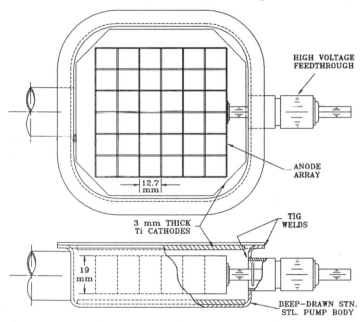

Figure 2.4.9. The first Varian, multi-cell sputter-ion pump.

Flanges of the earlier pumps had Alpert *step seals*, though three years later, pumps of all sizes were sealed with ConFlat® flanges.[79] This is a good example of the ratcheting effect of technology. The UHV sputter-ion pump came before the ConFlat® flange, but made the development of reliable, bakeable, take-apart flanges important. A large Alnico V® magnet provided the external magnetic field. The lighter ferrite magnets had not yet found wide industrial use. The speed of this pump was of the order of 8 ℓ/sec for N_2. Variations of this simple design are used as appendage pumps to this day on many of the microwave tubes produced throughout the world.

This first Varian pump proved to be an excellent *appendage* pump for use on large microwave tubes. At the time it was being developed, I was a tube technician at the G.E. Microwave Laboratory in Palo Alto. In 1957, we were developing a large, highly classified, super-power klystron for a radar system. We heard rumors of the *magical* electronic pumps which could be used to append our tubes. We obtained prototype pumps, and developed a battery-operated, high-voltage power supply so that appendage pumps, attached to klystrons, would pump on the tubes, even while in transit to the transmitter.

This appendage pump and even larger pumps under development at Varian were reported on that same year by Hall[80] and a year later by Jepsen.[24] In August of 1958, Lloyd and Huffman made patent application for a pump with modular elements which fit into *pockets* defined by the envelope of the pump body.[82] Figure 2.4.10 shows the concept of this modular design. Pump elements were fabricated so that the anode arrays were supported off of the cathode assembly frame by shielded ceramic insulators. The insulators had to be optically shielded from the cathodes to avoid electrical breakdown problems caused by sputtered cathode material. The design of these insulator assemblies became more sophisticated with time, as it was also important to minimize insulator contamination problems associated with the starting of pumps (*i.e.*, the *unconfined* discharge mode). The assembled elements were inserted into separate *pockets* in the pump body weldment.

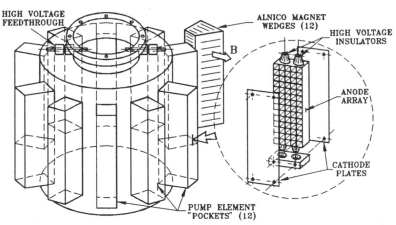

Figure 2.4.10. Representation of a variation of the first "pocket" pumps, patented by Lloyd and Huffman of Varian Associates.[82]

Thereafter, a variety of unique pump configurations began to appear on the market.[83,84] However, this development by Lloyd and Huffman repre-sented the next major breakthrough in sputter-ion pumping as: 1) it made possible manufacturing economies associated with the modular construction;

2) it suggested that the size of sputter-ion pumps was virtually unlimited; and 3) it incorporated innovative, and equally modular, magnetic *circuit* designs. All subsequent pump designs were minor variations of this modular design.

2.4.3 Noble Gas Instabilities

The diode pump proved to have a fundamental flaw. It was discovered that, after pumping on an air leak for an extended period of time, the pump would from time to time violently *belch* up gas in some sort of periodic fashion. The gas which was preferentially regurgitated by the pump was argon. These instabilities, called pump *memory* effects, would result when pumping on a gas mixture containing a partial pressure of an inert gas, or when pumping on a pure inert gas. Such instability problems were recognized early in the development of these pumps.[58,85,86] At times they were evidenced by gradual cyclical fluctuations in pressure; on other occasions, as very rapid increases in pressure. Brubaker was the first to report the cyclical, almost *sinuosoidal*, variations in pressure, sometimes characteristic of these instabilities.[56] Malter reported on this type of instability problem a year later.[86]

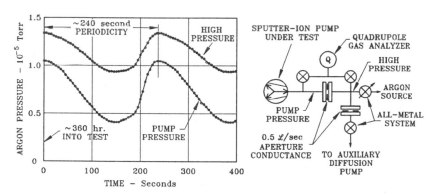

Figure 2.4.11. Example of sinusoidal–type argon instabilities observed with a sputter–ion pump, when simultaneously pumping on the system with a second stable auxiliary pump.

The sinusoidal-type instabilities are difficult to duplicate. They usually involve use of very controlled parameters and conditions, (*e.g.*, highly regulated power supplies, *etc.*) and the simultaneous pumping on the system with some form of stable, auxiliary pump. An example of this type of instability is shown in Fig. 2.4.11 (Welch, Stanford Linear Accelerator Center (SLAC), 1969). Extensive tests were conducted at that time on multi-cell pump elements, (32 cells, $d = 1.98$ cm ϕ, $\ell = 1.59$ cm) with Ti cathodes, which were inserted into custom-built pump pockets.

In other studies at SLAC, the more abrupt or *violent* type of instabilities were also duplicated under laboratory conditions. An example of this type of instability is shown in Fig. 2.4.12. In this case, in the absence of a stabilizing auxiliary pump, at the onset of the instability, the pressure gradually rises, over a period of hours. After this, abrupt, periodic peaks or *spikes* in pressure are observed, after which, the pump will again be stable for a period of several hours or even days. The pump and apparatus were identical to that of Fig. 2.4.11, except in this case the auxiliary pump was not used.

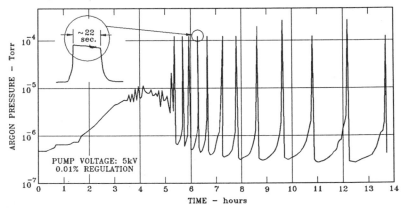

Figure 2.4.12. Example of instabilities in a system with one sputter-ion pump with titanium cathodes, after pumping argon at 5 x 10⁻⁷ Torr for a few hours.

The shape and periodicity of the instability *wave functions* will vary with any of the parameters of Table 2.2.1, with the volume of the vacuum system, and with the presence of auxiliary pumping on the system. In fact, it was noted in these studies that under highly controlled circumstances, including temperature, the pump elements described above could pump on Ar leaks for extended periods. In one instance, a new pump pumped on an Ar leak, at a pressure of $\sim 2 \times 10^{-7}$ Torr for only $5\frac{1}{2}$ hrs, having pumped only $\sim 2.4 \times 10^{-3}$ Torr-$\mathcal{L}$ Ar prior to the onset of instabilities. On readjusting the anode voltage from 5.0 kV to 4.2 kV, this same pump pumped on an Ar leak, at a pressure of $\sim 2 \times 10^{-6}$ Torr, for ~ 700 hrs, pumping ~ 3.0 Torr-$\mathcal{L}$ Ar, with no evidence of instability. Reducing the Ar pressure for several days, to $\sim 2 \times 10^{-7}$ Torr, did not prompt the onset of instabilities. These results suggested two things regarding sputter-ion pumps: 1) there are narrow voltage "bands" within which, under highly controlled conditions, a given pump may stably pump Ar; 2) in the absence of the reporting of all of the experimental data and test parameters, discretion should be used in interpreting published claims of instability or stability relating to a given pump configuration.

SLAC Pump Instability Problem

High vacuum systems which are frequently cycled back to air are less prone to argon instabilities.[85] However, systems with air leaks, yet operating for extended periods at pressures even $\leq 10^{-8}$ Torr, may evidence argon instabilities in a matter of a year of so. This became a serious problem in the SLAC Two-Mile accelerator, and prompted the above studies. Anomalous pressure bursts in sectors of the accelerator resulted in the interruption of operation of the machine. The vacuum system is described elsewhere.[87] For purposes of this discussion, it is divided into thirty, 100 m sectors. A simplified schematic of one of these sectors is given in Fig. 2.4.13.

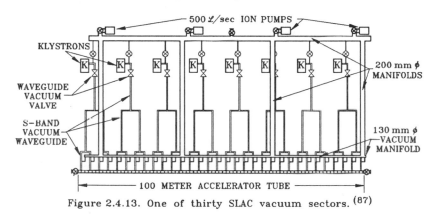

Figure 2.4.13. One of thirty SLAC vacuum sectors. [87]

The only vacuum connection between each of the sectors existed at the conductance-limited, disk loaded waveguide (*i.e.*, the accelerator beam pipe) located in the accelerator tunnel. Therefore, from a vacuum standpoint, the sectors were somewhat independent. Each sector was pumped by four, 500 $\mathcal{L}$/sec diode sputter-ion pumps, equally spaced along a 100 m, 200 mm ϕ manifold running the full length of the sector. In early 1967 we observed anomalous pressure *bursts* in many of these sectors. We suspected argon instabilities. R.S. Callin, of the SLAC Vacuum Group, substantiated this in a series of measurements with a partial pressure analyzer. He reported the following:

1) There were steady-state, pressure gradients in the 100 m manifolds of six sectors evidencing these periodic bursts. The average of pressures at one end of the manifolds was ~4 $\times$ 10^{-9} Torr, where the average of pressures at the other end of the troublesome sectors was ~3 $\times$ 10^{-8} Torr. We had long suspected there were air leaks in these sectors, but they could not be located.

2) The gas bursts, determined by Callin to be argon, *always* initiated in the pump at the lowest-pressure end of the 200 mm ϕ manifold. During a burst,

the pressure would increase to $\lesssim 5 \times 10^{-5}$ Torr in ~15-30 minutes, and then recover to the original pressure in an equivalent or less time.

3) Argon pressure bursts in the pump furthest removed from the air leak would cause one or two of the adjacent pumps to also become unstable.

4) Initially, occurrences of argon instability were random in time. Eventually, they occurred on a regular basis - usually in the mid-morning when the klystron gallery had warmed up from the night before (no heating of the gallery was required in "sunny" California).

It was at first perplexing that the pump furthest removed from the leak was the first to become unstable. We suspected it might have something to do with an argon enrichment phenomenon created by the distributed pumping system. Assume that the pumping speed of the sputter-ion pumps for Ar was ~2% of the speed for air.[85,88] Admittedly, this speed is very difficult to stably quantify in conventional diode pumps.[24] Assume that the air speed of the pumps is the rated 500 $\mathcal{L}$/sec; at the lower pressures, the speed is probably ~100 $\mathcal{L}$/sec; but, neglect this fall-off in I/P with pressure. Assume there is ~1.0 % Ar in the air, and that the inlet leak rate is proportional to the partial pressure of the atmospheric gases. To simplify the problem, assume that, excluding Ar, the remainder of the in-leaking gas is N_2. The pressure at the leaking end of the manifold is 3×10^{-8} Torr. With these assumptions, one can calculate the results of Fig. 2.4.14. That is: 1) There is an Ar enrichment process that occurs at the location of the leak. Where the partial pressure of Ar in the atmosphere is assumed to be ~1 %, the partial pressure of Ar near the leak is ~20 %. 2) The partial pressure of Ar at the pump furthest removed from the leak is ~23 % of the partial pressure of Ar at the pump nearest the leak. But, 3) the Ar pressure at the furthest removed pump is ~99.7 % of the total pressure, neglecting system background gases.

Each of these pumps had sixteen elements identical to the element described at the beginning of Section 2.4.3. Assuming the pumps operated two years prior to the onset of instabilities, calculations show that ~5 × 10⁻² Torr-$\mathcal{L}$ of argon was sufficient to induce instabilities in a single element. However, in this case, varying the pump voltage did not eliminate the instability problem. From this, we concluded that the pumping of chemically active gases mixed with the in-leaking Ar aided in the pumping and retention of Ar. The noble gas stabilization effect of pumping mixed gases has been reported elsewhere.[89,48,24] The enrichment phenomenon has not.

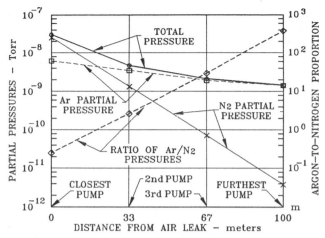

Figure 2.4.14. Argon enrichment associated with an air leak at one end of a manifold along which pumps are distributed.

Diodes with Slotted Cathodes

Several schemes were developed for the stable pumping of noble gases. It was found that if slots were machined into the cathode plates, stable Ar pumping could be achieved up to pressures of ~10^{-5} Torr.[85] A patent application was made by Jepsen on such a pump,[90] only two years after publication of results of the first multi-celled pump.[78] This *slotted cathode pump*, as described in Jepsen's patent, is shown in Fig. 2.4.15.

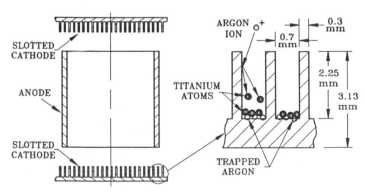

Figure 2.4.15. Slotted titanium cathodes for noble gas pumping. [90]

It was assumed that, because most of the ions striking the cathodes do so at large angles of incidence, the sputter-yield was greater. This had the effect of increasing η, the pumping efficiency. Also, it was believed that noble gas molecules, weakly physisorbed in the lower regions of the troughs formed by the slots, would be physically trapped by subsequently sputtered material.

The slotted-cathode pump did not prove to be the *perfect panacea* for noble gas pumping with sputter-ion pumps. For example, Vaumoron showed that these pumps would also become unstable when pumping on a 20% Ar, 80% N_2 mixture, at pressures in the 10^{-5} Torr range.[9 1] However, these pumps could pump on air leaks for extended periods without the evidence of noble gas instabilities.[8 5] This improved stability was later discovered as resulting from the burial (*i.e.*, physisorption) of high-energy, neutral atoms. But, this is discussed below. The slots had to be machined in each of the cathode plates. This was an expensive operation. The slotted diode pump was short-lived, as less expensive alternate solutions were soon discovered.

2.4.4 The Triode Pump

Vanderslice made patent application for the first triode sputter-ion pump in September 1958.[9 2] It was a sputter-ion pump with positive anode, negative cathode and slightly more negative *collector* plate. That is, the anode, cathode and collector, arrayed in this order, were operated at three different voltages. This triode pump effectively pumped noble gases. The cathode was somewhat transparent to bombarding ions and cathode material sputtered by these ions. Some of the ions could pass through the cathode mesh-structure and bombard the collector material, but with negligible sputtering. It was believed that noble gas ions, weakly physisorbed on the collector, were sputtered over with cathode material and trapped within the collector. Similar to the single-cell diode patent of Gurewitsch and Westendorp,[1 9] also of the General Electric Company, Vanderslice's patent described a single-cell, triode sputter-ion pump. Clarke, of the General Electric Company, later patented a form of multi-cell triode pump.[9 3]

In May of 1959 Brubaker made patent application on the first multi-cell, triode sputter-ion pump.[9 4] Later that year he presented a paper on the characteristics of this new triode pump *vs.* the conventional diode sputter-ion pump.[5 6] This was the first comprehensive paper on the noble gas instability problems of conventional diodes. He reported on various manifestations of these instabilities, from violent pressure bursts to, at times, almost sinusoidal variations in pressure. He postulated that Ar, shallowly buried in the cathodes of diode pumps, was liberated by subsequent sputtering of cathode material. He described an experiment where there was a factor of 30 increase in the Ar pumping speed of a single-cell diode as the result of the use of an auxiliary filament to sublime Ti over weakly trapped Ar atoms in a Ti cathode. This over-sputtering of material, as proposed by Herb in 1953,[1] was thought to be the main mechanism responsible for the pumping of noble gases. The effect is real as evidenced in the SLAC instability problem (*i.e.*, a mixture rich in N_2 will augment the stable pumping of Ar). Brubaker's triode was configured as shown in Fig. 2.4.16. In this pump configuration, the anode is operated at + 3 kV with respect to ground, the cathode at ground potential, and an interposing, somewhat transparent *auxiliary* cathode grid, at −3 kV with respect to ground.

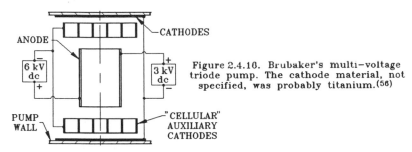

Figure 2.4.16. Brubaker's multi-voltage triode pump. The cathode material, not specified, was probably titanium.[56]

Brubaker reported triode speeds for Ar of about 25% that of air. He also reported that the triode had four times the pumping speed for air as that of the diode. Though the former results have proven to be the case, the latter has not for similarly sized pumps. In fact, for reasons to be discussed, triode pumps typically have less pumping speed for air, per unit pump volume, than do diode pumps. However, Brubaker was the first to note improved pumping of triode pumps for Ar. He reported improved pumping as "... ions incident at a grazing angle do more sputtering than those which strike the surfaces at normal incidence." He also pointed out the apparent ease with which argon instabilities could be initiated in a diode pump, with slight variations in voltage, where triode pumps seemed much more immune to this phenomenon. The triode pump was effective in pumping noble gases. But, the correct model for the pumping of these gases remained a mystery.

Two years after the publication of Brubaker's paper, Hamilton, of the Consolidated Vacuum Corporation, discovered a *strange* phenomenon.[47] He varied the electrode potentials of a triode pump, as shown in Fig. 2.4.17, and measured associated changes in pump speed.

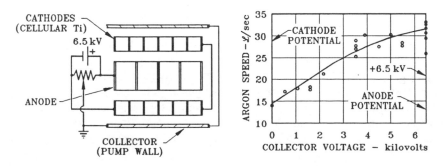

Figure 2.4.17. Hamilton's triode pump configuration and argon speed results as a function of electrode potentials.[47]

Hamilton determined that if the electrodes of the triode pump furthest removed from the anode (*i.e.*, Brubaker's *cathodes*) were operated at the same potential as the anode, pumping speed for the gases, including Ar, was comparable to that observed with Brubaker's potential settings. These results caused quite a stir in the industry. Of course, the industry immediately adopted the much simpler, single potential power supplies for use with triode pumps. But the perplexing aspect of Hamilton's results was that the proposed mechanism for triode pumping of noble gases *went out the window* with his findings. That is, it was believed that Ar ions were shallowly buried in the equivalent of Brubaker's cathodes (*i.e.*, see Fig. 2.4.16) - the pump walls, in Hamilton's case - and subsequently covered over by material sputtered from the auxiliary cathodes. Hamilton's results indicated ion burial in the cathodes (*i.e.*, pump walls) could not be occurring, as the kinetic energy of any ion reaching the cathodes had to be zero, assuming no significant rf energy was being generated in the discharge.

Through use of autoradiography, Andrew and others reported that a significant amount of the noble gases were pumped at the anodes.[89,95,96,97]. This too seemed energetically impossible, for how could gas ions reach the anode? Rutherford speculated that anode pumping was due to rf voltages present in the discharge.[88] The triode pump effectively pumped noble gases, and diodes to some extent, but an understanding of the mechanism was still not at hand.

The Varian StarCell® triode pump, first reported on by Pierini and Dolcino in 1983,[98] is the most recent development in triode pumps.[99,100,101] The conventional triode (*i.e.*, a pump with the Hamilton potentials, hereafter) was expensive to manufacture because of the high labor content associated with assembling the grid-cathode module. Each Ti cathode strip had to be individually mounted in the stn. stl. cathode frame. Also, after extended use at high pressures or the extended pumping of gases containing H_2 (*e.g.*, H_2O), one or more of the grid strips tended to burn through and short to the pump wall or anode. The StarCell® pump configuration, shown schematically in Fig. 2.4.18, eliminated the cathode strip shorting problems of the conventional triode pump. Also, the conventional triode pumps, after pumping H_2 for extended periods, are subject to thermal *run-away* problems.[102] This problem, discussed below, is somewhat ameliorated by the StarCell® cathode configuration (*i.e.*, due to the extended cathode surface area of this pump). However, the primary advantage of this pump configuration is in the reduction of triode manufacturing costs. Special tooling made possible the formation of the *stars* in the cathode plates. This investment in tooling eliminated the repetitive, high labor costs associated with assembling the grid assemblies of the conventional triode pumps. These labor savings can be passed on to the buyer. However, replacement cathodes must be purchased from the original equipment manufacturer.

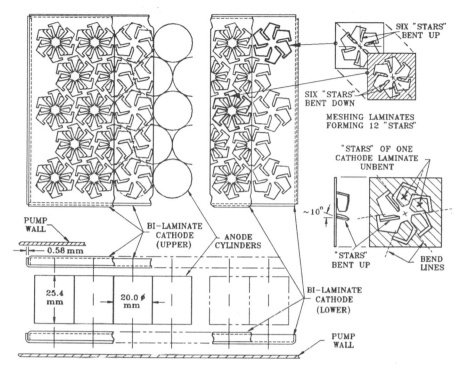

Figure 2.4.18. Pierini and Dolcino's StarCell® pump element.[98]

Magnetic Fields in Diode and Triode Pumps

Subsequent to Hamilton's discovery,[47] all commercial triode pumps were manufactured and sold with the cathodes operated negative with respect to ground, and the anodes and pump bodies at ground potential. Brubaker's *cathode* was eliminated, as the walls of the pump body served this function thereafter. For reasons which will become evident, eliminating the use of titanium pump *walls* may have been a fundamental error.

Operating the cathodes at a negative potential required that, for electrical isolation, additional spacings, g, had to be provided between the cathodes and pump body. Therefore, for the same cell dimensions (*i.e.*, ℓ and d) the magnetic field gap would have to be greater for the triode pump. A larger magnetic field gap, for the same amount of magnetic material, would result in a decrease in magnetic field for the triode.

Assume that the magnetic circuitry of the diode pump was designed so that the magnetic field was optimum for the maximum I/P (*i.e.*, near the *critical* magnetic field of Fig. 2.2.3). An increase in the magnetic field gap of the pump would result in a corresponding decrease in magnetic field and consequential decrease in triode pump speed. We see that because of the need for a larger magnetic field gap, the speed of the triode pump would be less than that of the diode.

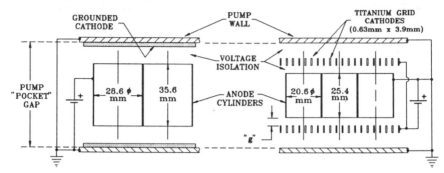

Figure 2.4.19. Conventional triode and diode sputter–ion pumps configured to be installed in pump "pockets" of the same size.

Some manufacturers sold both diode and triode pump configurations. The above considerations suggested that the size of the diode pump pockets would have to be different from pockets used for the triodes. Also, the fall-off in the triode magnetic field had to be addressed. Therefore, for reasons having to do with both the economics and the physics of sputter-ion pumps, these manufacturers decided to make all pumps have the same size pockets and, therefore, the same magnetic field. Anodes of triode pumps were shortened to compensate for the two additional electrical gaps and to make use of the same pockets and magnetic field circuits used with diode pumps. This resulted in a triode configuration comparable to that shown to the far right of Fig. 2.4.19.

The value of I/P is directly proportional to ℓ, the anode length. Therefore, a reduction in anode length in the triode pump resulted in a corresponding reduction in the speed of the pump. It was known that the fall-off in speed of sputter-ion pumps with a decrease in cell diameter becomes significant only at low pressures. This is shown in Fig. 2.2.7. Therefore, in order to increase the high-pressure speed of the triode pumps, the diameter of the cells was decreased so that more cells could be packed into the same volume. Because of this, at high pressures, the pumping speed per unit volume of the triode pump is very comparable to the diode pump. However, at lower pressures and with nominal magnetic fields, this is not the case.

Field Emission in Triode Pumps

There appears to be a tendency for triode pumps to have problems with field emission from the cathodes.[103,98,104] This is called a *problem* as in many applications pump current is used as an indication of system pressure. For example, a new 120 $\mathcal{L}$/sec triode pump will have of the order of ~100 μA field emission current on initial start-up. This corresponds to current equivalent to a pressure of ~5 × 10^{-8} Torr. It is believed that *whiskers* develop on the grids of the cathodes each time a triode pump is started, and that these whiskers cause the field emission. The whiskers may be *burned off* by a process called hi-potting - the momentary application of a ~20 kV, AC voltage. Unless the pump is hi-potted, field emission current would obscure indicated pressures inferred by pump current from the Penning discharge. The presence of field emission in a pump may be verified by removing the magnets, and while under high vacuum, measuring the pump current as a function of pressure. If field emission is present, a plot of the $\ln(IV^{-2})$ *vs.* V^{-1}, where V is the pump voltage and I the current, will result in a straight line which is called a *Fowler-Nordheim plot*.[105]

2.4.5 The "Differential" Sputtering Pump

In 1965, James and Tom, of Perkin-Elmer (Ultek), applied for patent on a new type of diode sputter-ion pump.[29] They mistakenly assumed that the current model for the pumping of inert gases, as postulated for triodes, was correct. Therefore, they constructed a pump having two types of cathode materials, as shown in Fig. 2.4.20. One cathode material was reported to be more readily sputtered than the other. In a paper which they published a year later, they assumed that noble gas ions, weakly buried in the cathode with the lowest sputter yield, would be preferentially buried by material sputtered from the cathode with highest sputter-yield.[106] They also suggested, in this early paper, that "... the high rate of deposition of atoms onto the anode surfaces (due to a high sputter-yield cathode) can substantially increase the pumping of fast neutrals that impinge on the anodes."

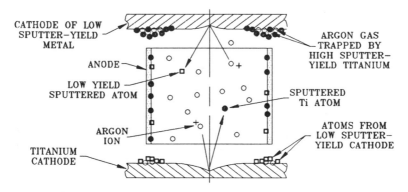

CATHODE OF LOW
SPUTTER-YIELD
METAL

ANODE

LOW YIELD
SPUTTERED ATOM

ARGON
ION

TITANIUM
CATHODE

ARGON GAS
TRAPPED BY
HIGH SPUTTER-
YIELD TITANIUM

SPUTTERED
Ti ATOM

ATOMS FROM
LOW SPUTTER-
YIELD CATHODE

Figure 2.4.20. The "differential ion" ("DI") sputtering model proposed by James and Tom to explain the pumping of argon. (29)

In their patent application James and Tom suggested that one cathode might be constructed of Cu and the other of Ti. They also proposed the use of Ta and Ti cathodes in that patent. Evidently, they assumed that both Cu or Ta had lower sputter-yields than that of Ti. In the first article which they published, no mention was made of the presumably proprietary Ta cathode material.[106] However, competitors and other researchers had no mercy, and soon the *cat was let out of the bag*. Three years later, in 1969, James and Tom published results of what came to be known as the *DI pump* (*i.e.*, Differential Ion), with Ta and Ti cathodes.[107]

These DI pumps very effectively pumped argon, and pumped chemically active gases with speeds only a few percent less than pumps with two Ti cathodes. For example, we at SLAC modified all of the pump elements of the two-mile linear accelerator, using Ti/Ta cathode pairs. Argon instabilities were never again observed at SLAC. However, there were some inconsistencies in the "differential sputtering" model of James and Tom. It was known that, for a broad range of ion energies, Ta and Ti have very comparable sputter-yields. Therefore, the differential sputter-yield model couldn't be correct. The correct model for noble gas pumping was still not fully understood. However, as first suggested by Lafferty in 1961,[95] people began to speculate that the burial of high energy neutrals must play an important role in the pumping of noble gases. Neither Lafferty nor Tom and James speculated on the origin of these neutrals. Lafferty concluded that some *charge exchange process* must exist to explain the presence of noble gases pumped in the anode assembly.

The High Energy Neutral Theory

The atomic structure of a metal comprises an ordered, three-dimensional array of positive ions of that material. These metal ions are *awash in a sea* of electrons, so that the net electrical charge of a piece of metal is zero. Some electrons are tightly bound to, and shared by, neighboring ions in the lattice. This gives the lattice a structure unique to the element. Other electrons, though *belonging* to particular metal ions at any moment, are free to meander about the bulk of the lattice structure. These are called *conduction electrons*. They stagger their way through the lattice in a preferred direction when an electric potential is applied to the metal.

On an atomic scale, the metal ionic sites have a positive charge. When a positive gas ion bombards the surface of a cathode plate, it is confronted by these positive point charges, and it is deflected at some angle, due to mutual coulomb repulsion. If the ion *bounces* back from the cathode surface, while at the same time *picking up* an electron from the metal, the ion is electrically neutralized. If, when bouncing back, it also retains much of its original kinetic energy, it is then said to be a *high energy neutral*. The reflected high energy neutral does not suffer coulomb repulsion on its next encounter with a solid surface. Therefore, as a high energy neutral, it will be more deeply buried into some solid surface in line-of-site with the primary collision. The higher the kinetic energy of the reflected neutral, the deeper the burial and permanent entrapment in the adjacent pump element, be it cathode or anode.

Jepsen was the first to demonstrate that the creation of high energy neutrals at the cathodes, and their subsequent burial on reflection, was the primary mechanism by which noble gases were pumped in all sputter-ion pumps.[108] The model was elegantly simple, appears in many freshman physics textbooks,[81] and has been verified by the work of a number of investigators (e.g., see: 109,110,70,74,111,112,97,90). The model had to explain the noble gas pumping of diodes with slotted cathodes, triodes, and the Ta/Ti DI pumps. Jepsen calculated the possible angles of deflection and energies of reflected neutrals for a given cathode material. He made the assumption that collisions between gas ions and ions in the metal lattice of the cathode were elastic (i.e., kinetic energy was conserved), and he treated the interaction of the two ions as a free-body problem. For example, assume the bombarding gas ion, represented in Fig. 2.4.21, has an initial kinetic energy of $\frac{1}{2}m_1\mathbf{u}^2$, where m_1 is its mass and $\mathbf{u}$ its velocity. Assume that a metal cathode ion, of mass m_2, has no initial velocity. In the collision process the gas ion will be deflected to an angle θ, and the metal ion to an angle ϕ, both depending on the impact distance, d, and the mass of each ion. Define the velocity vector of the gas ion after the collision as $\mathbf{v}$, and the velocity vector of the metal ion as $\mathbf{s}$. Conservation of kinetic energy and momentum results in the following equations:

$$0 = -m_1\mathbf{v}\,\sin\theta + m_2\mathbf{s}\sin\phi \qquad (2.4.4)$$

$$m_1\mathbf{u} = m_1\mathbf{v}\,\cos\theta + m_2\mathbf{s}\cos\phi \qquad (2.4.5)$$

$$\tfrac{1}{2}m_1\mathbf{u} = \tfrac{1}{2}m_1\mathbf{v}^2 + \tfrac{1}{2}m_2\mathbf{s}^2 \qquad (2.4.6)$$

The $\sin\phi$ and $\cos\phi$ terms may be eliminated by squaring and adding (2.4.4) and (2.4.5). Defining $R = m_2/m_1$, solution of $\mathbf{v}^2/\mathbf{u}^2$, i.e., the ratio of the energy of the reflected neutral vs. its energy as an incoming ion, is:

$$\frac{\mathbf{v}^2}{\mathbf{u}^2} = \frac{\{\cos\theta \pm (R^2 - \sin^2\theta)^{\frac{1}{2}}\}^2}{(R + 1)^2}. \qquad (2.4.7)$$

The table in Fig. 2.4.21 gives results of (2.4.7), as a function of θ, m_2 and m_1, for four gases scattered off Aℓ, Ti, Mo and Ta. We see that for normal incidence, neutrals of Ar scattered off Ti have $< 10\%$ of the energy of the primary ion. Reflected neutrals of Ar on Ta retain $41\% - 64\%$ of their original energy. Of course, when $m_2 < m_1$, then $m_2 < m_1\sin\theta$ implies that neutrals can only be scattered in the forward direction, as in the case of Ar on Aℓ. In this case, the maximum possible scattering angle, θ, is ~41°. This does not mean that Ar cannot be pumped by a pump with Aℓ cathodes.

It merely means that the pump must be designed so that most of the Ar ions strike the cathodes at grazing angles of incidence, as in the case of triode pumps.[72]

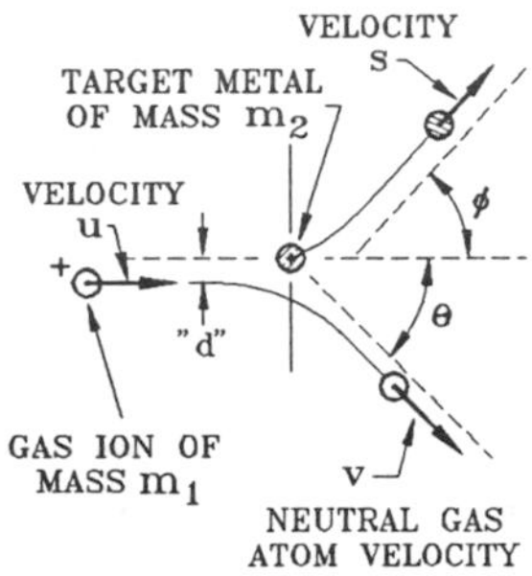

GAS		$\theta = 0$	$\pi/8$	$\pi/4$	$3\pi/8$	$\pi/2$	$5\pi/8$	$3\pi/4$	$7\pi/8$	π
Al	He	1.000	0.978	0.917	0.832	0.742	0.661	0.600	0.563	0.550
	Ne	1.000	0.890	0.621	0.333	0.144	0.062	0.033	0.023	0.021
	Ar	1.000	0.781	–	–	–	–	–	–	–
	Kr	1.000	–	–	–	–	–	–	–	–
Ti	He	1.000	0.987	0.952	0.902	0.846	0.793	0.751	0.724	0.715
	Ne	1.000	0.937	0.777	0.580	0.407	0.286	0.213	0.177	0.166
	Ar	1.000	0.878	0.581	0.272	0.091	0.030	0.014	0.009	0.008
	Kr	1.000	0.736	–	–	–	–	–	–	–
Ta	He	1.000	0.997	0.987	0.973	0.957	0.941	0.927	0.918	0.915
	Ne	1.000	0.983	0.937	0.871	0.799	0.734	0.682	0.650	0.639
	Ar	1.000	0.967	0.878	0.759	0.638	0.537	0.464	0.421	0.407
	Kr	1.000	0.931	0.756	0.546	0.367	0.247	0.178	0.145	0.135
Mo	He	1.000	0.994	0.976	0.950	0.920	0.891	0.867	0.852	0.846
	Ne	1.000	0.968	0.883	0.769	0.652	0.554	0.482	0.440	0.426
	Ar	1.000	0.938	0.779	0.584	0.412	0.291	0.218	0.181	0.170
	Kr	1.000	0.872	0.562	0.244	0.068	0.019	0.008	0.005	0.005

Figure 2.4.21. Kinetic energy ratio of reflected neutral atom to bombarding ion, as a function of the masses of the gas and target metal atoms.

The Ar pumping characteristics of the conventional diode, DI, slotted diode and triode pumps started to make sense to investigators. In the case of the conventional diode pump (i.e., pumps with two Ti cathodes), the reflected neutrals of Ar ions have very low energies and are therefore shallowly buried in the adjacent anode and cathode. A very slight change in one of the pump parameters, such as pressure, magnetic field (e.g., due to temperature changes), or voltage, can cause a slight modification in the space charge distribution. The magnetic field of a typical sputter-ion pump varies from 30 - 40% across the anode array (of identical cells). This suggests that the discharge modes (i.e., space charge distribution) in, for example, anode cells with high magnetic field could have one mode configuration, where the space charge distribution might differ markedly in cells with the lower magnetic field. Rudnitskii and Dallos have shown that such mode shifting occurs even at low pressures.[57,103] There may also be mode mixing, or unstable moding in cells between the two magnetic field extremes. A slight change in the space charge distribution in a few of the cells, due to some perturbation in a pump parameter, could result in ions bombarding previously unscathed portions of the cathodes. From these areas, weakly buried neutrals would be subsequently liberated. This is tantamount to a negative pumping speed for the gas.[113] The liberated neutrals in effect increase the local pressure in the cell. This gas in turn is ionized, resulting in an increase in the intensity of ion bombardment of the cathodes, etc. This is in essence a "positive feedback" situation, which, depending on the amount of weakly buried Ar, can result in pressure runaway. If the system is damped by the presence of a second stabilizing pump, the feedback may be slightly negative, and may result in a sinusoidal instability, or dithering effect in pressure. (I have also seen H_2 induced, sinusoidal pressure variations with a single pump, pumping on a closed system.[45] In this case, the pump self-stabilized due to changes in the discharge characteristics, pumping the H_2 elsewhere on the cathodes.)

In the case of diode pumps with Ta/Ti cathodes, neutrals reflected from the Ta cathodes have energies of the same order as the primary ions, and are deeply buried in the anode and adjacent cathode. Therefore, with slight changes in the discharge, there is less probability that previously buried neutrals will be desorbed.

In the case of triode pumps, the choice of cathode material is of less importance in the pumping of noble gases. This was demonstrated by Liu's experiments with $A\ell$ cathodes,[72] and in earlier work by Hall, when using this material as part of a triode cathode assembly.[114] Most of the ions impinging on the cathode grids do so at grazing incidence. Therefore, regardless of the cathode material used, a deflected, neutralized ion retains much of its original kinetic energy and becomes buried in the walls of the pump. A similar situation exists with the slotted diode pump. The grazing angle of incidence of ions on the slots of the cathodes makes possible the creation of high energy neutrals which in turn are buried and sputtered over at the bottom of the slot *trenches*. However, because it is possible in time to completely sputter through the cathode plates of a slotted diode pump, it is possible for Ar to be later liberated and cause instabilities. This is not possible with the triode pump (*i.e.*, a pump using Hamilton's potentials). In this case, the potential of the anode is the same as the walls of the pump body. This means that the kinetic energy of even the most energetic ion created in the discharge will be zero at the walls. Therefore, there is no ion sputtering of the pump walls, and it is impossible for ions to desorb previously pumped gas. The walls of the pump body apparently are not appreciably sputtered by energetic neutrals.

In 1970, Vaumoron and De Biasio conducted an extensive series of experiments which put Jepsen's fast neutral theory *to the test*.[115] They tested the theory with cathode materials including Ti, Cu, Mo, Ag and Ta, and for the gases He, Ne, Ar, Kr and Xe, in some 22 combinations of materials and gases. They found excellent correlation with the fast neutral theory, at pressures < 10^{-4} Torr, providing $R \gtrsim 2.4$. Baker and Laurenson conducted noble gas pumping experiments with cathode material pairs of $A\ell$/Ta, Ti/Ti and Ti/Ta and for the gases He, Ne, Ar and Kr.[116] These results also verified Jepsen's fast neutral theory, though intermittent pressure pulses were noted when pumping Ne, Ar and Kr with the $A\ell$/Ta cathode pair. Where Vaumoron's tests were conducted at steady-state pressures, Baker's tests involved deducing the speed of different cathode pair materials from the rate of pressure pump-down of the noble gas. The advent of the slotted diode, DI and triode sputter-ion pumps overcame one major limitation of these pumps; that is, the pumping of noble gases.

Beware of "Shortcuts"

A note of caution is given regarding noble gas pumping provisions. Because of the fall-off in speed noted for triode pumps at the lower pressures, some would prefer to use diode pumps for their low-pressure applications. However, because of the cost of Ta, it has been suggested that some of the elements of the diode pump might have the conventional Ti/Ti cathode pairs, where the others might be constructed with Ta/Ti elements. No data has been published which suggests that such *mixing* of elements in a given pump or system will afford immunity to instabilities in those elements with the Ti/Ti cathode assemblies. Granted, the system will be more tolerant to Ar leaks, as the preponderance of Ar will be pumped by the *noble elements*. But, the conventional diode elements (*i.e.*, elements with Ti/Ti cathodes) may still become unstable after pumping a given amount of Ar.

Also, the use of auxiliary pumping, such as a cryopump, DI or triode pump, to pump on a system once evidencing argon instabilities, may not cause the problem to go away. Such was the case when argon instabilities were noted in LINAC pumps at BNL (the Brookhaven National Laboratory). We first observed that argon instabilities existed in a number of conventional diode pumps used to pump accelerator modules (*tanks*). We sought to determine if instabilities would persist in some of the pumps, if Ti/Ta elements were substituted in other pumps on the same tank. We used a cryopump as an auxiliary pump. It was coupled to one of the tanks so as to afford a speed for Ar of $\sim 10^3$ $\mathcal{L}$/sec. The argon instabilities, verified with a quadrupole residual gas analyzer, persisted thereafter.

2.5 Hydrogen Pumping

Hydrogen is prominent in all UHV systems. Because of this, the pumping characteristics of sputter-ion pumps for this gas have been extensively studied. For example, microwave tube components are frequently brazed in hydrogen furnaces. This process results in large amounts of H_2 being liberated during the final bakeout of these tubes. Also, both fusion[117] and fission work gave further impetus to the study of the H_2 pumping capabilities of sputter-ion pumps.[118,119] Singleton has done the most definitive work on the sputter-ion pumping of H_2 to date.[120,102,121]

As noted in Fig. 2.4.8, H_2 is primarily pumped by diffusion into the cathodes of the pump. But, as you will soon see, the diffusion process is rather complicated. Excluding Li, Ca, and Na, titanium and alloys of this metal have the greatest sorption capacity for H_2,[123,124] and to first order, this is completely reversible.[125,71,102,123] I noted earlier that Gurewitsch and Westendorp, the first to report on use of Ti as a cathode material, used this material for this very reason.[19] In order for there to be substantive diffusion of gas in metals, three requirements must be met: 1) the metal must have a chemical affinity for the gas; 2) the gas must first dissociate into an atomic state; and 3) there must be a concentration gradient of the atomic species to promote its diffusion in the bulk material.[126] However, as will be shown, gases may be artificially implanted into materials for which they have a low chemical affinity, and diffusion "forced" by contrived concentration gradients.

Only ~2.5% of the ions created in a low-pressure H_2 Penning discharge are H^+ ions. Therefore, in order for H_2 to be significantly pumped, H_2^+ ions must first dissociate on or near the cathode surfaces. Also, if the surfaces are *clean*, H_2 molecules similarly dissociate at the cathodes. This requires that there be surface *sites* which will promote this dissociation. These are sometimes called *energy sites*. This merely means that there are locations on the cathode where the value of the activation energy, E_a, of Fig. 2.4.7, is sufficiently depressed so that molecular dissociation and chemisorption is promoted. If the surfaces of the cathodes are *contaminated* by the presence of, for example, TiN or oxides of Ti, E_a can become sufficiently positive where the surface dissociation of the H_2^+ ions will not occur.[49] Gas *contamination* will also inhibit the subsequent surface release of hydrogen in solution with a metal.[127] The energy sites are created by the sputter-cleaning of the cathode surface. However, even at high energies, the lighter H_2^+ ions have poor sputter-yield.[128,129,53] More specifically, at 7.0 keV, the sputter-yield of H_2^+ on Ti is only 0.01.[130] Because of this, when introducing pure H_2 into a pump, it may take a long time for the surfaces to be *scrubbed* clean by the bombarding H_2^+ ions. Therefore, the pump speed may initially be very low for H_2 and gradually increase with time. Eventually, as the cathode hydrogen concentration gradient equilibrates, a steady-state speed will be achieved which will be ~50% of the maximum speed (see Fig. 2.5.3). It is important to note that the high pumping speed for molecular H_2 is only temporary with Ti cathodes at high pressures. However, this effect can be long term with cathodes of Ti alloys.

Rutherford and Jepsen used noble gas ions as a *scrubbing agent* to enhance the pumping of H_2, reporting this effect in 1961.[49,131] They used Ar for this purpose, though this has been known to have disadvantageous side effects, as reported by Dean.[132] Dean used this technique to reduce the total system pressure, comprising mostly H_2, in a low-energy, sputter-ion pumped electron storage ring. However, he recanted doing so, as the beam decay times of an electron storage ring vary to first order with Z^2, where Z is the atomic number of the residual gas species. Though the total pressure in the ring was decreased by *scrubbing* the pump with Ar, because of its higher Z, the beam decay effects from the residual Ar proved to outweigh the beam lifetime benefits of the reduced pressure. Dean thereafter used He as a *scrubbing agent*. Though it has a much reduced sputter-yield, the net benefits were reported to be positive. Rutherford reported on the sustained

H2 pumping of scrubbed Ti cathodes even after removing the high voltage to the pump,[49] observing evidence of pumping almost two hours later. In light of the negligible benefits which might have stemmed from the prior sputtering of the cathodes by H2, this was conclusive evidence of the pumping of H2 by a diffusion process. Singleton noted that this sustained pumping effect did not exist when *scrubbing* the cathodes with N2 ions.[121] He, and others, believed H2 pumping probably ceased, on turning off the power, only because the cathodes became contaminated with N2 or other gases.[49,102] Also, he noted that with the simultaneous pumping of CO and H2, the speed for H2 decreased over that observed when pumping pure H2. Audi reported a similar effect, in certain pressure intervals, with the simultaneous pumping of N2 and H2 with a triode pump.[50] Cummings also noted that the H2 speed of a distributed ion pump, when pumping H2 and CO, was significantly reduced.[133]

Argon not only has a high sputter-yield, but it creates ruptures or dislocations in the cathode metal lattice. This *damage* tends to increase the effective surface area of the cathode, and also creates conduits for the diffusion of hydrogen into the bulk of the material. The pumping of pure H2 will also cause cracks and rifts in Ti cathodes. But, this is for different reasons.[131] In this case, changes in the crystalographic structure occur with differing H2 concentrations in the metal. In other words, there are volume changes in the cathode material, which take place on a macroscopic scale, as the result of variations in the atomic hydrogen concentrations in the Ti. During the sustained pumping of H2, these surface rifts increase the effective surface area of the cathodes. However, Singleton and Audi have shown that the gains in H2 pumping will vanish on the subsequent and sometimes simultaneous exposure to other active gases.[101,102]

Diffusion into the cathodes is a surface area and diffusivity-limited process. In an effort to improve on both, Jepsen constructed cathodes of sintered Ti granules, with particle dimensions ≤ 0.2 mm ϕ.[134] The speed per unit area of this sintered Ti cathode was $\times 7$ greater than pumps with conventional Ti cathodes, and decayed to a steady-state speed of half this value, 25 minutes after removal of pump power. If the sintered granules formed hemispheres, the maximum effective increase in cathode surface area would be $\sim\times 2$. Therefore, diffusion effects, into the bulk of the sintered matrix, must have played a significant role in the enhanced H2 pumping. This technology, then primarily having application in fusion-related programs, was not pursued commercially. New UHV applications have encouraged the revisiting of this technology in the form of NEG (nonevaporable getter) pumps.

2.5.1 Hydrogen Pumping in Diode Pumps

In the early 1970's, Vissers experimented with various schemes for detecting the presence of hydrogen in molten Na heat exchanger loops of reactors.[118] These Na loops interfaced with secondary water cooling loops. If a water leak developed at the interface - a serious problem in these systems - the concentration of nascent hydrogen would increase in the Na-cooled loop. Vissers developed a *water leak* monitor to detect changes in the concentration of hydrogen in these liquid Na loops. It comprised a sputter-ion pump attached to a reinforced, Ni diaphragm which in turn was inserted into the molten Na stream. Nascent hydrogen, stemming from a water leak, would diffuse through the Ni diaphragm and into the sputter-ion pump. One could therefore pump the hydrogen, while at the same time use the pump current as an alarm, to indicate a water leak into the Na loop.

One problem with this scheme was that the speed of the pump for H_2 varied so widely, that it was difficult to do quantitative measurements. For example, if the speed was initially low, the H_2 pressure indicated by pump current would have the value I. If, however, the speed increased in time by ×3, for the same H_2 throughput, the current for the new pressure would be $I/3$. Variations in H_2 speed of this magnitude are quite possible.[120] This implied that one could not rely on pump current readings as an accurate indication of the magnitude of a water leak into the Na loop. A second problem had to do with increasing the capacity of the pump for H_2, so as to prolong the life of the instrument. Though the idea for the original instrument was Vissers', people at Varian later helped in this effort by continuing research on H_2 pumping. Hill experimented with and made patent application on the use of various alloys of Ti in sputter-ion pumps, specifically aimed at the *water-leak* monitor application. In a patent application claim in 1975, he noted that greater hydrogen capacity is to be achieved by the use of one of the Ti alloys, "... stabilized to maintain it in the body-centered cubic (bcc) crystal lattice form" (presumably, rather than the hexagonal, close-packed (hcp) form of pure Ti).[135]

In 1967 Lamont *reported* on the enhanced sputtering of bcc cathode materials, when he made patent application on what became known as the diode *post pump*.[136] (He gave credit to Brubaker, who first used *posts* in a triode configuration.[56]) The *posts* in Lamont's pump were grounded to the cathodes and protruded part way into the anode volumes. He claimed that the use of bcc materials enhanced cathode (*i.e.*, post) sputtering over that of the pure hcp Ti.

I experimented with variations of Vissers' H_2 monitor, and developed an instrument which was more immune to variations in pump speed.[119] The monitor was formed by putting two Penning cell assemblies in series, separated by a small aperture. Cells close to the H_2 source had Aℓ cathodes, with reduced pumping,[73,120,137] where the cell furthest removed from the source had Ti cathodes. If the current to the two *pumps* was summed, there would be an amplification effect which nulled out, for the greater part, current changes stemming from changes in *pump* speed. However, quantitative tests still had to be conducted on a variety of Ti alloys to address the H_2 capacity problem of the *water-leak* monitor. I was then managing the Pump and Instrument R&D Department at Varian. We were all *working* managers, and I took on the cathode materials study program. We had to construct some sort of materials test vehicle with the following features: 1) it had to be bakeable; 2) it had to be a diode configuration (triodes have problems in pumping large quantities of H_2); 3) it had to facilitate measuring and controlling cathode temperature; 4) replacement of the cathodes had to be possible without welding or machining; and, 5) we needed the assurance that all boundary conditions would remain the same except the one variable, the choice of cathode material. The above conditions prompted the development of the Varian HiQ® pump as the test vehicle, and for which I submitted a patent application shortly thereafter.[138] The pump was constructed as shown in Fig. 2.5.1. The cathodes were sealed to the pump body with ConFlat®-type seals, as shown in the blow-up in the figure. Titanium water-cooling tubes were vacuum brazed on the outside surfaces of the 4.7 mm thick cathode plates (1). The magnetic circuitry was formed by the pump body (2), the cathode clamp rings (3) and the steel end-caps (4), which captured the ferrite magnet disks (5). The total surface area of both cathodes under the projected area of the total anode assembly (*i.e.*, not just the anode cell diameters) was ~190 cm². At a pressure of ~2 × 10^{-4} Torr, the total *pump sensitivity* for N_2 was ~400 A/Torr, dropping to ~70% of this value at ~3 × 10^{-8} Torr. The pump sensitivity was ~120 A/Torr for H_2 at ~2 × 10^{-4} Torr. The magnetic field varied between 0.09 - 0.13 T across the 19, 21 mm ϕ × 28 mm long anode cells. If you are a *stickler for detail*, you will note that the pump shown in Fig. 2.5.1 has 19 anode cells. In the final design, one cell was removed to make room for an anode support bracket. The anode structure shown in the figure had a mechanical resonance which caused it to self-destruct when in transport (*i.e.*, while in the trunk of a salesman's automobile). Data apply to the 19-cell configuration.

The pump operated at a voltage of 7.5 kV, decreasing by only ~7% at a N_2 pressure of 10^{-3} Torr (i.e., when used with a robust, matching power supply). We initially experimented with anodes comprising several plates, with aligned holes. Anodes, constructed in this manner, were first suggested by Hall, et al., in 1957.[4] Though such anodes have found wide acceptance in *distributed* ion pumps, used in electron-positron and other storage rings,[75,28,139,70,140] we found them unacceptable for the HiQ® pump applications, as the anode plates could be melted, due to electron bombardment, during the glow discharge starting of the pump.

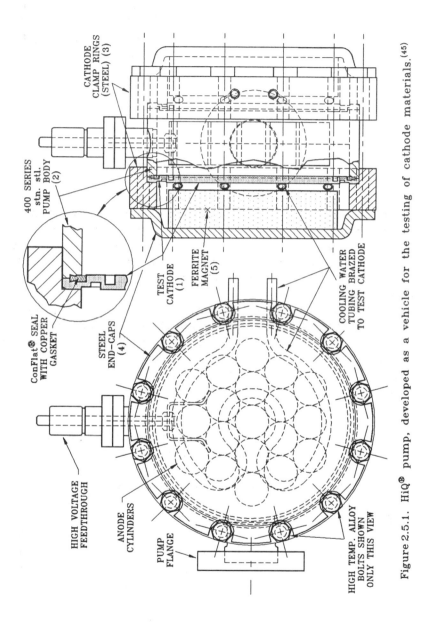

Figure 2.5.1. HiQ® pump, developed as a vehicle for the testing of cathode materials.[45]

CATHODE
CLAMP RINGS
(STEEL) (3)

400 SERIES
stn. stl.
PUMP BODY
(2)

ConFlat® SEAL
WITH COPPER
GASKET

TEST
CATHODE
(1)

FERRITE
MAGNET
(5)

STEEL
END-CAPS
(4)

COOLING WATER
TUBING BRAZED
TO TEST CATHODE

HIGH VOLTAGE
FEEDTHROUGH

ANODE
CYLINDERS

PUMP
FLANGE

HIGH TEMP. ALLOY
BOLTS SHOWN
ONLY THIS VIEW

With this test vehicle, we conducted tests on a number of cathode materials. The first test was with an alloy comprising Ti-10V-11Cr-3Aℓ. It was a variation of the more common alloy, Ti-13V-11Cr-3Aℓ, reported on by Hill.[135] It proved to be the most promising of the materials tested in terms of H$_2$ capacity. We were able to pump ~16,000 Torr-$\mathcal{L}$ of H$_2$ with this cathode material, at a steady-state pressure of ~2.0×10^{-4} Torr, and with the near-constant speed of 50 $\mathcal{L}$/sec.

To appreciate the significance of this number, tests of others are cited. For example, Singleton reported pumping, with some decay in speed, the equivalent of ~10^3 monolayers of H$_2$, with a single-cell pump with Ti cathodes.[120] The Ti-10V-11Cr-3Aℓ tests corresponded to pumping the equivalent of ~1.8×10^6 monolayers of H$_2$. In 1980, Lan Zeng-Rui and Lu Ming reported on the H$_2$ pumping of a pump having one cathode of Ti and the second of an unspecified Ti-Mo alloy.[141] They reported pumping ~6.7 Torr-$\mathcal{L}$ of H$_2$/cm^2 prior to discontinuing their tests, while noting that the speed of their pump did not "... decrease." (The Mo serves as a β-stabilizer in Ti alloys.[142]) In tests with the Ti-10V-11Cr-3Aℓ alloy cathodes, a total of ~84.2 Torr-$\mathcal{L}$ of H$_2$/cm^2 was pumped prior to end of life. McCracken and Maple measured trapping efficiencies of 10-30 keV H$^+$ ions on 0.25 mm thick Ti, Ta, Zr and Mo "cathodes".[71] They noted negligible change in trapping efficiencies after accumulating the equivalent of ~10^{19} H$^+$ ions/cm^2 on Ta, Zr and Ti. Tests with the Ti-10V-11Cr-3Aℓ alloy led to *trapping* the equivalent of ~5.5×10^{21} H$^+$ ions/cm^2, prior to end of life. In all tests, the quantity of H$_2$ pumped was determined by measuring the value of VdP/dt of a known volume, pressurized with H$_2$, and attached to the system through a variable leak.

One would think these results with the Ti-10V-11Cr-3Aℓ material to be a *smashing* success. Rather, they were a *shattering* failure, as the restrained cathodes, because of hydrogen embrittlement, tended to spontaneously shatter into a dozen or so pieces at the end of life. This presented some difficulties, in light of the HiQ® pump configuration. Also, though we first thought the Ti-10V-11Cr-3Aℓ material to be a common, readily available alloy, about this same time we learned that it was the left-overs of a mill-run made for an air-frame manufacturer; we thereafter called this material *unobtanium*. We speculated that the more common Ti-13V-11Cr-3Aℓ alloy would be no more promising (*i.e.*, less shattering). This was a double disappointment. But, in truth, this is the way we all muddle our way through development programs.

Being optimists, we started to view the test vehicle (*i.e.*, the pump) as having as much, if not more, market potential than the materials thus far tested. Therefore, we continued to evaluate the high throughput pumping capabilities of the pump, for various gases, and while using cathodes made of pure Ti. In the meantime, we explored the possibility of using other materials.

One material looked particularly attractive, from an availability standpoint. This material, Ti-6Aℓ-4V (*6-4* material, hereafter), was likened to 1020C carbon steel used in the farm implement business (*i.e.*, in terms of its availability). It was a desperate *stab in the dark*, but cathodes of this material were fabricated and installed in several test vehicles. The results proved most promising. The H_2 speed of this *6-4* material, compared with pure Ti, is given in Fig. 2.5.2.

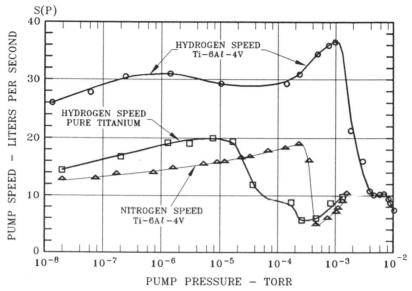

Figure 2.5.2. Pumping speed for hydrogen and nitrogen with Ti−alloy cathodes and for hydrogen with pure Ti cathodes.

When pumping H_2 with the pure Ti cathodes, end of life was manifest as a sinusoidal instability. The onset of these instabilities occurred after pumping only ~10^3 Torr-$\mathcal{L}$ of H_2 at a pressure of ~1.0×10^{-4} Torr. By reducing the pressure to ~7×10^{-6} Torr , additional H_2 could be pumped before the cathodes became saturated. This suggested, as indicated in Fig. 2.5.2, that H_2 pumping was being diffusion limited at the higher pressures. Note that at H_2 pressures ≳ 1.7×10^{-5} Torr, the speed of pumps with Ti cathodes starts to decrease with increasing pressure. The test vehicle precluded this reduction in speed as being due to either thermal effects or changes in the discharge characteristics. Therefore, I conclude that this must be the onset of diffusion-limitations of the pure Ti. The maximum H_2 capacity observed with Ti cathodes, for operating pressures of 7.0×10^{-6} to ~1.0×10^{-4} Torr, and cathode temperatures ranging from 70 - 200° C, was ~3×10^3 Torr-$\mathcal{L}$.

Figure 2.5.3 shows the change in speed of the *6-4* material as a function of the quantity of H_2 pumped. The H_2 capacity of pumps with the *6-4* cathodes, ranged from ~1.05×10^4 (at a pressure of ~2.5×10^{-4} Torr, and with water cooling) to 1.3×10^4 Torr-$\mathscr{L}$ (at a pressure of ~4.0×10^{-5} Torr, and without cooling).

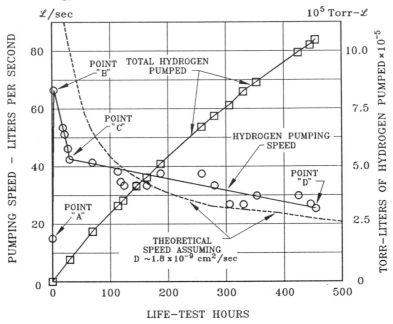

Figure 2.5.3. Speed and quantity of hydrogen pumped with Ti–6Al–4V cathode material.

The onset of the sharp decrease in H_2 speed at ~10^{-3} Torr, with the *6-4* material is not due to diffusion-limitation effects. Rather, it is due to a mode change in the discharge characteristics of the Penning cells.[143] The onset of this mode transformation was at a pressure of ~2.3×10^{-4} Torr when pumping pure N_2. An hysteresis effect was not observed. The H_2 speed of the pump with the *6-4* material at pressures $\gtrsim 3 \times 10^{-4}$ Torr was probably not steady-state, as the capacity of the *6-4* cathodes could be increased slightly by reducing the operating pressure to ~4×10^{-5} Torr (an arbitrary reduction in operating pressure). A variation in apparent H_2 speed, with operating pressure, was not observed when using the *unobtanium* cathodes, suggesting a much higher diffusivity in this totally, β-stabilized, bcc material. In pumps with the *6-4* cathodes, end of H_2 life was manifest as an abrupt decrease in pump speed, rather than the dramatic implosion with the *unobtanium* cathodes. A summary of transient and steady-state pumping results with the pure Ti and *6-4* cathodes is given in Table 2.5.1.

Table 2.5.1. Pump speed for molecular hydrogen for cathode materials of Ti–6Al–4V and pure titanium.

H_2 PRESS. Torr	ION DENSITY H_2^+/cm^2-sec (note "a")	H_2 SPEED Liters/sec		H_2 MOLECULES PUMPED/cm^2-sec ("b")		H_2^0 PUMPED /H_2^+ ION	
		Ti	"6-4"	Ti	"6-4"	Ti	"6-4"
1.4×10^{-3}	5.53×10^{15}	10.0	34.0	2.42×10^{15}	8.24×10^{15}	0.44	1.50
1.0×10^{-3}	3.95×10^{15}	8.7	36.6	1.51×10^{15}	6.34×10^{15}	0.38	1.60
2.6×10^{-4}	1.03×10^{15}	5.9	31.5	2.66×10^{14}	1.42×10^{15}	0.26	1.38
1.0×10^{-4}	3.95×10^{14}	9.4	28.5	1.63×10^{14}	4.94×10^{15}	0.41	1.25
1.7×10^{-5}	6.70×10^{13}	19.4	29.0	5.71×10^{13}	8.54×10^{13}	0.85	1.28
1.0×10^{-5}	3.95×10^{13}	20.0	29.5	3.46×10^{13}	5.11×10^{13}	0.88	1.29
1.0×10^{-6}	3.95×10^{12}	18.8	31.2	3.26×10^{12}	5.40×10^{12}	0.83	1.37
"c" 2.3×10^{-4}	9.08×10^{14}	–	67.0	–	2.67×10^{15}	–	2.94
"d" 2.3×10^{-4}	9.08×10^{14}	–	25.0	–	9.96×10^{14}	–	0.91

NOTES: "a": Based on assumed constant "pump sensitivity" of 120 A/Torr.

"b": Total cathode area ~190 cm^2.

"c": After pumping ~0.5 Torr-l /cm^2 (Point "B" of Fig. 2.5.3.).

"d": After pumping ~55.3 Torr-l /cm^2 (end of life, Fig. 2.5.3.).

Data on the N_2 speed and life of pumps with *6-4* cathodes was reported elsewhere.[45] Briefly, the end of life occurred after pumping ~6.2 × 10^3 Torr-l of N_2, at a pressure of ~1.3 × 10^{-4} Torr. The *failure*, after pumping N_2 at this pressure for ~10^3 hours, was caused by N_2^+ ions *drilling* small holes through the 4.7 mm thick cathodes. After the N_2 life tests, it was evident, from the comparative cleanliness of the cathode plates, that most of the nitrogen was pumped on the anodes. Also, a *crusted* build-up of Ti and pumped gas on the anodes was colored the golden brown of TiN.

Diffusion in the Cathodes

The cathodes are bombarded by ions ranging in energy from 0 - 7.5 keV. KenKnight and Wehner report that: 1) there is little difference in the H_2^+ sputter-yield of metals *vs.* oxides or nitrides of the same metals; 2) in the range of 1.0 - 7.5 keV there is, in fact, an increase in sputter yield with decreasing energy (*e.g.*, H_2^+ on Cu); and, 3) the sputter-yield of 7.5 keV H_2^+ ions on Ti is 0.011.[130] I will make the assumptions that the energy distribution of ions in the pump is approximately linear; that the sputter-yield of all ions is comparable; and, that the surface contaminants have a thickness ≤ 100 Å. For the given sputtering rate, and an operating pressure of ~2.3 × 10^{-4} Torr, the surfaces of the cathodes will be sputtered clean in ≤ 1.5 hours.

The range of penetration of the bombarding ions will differ according to energy of the ions and density of the material. In work with ion implanters, the beams are monoergic and, prior to annealing, when bombarding surfaces with high doses (*i.e.*, $\gtrsim 10^{15}$ implants/cm^2), the dopant tends to stratify about an average range.[144] In our case, the energy of the *beam* varies from 0 - 7.5 keV. Young experimented with the range of electrons, H^+ and H_2^+ ions in $A\ell$.[145] He determined that the range, R, in μm, of H_2^+ ions in $A\ell$ is: R = 0.015 $E^{0.83}$, where E is the energy of the ion in keV. Therefore, because of the difference in densities, we would expect the range in Ti to be ~0.01 × $E^{0.83}$. We then calculate that the maximum range of the H_2^+ ions in Ti to be ~500 Å. This concurs with later findings reported by Andersen and Ziegler.[225] Of course range, R, is distributed about an average value.

The macro-fissures and cracking often observed when pumping large quantities of H_2 with Ti,[131] and very evident in the *6-4* material at end of life, were not at all evident when pumping H_2 with the *unobtanium*. The mechanisms for cathode rift formation, with the adsorption of H_2, are extremely complex, beyond the scope of this book, and certainly my complete understanding. However, qualitatively, the observed cracking probably occurs for the following reasons. At room temperature the α-phase of Ti (*i.e.*, the crystalographic structure of pure Ti) is hcp. Pure α-Ti, at room temperature, has a low H_2 solubility limit of ~20 ppm (by weight).[142] As H_1 goes into solution with the metal, some of the Ti must convert to a γ-phase, which is fcc (face centered cubic). *Going into solution* amounts to the H_1 atom *jumping* into an interstitial site, when given the opportunity by lattice vibrations. Once the H_1 atom *squeezes in*, the lattice is distorted, and seeks the lower energy, fcc configuration (*i.e.*, with the H_1 atom interstitially in place). Such lattice distortions are carried to an extreme with the formation of titanium hydride, and ranging from $TiH_{1.53}$ to $TiH_{1.99}$ in composition.[146] The hydride is also a fcc arrangement, with the H_1 atom residing in the tetrahedral interstices. As an extreme example, the density of TiH_2 is ~86.5% that of pure Ti.[123] Therefore, each dimension of a cube of pure Ti would have to increase by ~6.4% in accommodating sufficient H_2 to form TiH_2. These local volume changes cause stresses between nearest-neighbor volumes, *i.e.*, unit cells in the lattice structure, inducing rifts and cracks in the cathodes. These induced rifts are called interfacial defects.[147]

The Ti-6$A\ell$-4V material is referred to as an α-β alloy. The $A\ell$ tends to substitutionally stabilize its neighbors in the α (hcp) form, where the addition of V acts as a β-stabilizer.[148] This means that in the absence of H_2, some of the alloy is in an α-phase (hcp) form, where some of the material is in the bcc form (*i.e.*, the β-phase). The increase in H_2 capacity of the *6-4* material is probably not due to the presence of $A\ell$. In fact, Paton and Williams, in a review paper, suggest to the contrary.[146] They report that only a metastable, supersaturated H_1 solution can exist in the α-stabilized material, resulting in high internal stresses. These stresses are finally manifest in the formation of titanium hydrides which precipitate along the α-β interface, causing fractures or cleavages. The presence of vanadium, a β-stabilizer, is probably responsible for the higher solubility and diffusivity of the 6-4 material. The H_2 solubility of completely β-stabilized materials (e.g., the unobtainium) ranges from $\times 40$ to $\times 100$ that of pure, supersaturated Ti, with the higher concentration resulting in catastrophic embrittlement.[146]

It is apparent that, because of crystolographic changes with H_1 concentration, the Ti and 6-4 materials become anisotropic, and a simple diffusion model does not apply.[149] Also, the spalling and cracking noted on the surfaces of the Ti and 6-4 cathode materials, suggests that late in life, zones on the surface of the cathodes may have operated at much higher temperatures than would be predicted by simple heat transfer calculations, dealing with an isotropic media. A thermal jump condition would exist at the interfaces of the spallated zones. A diffusion barrier also would be created at these interfaces. With substantial H_2 pumping, the concentration of H_1 in the Ti results in the formation of titanium hydride. At this time, the isotherm of titanium hydride comes into play, which, with excessive extrapolation (noted by Singleton[102]), is $\sim 10^{-5}$ Torr at 170° C.

Temperature also plays an important role in the diffusion process. It probably was in some part responsible for McCracken's results,[71] and the sharp increase in the speed of pumps with the 6-4 material at pressures $\geq 10^{-4}$ Torr. From ambient temperatures to $\sim 500°$ C, the solubility of H_1 in Ti is approximately constant.[123] However, the diffusivity, D, will vary markedly with temperature. The expression for diffusivity in three dimensions is:[147]

$$D \quad = \quad (1/6)\, a^2 \, \nu \, \exp(\Delta S_d/R)\exp(-\Delta E_d/RT)$$

$$= D_0 \exp(-\Delta E_d/RT) \qquad\qquad (2.5.1)$$

where a = the distance between interstitial jump sites encountered by the diffusing atom,

ν = the frequency of oscillation of the atom (i.e., the number of attempts per unit time at changing sites),

R = the universal gas constant in calories per mole -° K,

and ΔE_d and ΔS_d are the activation energy and activation entropy for diffusion per mole of diffusing species, respectively, in units of calories/mole. The diffusivity term appears in the equation of Fick's first law of diffusion, in an infinite slab (i.e., a flat sheet of material of finite thickness d in the x-direction, but unbounded in the y and z-directions), as:

$$Q \quad = \; - \; D \, \partial \, C(x,t) \, / \, \partial x \tag{2.5.2}$$

where Q is the throughput or transport of H_1 within the slab, and $C(x,t)$ is the concentration gradient of H_1 in the slab. *Fick's second law of diffusion* is given in the less familiar form:

$$\partial C / \partial t \quad = \; \partial / \partial x \; (D \, \partial C / \partial x) \tag{2.5.3}$$

Because of the hodge-podge gradient of crystalographic species, the diffusivity becomes a function of depth in the material; or, $D = f(x)$. Secondly, where temperature gradients exist (*e.g.*, exacerbated by spalling), then the diffusivity will take the form $D = f(D_0(x), T(x))$. Then, (2.5.3) has the form:

$$\partial C / \partial t \; = \; \partial / \partial x \; D_0(x)((\exp(\Delta E_d / RT(x))) \, \partial C / \partial x, \tag{2.5.4}$$

where, $T(x)$ means that the temperature, T, is a function of x, the distance within the slab. Assuming we were able to express $T(x)$ exactly, expressing $D_0(x)$ is another matter.

At a constant pressure, there is qualitative similarity in the speed-vs.-time characteristics of pumps with cathodes of pure Ti and the *6-4* materials. Let us examine these results more closely. Note that the speed of the pump with the *6-4* material, shown in Fig. 2.5.3, initially appears to be high (*i.e.*, ~66 $\mathcal{L}$/sec). Actually, this is the speed of the pump ~2 hours into the test. It is difficult to measure the initial transient speed of any pump at time $t = 0$, but, based on H_2^+ *implant* considerations, and assuming an initial contaminant layer of ~100 Å, we calculate a speed, at $t = 0$, of ~15 $\mathcal{L}$/sec. In fact, there appear to be *seven* phases involved in the pumping of hydrogen in the *6-4* material - many of which occur simultaneously. These are:

1) A process of the gradual sputter-cleaning of the surfaces by the bombarding ions (*i.e.*, interval "A"-"B" of Fig. 2.5.3).

2) A process of the gradual restructuring of the material, to a finite depth, δ, the *implantation laminate* (see Fig. 2.5.5), as a consequence of H_1 implantation and the pumping of ionized and molecular hydrogen (*i.e.*, interval "A"-"B" of Fig. 2.5.3). At point "B", the pumping ratio of H_2 / H_2^+, given in Table 2.5.1, suggests that the rate at which molecular hydrogen is being removed from the system is ~x3 that of the impingement rate of the H_2^+ ions. At the end of the life-test this ratio was ~ 0.91.

3) The creation of a second *bulk laminate* of thickness $\lambda(t) - \delta$, where the high H_1 concentration at $x = \delta$ forces the diffusion into and restructuring of the material in the range $\delta \le x \le \lambda(t)$ (*i.e.*, interval "A"-"C").

4) A decrease in pumping speed as $\lambda(t)$ increases (*i.e.*, the bulk laminate grows in depth) to the point of $\lambda(t_2)$, and the incoming flux is in equilibrium with the gradient in the bulk laminate (*i.e.*, interval "B"-"C" of Fig. 2.5.3).

5) A steady-state equilibrium of H_2 flux entering the implantation laminate and the H_1 flux leaving this laminate and entering the region $x > \delta$ (*i.e.*, the point "C" of Fig. 2.5.3, which is ~29 hours into the test).

6) The steady-state growth of the bulk laminate by diffusion of implanted hydrogen from the initial implant depth, through the interface $x = \delta$ and into the bulk laminate (*i.e.*, interval "C"-"D" of Fig. 2.5.3, and $\lambda(t_s)$ of Fig. 2.5.5).

7) The point where the surface concentration needed to cause diffusion from the implantation laminate and into the bulk laminate (further increasing its thickness) exceeds TiH_2, resulting in a net speed of zero (or the onset of thermal problems initiated by spallation of the material; *i.e.*, point "D").

The above model suggests there are three diffusion processes going on: 1) the diffusion of implanted H_1 in the implantation laminate $0 \leq x \leq \delta$; 2) the diffusion of H_1 into the bulk laminate $\delta \leq x \leq \lambda(t)$; and 3) diffusion into the bulk material (of negligible consequence in Ti, as will be shown). Crank refers to such processes as either "pseudo-Fickian" or "non-Fickian" in nature.[150] This merely means that the presence of the diffusing material alters the diffusing medium so as to modify the diffusivity of the material in the medium (*i.e.*, $D = f(C(x))$, as we suspected). I will leave solution of such equations to the mathematicians. What, then, can we conclude from the data in the two figures? A great deal, if we are willing to risk assuming the seven perceived steps apply in the hydrogen pumping model.

Titanium Cathode Material, a Model

Let us assume that an implantation laminate is created, of thickness δ, and with a concentration of ~TiH_2, which remains constant after a short time. A 500 Å laminate would accommodate only 1.6 Torr-$\mathcal{L}$ of H_2 (i.e., 8.5×10^{-3} Torr-$\mathcal{L}/cm^2$) prior to *filling* to a state of the creation of TiH_2. Let us assume the H_1 concentration in the laminate reservoir is C_δ ~1.14×10^{23} and see where this leads. For the time being we will assume that $D \neq f(x,t)$ at $x > \lambda(t)$. Assuming the bulk of the cathodes is initially void of H_1, the H_1 concentration in the cathodes, of thickness d, is then given by:[151]

$$C(x,t) = C_\delta \left[1 - \frac{x}{d} - \frac{2}{\pi} \sum_1^\infty \frac{1}{n} \sin \frac{n\pi x}{d} \exp(-D(n\pi/d)^2 t) \right] \quad (2.5.5)$$

This equation stems from the solution of Fick's first and second laws of diffusion, assuming zero initial concentration of H_1 in the material and the above assumptions, including $D \neq f(x)$. Room temperature (RT) diffusivity data were not found for pure Ti. Available data, taken at temperatures $\geq$ 500° C, suggests a RT Ti diffusivity of $D \sim 4.25 \times 10^{-12}$ cm^2/sec.[152] Also, the RT diffusivity of H_1 in TiH_κ is $D_\kappa \sim (2 - \kappa)(3.0 \times 10^{-12})$ cm^2/sec, $1.96 \geq \kappa \geq 1.35$, as given in this same reference. Two other references in this same work place the RT diffusivity of H_1 in $TiH_{1.55}$ at 0.48 to 3.4×10^{-12} cm^2/sec. A summary of these data is given in Fig. 2.5.4.

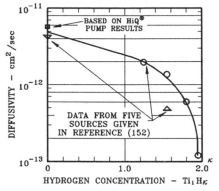

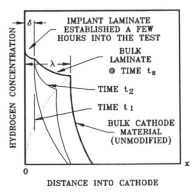

Figure 2.5.4. Hydrogen diffusivity vs. concentration in titanium.

Figure 2.5.5. Proposed model for hydrogen diffusion in cathodes.

Diffusion processes in the cathodes result in concentration gradients, in time, as shown in Fig. 2.5.5. A time t_1, the cathodes are sorbing H_1 and an implant laminate of depth δ is established. At time t_2, the edge of the bulk laminate, $x = \lambda$, starts to grow away from the surface such that $\lambda(t) > \delta$, etc. Early in this process, the diffusivities in the implant laminate, bulk laminate, and bulk cathode material are all similar, so that the distinction between δ and λ is at best *fuzzy*. This is also true for all times at low operating pressures. Let us first assume two things: 1) $D_\lambda = 4.25 \times 10^{-12}$ cm^2/sec (*i.e.*, the diffusivity at the edge of the bulk laminate); and, 2) a steady state speed, at 10^{-4} Torr, of ~ 9 ℓ/sec, for the pump with pure Ti cathodes. This corresponds to a flux 3.12×10^{14} H_1 atoms/sec-cm^2 passing the plane at $x = \lambda$. Evaluating (2.5.2) and (2.5.5) at $x = 0 \sim \lambda$, and solving for C_λ *vs.* time, results in an increasing hydrogen concentration in time as shown in Fig. 2.5.6. We see that pumping ~ 34 hours at this H_1 flux, the concentration, at $x = \lambda$, necessary to sustain pumping at this flux, exceeds that of TiH_2 - a physical impossibility.

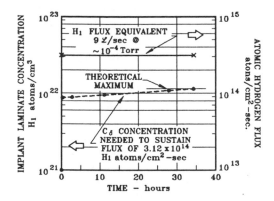

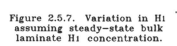

Figure 2.5.6. Maximum possible concentration exceeded in Ti at H₁ flux at ~10⁻⁴ Torr.

Let us try another approach; let us arbitrarily assume that C_λ stabilizes at 8.8 × 10²² H₁ atoms/cm³, and calculate how the flux through the plane $x = \lambda$, i.e., the speed of the pump, varies in time. This results in a change in hydrogen pumping speed as shown in Fig. 2.5.7. This result more closely corresponds to what we observe in the laboratory.

Figure 2.5.7. Variation in H₁ assuming steady-state bulk laminate H₁ concentration.

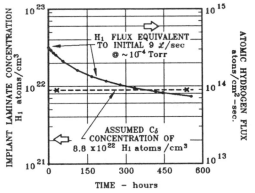

For a constant value of C_δ we see that the speed of the pump has decreased after ~560 hours, to ~23 % of the initial speed. With this model, we calculate that a total of ~720 Torr-$\mathcal{L}$ of H₂ was pumped into the bulk of the cathode, with 280 Torr-$\mathcal{L}$ unaccounted for (i.e., the experimentally observed 1000 Torr-$\mathcal{L}$). The values of C_λ ~8.8 × 10²² H₁ atoms/cm³, and D_λ ~5.0 × 10⁻¹² cm²/sec were established by iterative computer calculations of the best fit for the H₁ flux in the Ti cathodes at a quasi, steady-state pressure of 1.6 × 10⁻⁵ Torr. You will recall that this is the pressure above which the speed decreased markedly.

Because $D = f(x)$, we may not venture into the material zone $x \lesssim \lambda$, with (2.5.2) and (2.5.5). However, let us look in the region $\lambda \leq x$, assuming C_λ 1.14 × 10²³ H₁/cm³, and observe the total flux passing a plane $x = \ell$, for various values of ℓ, while approaching the bulk laminate *from the right*. The purpose of this exercise is to determine how much H_1 has diffused into the bulk of the cathode as a function of time. Invoking (2.5.2), and *counting the atoms* passing the plane $x = \ell$, for time t, amounts to solving the following equation:

$$\int_0^t [-D \, \partial/\partial x \, (C(x,t),|_{x=\ell}] \, dt = \text{number of } H_1 \text{ atoms/cm}^2 \quad (2.5.6)$$

Evaluating (2.5.6) at $\ell \sim 10^{-2}$ cm, we find that after 560 hours, a total of only ~12 Torr-$\mathcal{L}$ of H_2 has diffused into the cathodes. This conclusively establishes that the mechanism for the pumping of H_2, in Ti, at high throughputs, is almost exclusively through the creation and growth of the modeled bulk laminate. The speed of the pump becomes zero when the concentration of H_1 in the implant laminate must exceed TiH_2 in order to *drive* H_1 through the bulk laminate. Because of limits on the diffusivity out of the bulk laminate (*i.e.*, its growth), the hydrogen *capacity* of the pump depends on the influx rate. Had we operated at lower pressures, or modestly higher temperatures, pump *capacity* would have increased (as was observed).

From the above findings, because of the small diffusivity value of pure Ti, we might have assumed that the cathodes were semi-infinite mediums to the diffusion of hydrogen. This problem has the simple solution:[1 5 3]

$$C(x,t) = C_\delta \, [1 - \text{erf}(x/2(Dt)^{\frac{1}{2}})] \quad (2.5.7)$$

In using (2.5.7), we still have to assume that $D \neq f(x)$, which we know is incorrect. However, assuming $C_\delta = 1.14 \times 10^{2\,3}$ H_1 atoms/cm^3, what would D have to be at $x = \delta \sim 0$ to support the observed flux of $3.12 \times 10^{1\,4}$ H_1 atoms/cm^2-sec (*i.e.*, a speed of ~9 $\mathcal{L}$/sec at ~10^{-4} Torr), at the edge of the implant laminate? Using (2.5.2) and (2.5.7), we calculate that after 34 hours of pumping, $D \sim 2.5 \times 10^{-1\,2}$ cm^2/sec, a reasonable result.

Ti-6Aℓ-4V Cathode Material

Let us assume the model which we used in the case of pure Ti applies to the *6-4* material. Note that the speed of the pump, as shown in Fig. 2.5.3, decreases almost linearly in time. I suspect that end of life was due to thermal problems stemming from spalling of the cathodes. If this is the case, we may express speed of the pump as: $S \sim 44 \, \mathcal{L}$/sec $- (1.16 \times 10^{-5} \mathcal{L}/\text{sec}^2) \times t$. From this we calculate that the speed of the pump is zero at $t_e \sim 3.79 \times 10^6$ seconds. Extrapolating linearly to $S = 0$, the average speed over the total life would have been ~22 $\mathcal{L}$/sec. From this, we calculate that the cumulative H_1 in the material per unit area, at $S \sim 0$, is ~$6.64 \times 10^{2\,1}$ H_1 atoms/cm^2, assuming that no H_1 *leaked* out the back of the cathodes. With this accumulated flux, we may establish a lower limit on the probable diffusivity of the 6-4 material, by assuming the maximum possible H_1 concentration is TiH_2. (Note that the depth of spalling was of the order of 1.0 - 1.5 mm.)

We make these calculations by using the equivalent of (2.5.5), but at $x = \lambda \sim 0$, and assuming $C_\lambda \sim 1.1 \times 10^{2\ 3}$ H_1 atoms/cm^3. This simple approach suggests that in the absence of H_1 in the 6-4 material, $D \gtrsim 1.8 \times 10^{-9}$ cm^2/sec. Results of this calculation are shown as the dashed curve in Fig. 2.5.3. On the other hand, using (2.5.2) and (2.5.7), we calculate that the maximum possible diffusivity of the 6-4 material is $D \sim 3 \times 10^{1\ 0}$ cm^2/sec at, for example, time, $t = 29$ hours.

Two questions remain: 1) If speed at the higher pressures is diffusion limited, why does it not increase markedly with decreasing pressures, assuming activation sites are continuously being generated by ion bombardment? 2) Why is there a disparity of almost a factor of two in the early speed data of pumps with the 6-4 vs. Ti materials? You will note from Table 2.5.1, that for all pressures, the rate of pumping by the pure Ti was less than the impingement rate of $H_2{}^+$ ions. This was not the case with the 6-4 material. This can be explained by differences in solubilities of the two materials in conjunction with probable differences of $\sim \times 100$ in diffusivities. Remember that the impingement rate of molecular hydrogen is $\sim \times 370$ that of the $H_2{}^+$ ions, for all pressures. This impinging flux of gas creates the equivalent of a concentration gradient at the surface. If the concentration at the surface is sufficiently low, H_2 will impinge on the surface, be dissociated and diffuse into the bulk. If the solubility limit at the surface of the material is at a maximum, and the diffusivity is insufficient to deplete this H_1 concentration, no molecular H_2 will be pumped.

Beta-Stabilized Cathode Materials

These materials have very high diffusivities. For example, Zeilinger and Pochman, using neutron radiography, established that the diffusivity of β-titanium (i.e., Ti-13V-11Cr-3Aℓ), is $\sim 2.8 \times 10^{-7}$ cm^2/sec[1 8 3] at RT. With such high diffusivities, a bulk laminate does not have an opportunity to build. Also, the *surface* implant concentration varies in time, but never achieves the TiH_2 level (i.e., at the time of catastrophic failure). There are two possible models for the pumping of hydrogen; one assuming the escape of H_2 out the back side of the cathodes, the other assuming negligible H_2 loss through the cathodes. Assuming the latter to be the case, the following equation approximates the concentration as a function of time:[1 5 4]

$$C(x,t) = \frac{qd}{D} \left[\frac{Dt}{d^2} - \frac{3x^2 - d^2}{6d^2} - \frac{2}{\pi^2} \sum_1^\infty \frac{-1^n}{n} \cos \frac{n\pi x}{d} \exp(-D(n\pi/d)^2 t) \right], \quad (2.5.8)$$

where q is the H_1 implantation rate per cm^2, on the face of the cathode. (Note that this equation is not applicable for small values of t.) With a speed of ~50 $\mathcal{L}$/sec at 10^{-4} Torr, 16,000 Torr-$\mathcal{L}$ would be pumped in ~3.2 × 10^6 sec. The above equation predicts a final H_1 concentration at $x = 0$ of ~1.27 × 10^{22} H_1 atoms/cm^3. At the back side of the cathode, the concentration would be ~1.17 × 10^{22}. Assuming a linear concentration gradient at failure, the total calculated average H_1 concentration is ~1.2 × 10^{22} H_1 atoms/cm^3, which, assuming all the gas was retained in the cathodes, leads to a final measured average concentration of ~1.17 × 10^{22} H_1 atoms/cm^3. This closely corresponds to the H_1 concentration which Paton indicates is sufficient to cause catastrophic failure in β-alloys of Ti.[146]

Conclusions on Hydrogen Pumping

The model for the pumping of H_2 in sputter-ion pumps depends on both the solubility and the diffusivity of the gas in the cathode material. Depending on the diffusivity, there appear to be two extreme models in the pumping of hydrogen. The first deals with materials with moderately low diffusivities, such as Ti. In this case, the hydrogen is implanted to a limited depth in the material. Thereafter, a bulk laminate is established, the depth of which depends on two diffusivities - the diffusivity in the parent material, prior to reaching a maximum solubility, and the diffusivity in the bulk laminate. The depth of the bulk laminate grows after the H_1 concentration at the inner edge exceeds the solubility of the material for a given crystal lattice structure. If the crystal lattice is able to reorder itself to accommodate more H_1, the depth of the bulk laminate will grow. Eventually, equilibrium exists between diffusion at $x > \lambda$ (*i.e.*, further growth in the laminate) and the concentration at the surface of the cathode, and no additional H_2 is pumped, for the given flux. The degree of *recovery*, in further pumping of H_2, depends on rate of diffusion into the bulk material in the absence of further implantation of H_1 into the material. This model is very similar to that proposed by Singleton.[102,120] The other extreme of the model is when the diffusivity of the material is so large that for the given flux an equilibrium surface concentration is never established.

Therefore, three elements are essential to the extended pumping of H_2: 1) a high, inherent solubility for hydrogen (*e.g.*, the β-alloy); 2) facility of the material to reorder its crystal lattice structure to accommodate additional H_1 once the solubility limit is exceeded (*e.g.*, pure Ti, and the *6-4* material); and 3) a high diffusivity (*e.g.*, the *6-4* and β-alloy materials). Note that a high inherent solubility corresponds to there being a low equilibrium pressure over the metal when the gas is in solution with the metal. Also, increasing the cathode surface area has the obvious advantage of decreasing the H_2 flux.

2.6 Triode Pumping

Two primary forms of triode pumps are in use at this time. One configuration features gridded cathodes, as shown in Fig. 2.6.1. Hereafter, it is called the GCT pump (*i.e.*, gridded cathode triode). The second is the StarCell® pump, marketed by Varian Associates, and shown to the right in Fig. 2.4.18. Use of the former pump configuration continues to date primarily as *distributed ion pumps* (DIPs), used in particle storage rings throughout the world (see Section 2.10).

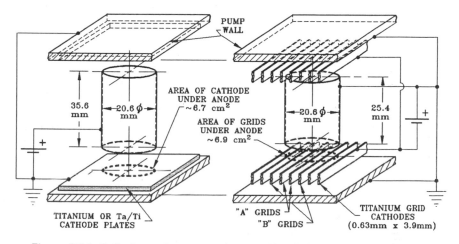

Figure 2.6.1. Cathode surface area of conventional diode and triode pumps compared with the effective ~13.3 cm^2 area of the StarCell® triode pump.

Solutions of ion trajectories proximate to the GCT pump grids are complex, but they may be found using a method described by Spangenberg, for gridded tubes.[155] This approach does not yield exact solutions, and requires iterative computer calculations. In the absence of a significant RF component in the discharge, all of the ions impinge on the cathodes. Also, though grids from time to time will be either melted or sputtered through, there is little evidence of back-side sputtering. Hamilton did note the back-side pumping of Ar in an early triode pump configuration.[47] However, most of the Ar ions strike the grids the first pass through, and many are reflected off these grids and pumped in the walls of the pump body as energetic neutrals.

The surface area of the cathodes plays an important role in the pumping of H$_2$. An enlargement of the GCT pump element is shown in Fig. 2.6.1, where one cell of the StarCell® element is shown in Fig. 2.4.18. Assuming that ions impinge on both sides of the A and B grids of Fig. 2.6.1, the equivalent cathode surface area, under a single cell, is ~6.9 cm^2. With a similarly sized anode and assuming both sides of the radial vanes accommodate gas - a reasonable assumption with neutral reflections - the StarCell® has an equiva lent cathode surface area ~13.3 cm^2. A diode cathode, under the same anode cell, would have a surface area of ~6.7 cm^2. Therefore, from a cathode surface area standpoint, the StarCell® pump should have a decided advantage over conventional diode and GCT pumps in the low pressure pumping of H$_2$. Assuming the same I/P characteristics of diode vs. triode pumps, and the absence of other variables, this would certainly be the case.

As the H$_2$ pressure and associated throughput is increased, triodes evidence problems of thermal run-away much sooner than diodes. The cathodes of diode pumps are operated at ground potential. Because of this, there is some thermal contact of these cathodes with the walls of the pump. Though a high thermal *jump* condition may exist between the cathode and the pump wall, there is some exchange of heat energy to the wall. Because the cathodes of triode pumps must be electrically isolated from the walls of the pump, they are also thermally isolated from the walls. This means that the primary mechanism for the dissipation of the ion beam energy is through radiation. The thermal problem is exacerbated by the fact that the construction of all triode cathodes makes the conduction of heat, in the plane of the cathode, negligible. Assuming a cell sensitivity for H$_2$ of ~5 A/Torr-cell, and a pump operating voltage of ~5 kV, power dissipated in each cell would be 2.5×10^4 W/Torr-cell. At 4×10^{-5} Torr, 0.5 Watts would be dissipated on each cathode of a given cell. Depending on the configuration of the cathode, this power would be sufficient to raise the temperature of a thermally isolated cathode to as much ~300° C. Snouse reported cathode temperatures as high as 500° C during the starting of triode pumps.[156] Baker reported problems with thermal runaway when pumping D$_2$ with triode pumps.[157] His solution was to reduce the voltage so as to decrease the power dissipated in the triode elements. For the above reasons, triode pumps have very low throughput limitations for H$_2$. For that matter, in the absence of water cooling, this is also true, to a lesser extent, of diode pumps. They too are not immune H$_2$ thermal runaway problems.[158]

In a recent paper, Liu, et al., reported on the pumping speed of a GCT pump, with Aℓ cathodes, and for the gases H$_2$, O$_2$, N$_2$ and Ar.[72] They modified a Varian 60 ℒ/sec GCT pump, by removing the Ti grids (see Fig. 2.6.1) and simply replacing them with Aℓ grids. Speed findings of this paper are important to the understanding of active-gas pumping mechanisms of triode pumps, having the conventional Ti cathodes. They reported pumping speeds *vs.* pressure, at saturation conditions, for the above four gases. In this instance, I interpret *saturation* as being comparable to Point "C" of Fig. 2.5.3, for the given gas.

Liu observed only moderate differences in the speed of the pump, with Aℓ vs. Ti cathodes, for three of the four gases, H$_2$ being the exception (*i.e.*, comparing his data with speeds reported by the manufacturer, with Ti cathodes). Denison measured the speed of a new triode pump for Ar and determined that it was ~39% of the speed of the pump for N_2.[1 5 9] Late in life, the speed reduced to ~20% of the rated N_2 speed. Therefore, Liu might have expected an initial speed for Ar of ~24 $\mathcal{L}$/sec, and a decrease to ~12 $\mathcal{L}$/sec late in life. Liu reported a maximum speed for Ar of ~18 $\mathcal{L}$/sec after *saturation*.

Liu reported a peak speed of ~95 $\mathcal{L}$/sec for N_2 with the $A\ell$ cathodes. This corresponds to an ~50% increase over that which would be observed with Ti cathodes. These results probably in part stem from the high sputter-yield of $A\ell$, compared to Ti. Comparative sputter-yield data for $N_2{}^+$ on $A\ell$ and Ti were not found. However, one may draw some conclusions from published data for other gases. Based on the sputter yields of Ne^+ and Ar^+ on Ti and $A\ell$,[5 4] the sputter-yield of $N_2{}^+$ on $A\ell$ is probably ~×2 that of Ti. The sputter-yield of 1.0 keV Ar^+ ions, normal to an $A\ell$ target, is ~2. Also, you will recall that in all triode pumps, the ion bombards the cathode with a high angle of incidence. Based on 1.05 keV Ar^+ sputtering data, we would expect the increase in the sputter-yield of $A\ell$ and Ti, with angle of incidence, to be comparable (*i.e.*, maximizing at ~×2 that of normal incidence).[5 3] However, we should not give too much significance to sputter-yields of triode configurations *vs.* those observed in diodes. For example, the nitrogen pumping efficiency (η, as defined in (2.4.6)) for the HiQ® pump is ~25%, where for triodes it is ~26%; *i.e.*, very comparable. However, I have observed only a ~25% gain in N_2 speed in a diode pump with $A\ell$ cathodes, suggesting some gains from the sputter-yield of triode configurations.[7 3]

At an incidence of $\pi/3$, the lower the energy of the bombarding ion, the further the departure from the cosine law (for example, for $\pi/3$, 5.0 keV Kr^+ on W, most of the sputtered W atoms are at $-\pi/3$).[5 3] Lastly, for Ar^+ ions with energies of ~1 keV, ~30% of the sputtered $A\ell$ neutrals and ~40% of the sputtered Ti neutrals will have energies ≥ 7.4 eV (*i.e.*, the dissociation energy of N_2).[5 3] From the above, we can envision a model for the pumping of N_2 with $A\ell$ cathodes, as follows: 1) For every N_2 ion bombarding the cathode, ~4 $A\ell$ atoms are sputtered off the cathodes. 2) A deposit of $A\ell$ builds up on the walls of the pump. The deposit is always rich in weakly physisorbed N_2 and some $A\ell N$. 3) Weakly physisorbed N_2, in the $A\ell$ deposit, may be dissociated by both bombarding N_2 and $A\ell$ neutrals, to promote further formation of $A\ell N$ within the bulk and on the surface, respectively. Therefore, the increased pumping of N_2 with $A\ell$ vs. Ti cathodes is essentially due to the higher sputter-yields of the $A\ell$ cathodes, and perhaps the energetic neutral dissociation of N_2 in and on the sputtered $A\ell$.

Liu used depth profiling, Auger analysis to determine the concentrations of gas and metal on various surfaces in the pump after pumping air for 3000 hours. Okano made earlier use of this technique in assessing the regions of pumping in a diode pump with juxtaposed $A\ell$ and Zr cathodes.[7 4]

Analysis of a metal plate, substituted for the wall of Liu's pump, indicated uniform concentrations of N_1 and $A\ell$, in equal proportions. Also, a uniform 40% nitrogen concentration was found in the cathodes, off anode center, at depths of ~450 Å, the limit of the profiling. However, based on earlier work of Audi, we would expect the N_1 in the cathodes to decrease with further depth profiling.[99] Also, Hamilton observed that N_2 is pumped by both physisorption and chemisorption in sputter-ion pumps. He noted that physisorbed N_2 (probably implanted in the cathodes) could be later desorbed on the introduction of Ar.[47]

Regarding H_2 pumping, Singleton reported that *saturation* (*i.e.*, in this case meaning zero speed) occurred in $A\ell$ cathodes after pumping only ~10^{-2} Torr-$\mathcal{L}$ H_1/cm^2 of cathode.[120] (In a similar experiment, I found close agreement with Singleton, and negligible capacity for H_2 when using $A\ell$ cathodes in a multi-celled, diode pump.[73]) This corresponds to *filling up* a beam-fabricated, implant laminate to a point of saturation. However, an important distinction is that in the case of Ti, the crystal lattice of the implant laminate (or, bulk laminate) will reorder itself to permanently accommodate more H_1 up to the point of formation of TiH_2. This is not the case with $A\ell$, which, as reported below, has very low solubility and negligible diffusivity for H_1. Therefore, we might expect that on turning off the pump with $A\ell$ cathodes, after *filling* the implant laminate, a great deal of the hydrogen would quickly diffuse back into the vacuum system.

Liu reported a steady-state H_2 speed of ~32 $\mathcal{L}$/sec at a pressure of 10^{-7} Torr, decreasing to ~20 $\mathcal{L}$/sec at ~10^{-6} Torr. We note that hydrogen sorption in $A\ell$ is endothermic. Work reported by Hatch establishes RT solubility limits of H_2 in $A\ell$ from 0.56 to 4.9 × 10^{-6} Torr-cm^3/cm^3 $A\ell$.[160] Further, up to a point, H_2 solubility in $A\ell$ tends to decrease with increasing amounts of trace elements.[161] Because of this, we would expect the solubility of H_2 in 6061 $A\ell$ to be less than that of 1100 $A\ell$. Once the H_1 concentration exceeds the very low solubility limit, perhaps as the result of quick-quenching or implantation, there is some evidence that H_1 in $A\ell$ forms small bubbles which diffuse and coalesce into larger bubbles, causing internal blistering,[162] behaving similar to inert gases in metals.[163,164] The solubility of H_1 in Ti, at one atmosphere and RT, is ~1.8 × 10^6 Torr-cm^3/cm^3 Ti; this is ×10^{12} higher than in $A\ell$, *give or take* a few orders of magnitude. Therefore, diffusion into the $A\ell$ cathodes was of limited significance in the H_2 pumping observed by Liu.

We expect the steady-state N_2 and H_2 speeds of any pump with Ti cathodes to be comparable at, for example, 10^{-7} Torr. Therefore, the fact that Liu reports only a 50% reduction in H_2 speed with $A\ell$ cathodes, vs. the speed of the same pump with Ti cathodes, suggests the pumping of H_2 in Liu's experiment stemmed in part from the burial of high-energy neutrals in the stainless steel walls and anodes of the pump. After *filling* the implant laminate in the $A\ell$ cathodes, the H_2 population is sustained in the $A\ell$

cathode only as the result of further ion impingement, but the net speed of the $A\ell$ cathodes would very quickly become zero. After this the walls of the pump consumed the H_2. We are also led to this conclusion by results reported by Norton.[165] He reports that the permeability, *i.e.*, the product of $D \times s$ (s is the solubility) when extrapolated to room temperature, is ~$\times 10^7$ greater in stn. stl. than in $A\ell$. Were the walls of the pump constructed of $A\ell$, we would expect the speed of the pump to approach zero on the pumping of ~$2 - 3 \times 10^{-2}$ Torr-ℓ H_2/cm^2. Of course, the presence of $A\ell_2O_3$ on the walls would cause further decrease in the GCT pumps capacity for H_2.[70]

Calder and Lewin report measured values of H_1 concentration in stn. stl., at RT, of ≥300 Torr-cm^3/cm^3 metal (*i.e.*, ≳ 2×10^{19} H_1 atoms per cm^3).[208] Eschbach notes that the RT, H_1 diffusivity of stn. stl. to be ~8.2 $\times 10^{-14}$ cm^2/sec.[166] This limited diffusivity may in part account for the reduction in apparent H_2 pump speed at pressures ≳ 10^{-7} Torr in Liu's experiment. Using data for the solubility of H_1 in stn. stl., we are still unable to account for the steady-state speed at ~10^{-7} Torr. For example, the 60 ℓ/sec triode pump has a stn. stl. *cathode* surface area (*i.e.*, the stn. stl. area under the anode cylinders) of ~500 cm^2. Using Calder's number for the solubility, and Eschbach's number for diffusivity, we predict with (2.5.7) the maximum possible flux of H_1 of ~2×10^6 atoms/cm^2-sec. Liu's H_2 data suggests a flux of ~4×10^{11} H_1 atoms/cm^2-sec. Assuming neutrals were implanted in the anodes with equal facility would only reduce this value by ~$\times 3$. Therefore, an implant laminate, similar to that in the case of Ti, must be "constructed" by neutrals impinging on the walls of the pump.

McCracken, using 18 keV H^+ ions, bombarded a number of materials, including stn. stl. targets, and measured the reflection coefficient as a function of time for beams of intensities of ~0.9 mA/cm^2.[167] He determined reflection coefficients by measuring the H_2 pressure build-up as a function of time. Interpreting his data, a beam reflection function, R, for stn. stl. can be approximated by R ~$k_0 \sin\alpha t$, $0 ≲ t ≲ 5$ seconds; R = k_0, $t ≥ 5$ sec, where α = $\pi/10$ sec. and k_0 = 0.9 mA H_1^+/cm^2 (i.e., 5.6×10^{15} H_1^+ per cm^2-sec). The gas accommodated is then merely $_0\int^5 (1-R)dt$. This has the value of ~10^{16} H_1^+ ions/cm^2. With Young's $A\ell$ range data,[145] one calculates the maximum range for 18 keV H_1^+ ions in stn. stl. to be ~600 Å. Assuming that the H_1 could readily diffuse out from this maximum range, a laminate "filling" would occur with an H_1 concentration of ~1.7×10^{21} H_1 atoms/cm^3, a concentration $\times 100$ higher than reported by Calder and Lewin. Using these energy-promoted solubility limits, and compensating for range, we would predict that Liu's pump would cease to pump H_2 at 10^{-7} Torr after a few hours. It is probable that a steady-state cathode implant laminate had not yet been *fabricated* at the time of the measurements. Also, neutral burial of H_1 in the anodes would extend this to ~10 hours. However, we should avoid taking any of these calculations too seriously, as the *laws of physics* always seem to be challenged by the quantities of gas which are consumed by sputter-ion pumps - at least our understanding of these laws.

Recent work substantiates the above model - that is, a significant amount of the H_2 in Liu's experiments was pumped in the walls of the pump.[73,246] It is possible that the effects of pumping substantive quantities of neutral H_2 in the stn. stl. walls and anodes of pumps may explain why the hydrogen speed of pumps, at low pressures, *falls off* faster than predicted by *I/P* considerations. On reducing the H_1 flux into the stn. stl. walls and anodes, the hydrogen therein, formerly in equilibrium with the higher flux intensities, is liberated into the vacuum system. This has the effect of temporarily reducing the net pumping speed at lower pressures. Of course, cathode contamination by other active gases at low pressures causes further reduction in speed with partial pressures comparable to that of H_2.

2.7 Transient Speed Effects

Early publications tended to overstate the relative speed of pumps for H_2 vs. the other gases (*e.g.*, N_2). This was because few had full appreciation for the transient speed characteristics of these pumps. The speed of sputter-ion pumps will vary widely depending on the history of pump use. For example, if you bake a new pump overnight at say 300° C and then conduct N_2 speed measurements, starting at 5×10^{-9} Torr, you will initially observe speeds ×4 higher than will subsequently be noted at this same pressure after a week of pumping. If the initial measurement is conducted at 10^{-8} Torr, it will take about 8-10 hours for the speed to decay to a steady-state level. It will take progressively less time to reach steady-state speed values, the higher the initial pressure at which measurements are conducted. However, my experience has been that the time required to achieve a steady-state speed is not a linear function of pressure. The steady-state speed ultimately achieved has been coined the *saturation* speed. This does not mean that the pump has reached its capacity for the gas in question. Rather, it means that at the given pressure, the speed of the pump does not significantly change with time. If speed data are taken while progressively increasing pump pressure, the measured speed will initially be higher than the saturation speed, and decay according to the above time-table. If speed measurements are taken, starting at higher pressures, and then at progressively lower pressures, the initial speed measured will be low, and increase to the saturation speed level according to the above time-table. This effect was first reported by Jepsen.[24] Up to this point, I have stressed that the speed of a cell is proportional to *I/P*. However, because of transient effects, and pump saturation effects (*e.g.*, see Section 2.5), this, as noted by Dallos, proves to be an over-simplification.[103]

When making meaningful speed measurements, it is necessary, but not sufficient, to have pumped a prescribed amount of gas prior to taking data. For example, the 1976 edition of the PNEUROP standard requires that prior to taking speed measurements, the equivalent of $\sim 4 \times 10^{-2} \times S$ Torr-$\mathcal{L}$ of the test gas must be pumped where S is the rated speed of the pump.[1 6 8] The pump is presumably *saturated* at this point. However, were one to pump this amount of the test gas, at say a pressure of 10^{-5} Torr, and by reducing the leak rate, subsequently establish a pressure of 10^{-8} Torr, the observed speed thereafter would be excessively low for several hours. Conversely, were one to achieve an equilibrium speed at 10^{-9} Torr some time after having pumped the required saturation quota, on varying the leak rate so as to increase the pressure to 10^{-6} Torr, the initial measured speeds would be far higher than the subsequent steady-state speed. Reaching a *steady-state* speed - I prefer this to the term *saturated* - is primarily a time-dependant function, and extremely pronounced when making speed measurements with sputter-ion pumps. For example, these long-term, transient speed changes do not exist in cryopumps, or the momentum transfer pumps. With these pumps, speed-dome outgassing and gauge pumping effects, due to pressure changes, are more significant than actual changes in pump speed. Because of these effects in sputter-ion pumps, I suggest that after first satisfying some saturation and *scrubbing* quota, meaningful data can only be obtained by dwelling at a given pressure for a minimum time, t_m, where,

$$t_m = k_1 \times \exp(k_2 \times (\log P_t/k_3)), \qquad\qquad (2.7.1)$$
$$k_1 = 24 \text{ hours},$$
$$k_2 = -0.96,$$
$$k_3 = 10^{-9} \text{ Torr, and}$$
$$P_t = \text{the test pressure, in Torr}, P_t \geq 10^{-9} \text{ Torr}.$$

Equation (2.7.1) is not founded on some complex theoretical analysis. Rather, it is based on decades of experience in the speed testing of sputter-ion pumps. The above expression yields minimum test durations for steady-state speed results at 10^{-9} Torr of ~24 hours, and 30 minutes at a pressure of 10^{-5} Torr. Equation (2.7.1) applies to each reading. For example, should the pressure be altered by $\times 2$, it is necessary that (2.7.1) be invoked for the new change in pressure, if meaningful data are to be obtained. The above equation suggests that it will take weeks of elapsed time to make meaningful measurements of pump speed for various gases, and pressures varying orders in magnitude. This proves to be the case. Neglecting set-up time, similar speed tests with a cryopump might take only 2-3 days.

2.8 Pumping Gas Mixtures

The effects of pumping mixtures of gases appears straight forward. Singleton reported the changes in the H_2 speed of a single-cell, diode pump as a consequence of increasing the partial pressure of N_2.[102] At an H_2 partial pressure of $\sim 2 \times 10^{-7}$ Torr, when increasing the partial pressure of N_2 from $\sim 10^{-9}$ to 10^{-5} Torr, the H_2 speed of the pump increased by $\sim \times 2.5$. As the partial pressure of N_2 was decreased, the H_2 speed correspondingly decreased. Singleton concluded that the increase in H_2 speed with increasing N_2 pressure stemmed from the chemisorption of H_2 on *free* Ti sputtered by the heavier N_2 ions. On turning off the voltage, there was no sustained H_2 pumping, as in the case of pumping Ar and H_2 mixtures. This indicated that the sorption of N_2 on the cathodes saturated activation energy sites, and precluded the dissociation molecular H_2.

Audi has done the most comprehensive work in the pumping of gas mixtures in triode pumps. In 1988 he published results of pumping N_2 and H_2 in a StarCell® triode.[50] That same year he reported results of pumping He and N_2, Xe and N_2, Ar and H_2, and N_2 and H_2 mixtures in a StarCell® pump.[101] The symbolism "X→Y", used below, is defined as: "measuring the pumping speed of gas X while progressively increasing the partial pressure of gas Y and holding the total pressure constant" ... In interpreting Audi's results, we must keep in mind general observations regarding ion pumps: 1) The I/P of sputter-ion pumps varies as a function of pressure. 2) The I/P of the StarCell® triode pump for N_2 is a maximum at $\sim 10^{-6}$ Torr, decreasing as pressure is either increased or decreased. 3) The I/P for different gases varies as a function of the sensitivity of the Penning cells for the gas species. This results in a translation of the I/P curve along the pressure-axis. 4) The I/P for gas species X, in a mixture of gases X and Y, varies as a function of the partial pressure of gas X, and is independent of the partial pressure of gas Y. Conversely, the I/P of gas Y is independent of the partial pressure of gas X. 5) If X→Y, and the sputter-yield of gas Y » X, even with the steady-state implant of gas X in the cathodes, as the pressure of gas Y is increased, eventually gas X will be sputtered away from the cathodes at the rate exceeding that of cathode implantation. Therefore, at that point, the measured speed of gas X would be solely due to pumping on surfaces other than the cathode. This is somewhat ameliorated by the unique configuration of the StarCell® cathodes. Some gas may be back-scattered, and implanted as neutrals on the back side of the slanted cathode vanes. Neutrals would also less readily sputter-desorb gas which has been implanted on these back-side surfaces.

The I/P (*i.e.*, speed) of air, H_2, and He are shown in Fig. 2.8.1. The I/P data for He were calculated assuming speeds are proportional to I/P, He speed is $\sim \times 0.25$ the air speed, and the speed maximizes at $\sim \times 4$ the pressure of the air maximum (*i.e.*, another I/P consideration). Ionization cross-section data of Rapp and Englander-Golden, for 300 eV electrons, were used to define the I/P peaks relative to N_2 (*i.e.*, air).[169]

Audi's results for $H_2 \rightarrow N_2$ are given in Fig. 2.8.2. Audi noted that for all pressures when He→N₂, the speed for He decreases. Therefore, results for He→N₂ are explained by Observations 3, 4 and 5. The difference between the He speed for a saturated *vs.* unsaturated pump is probably attributable to saturation effects of neutral He pumping on the anodes and walls of the pump, and the sputtering away of He implanted in the cathodes. The fall-off in He speed of an unsaturated pump, with He→N₂, is probably due to the decrease in I/P of He with the increasing partial pressure of N₂.

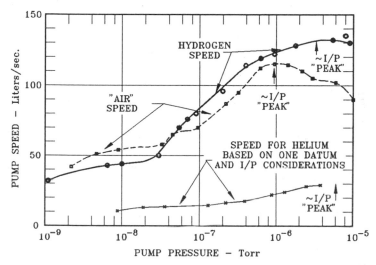

Figure 2.8.1. Shift in pump speed "profiles" of a 120 ℓ/sec. StarCell® pump with pressure, due to changes in I/P for different gases.

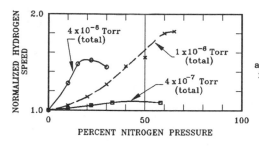

Figure 2.8.2. Variation in H₂ speed of a StarCell® pump at constant total pressure, with increasing N₂ proportion.[101]

Audi's H₂→N₂ data were more revealing (*i.e.*, Fig. 2.8.2). There are two *competing* mechanisms when H₂→N₂. One mechanism deals with the increase in H₂ speed as a consequence of an increase in the I/P of N₂, as the N₂ pressure is increased in proportion to the total pressure. The second deals with the H₂ liberated from the cathodes, (*i.e.*, the net flux), as a consequence of cathode sputtering by the N₂⁺ ions. Regarding the first mechanism, when H₂→N₂ at a total pressure of 10^{-6} Torr, and the N₂ partial pressure is 10^{-7} Torr, the N_2 speed of the pump is ~70 l/sec (*i.e.*, Fig. 2.8.1), and the rate at which N_2 molecules are being pumped is ~2.3 × 10^{14} N_2 molecules/sec. However, assuming unity sputter-yield, the rate at which *free* Ti atoms are being created is twice this number (*i.e.*, η ~0.25). If when $H_2 \rightarrow N_2$ the total pressure is the same, but the partial pressure of N_2 is ~6.5 × 10^{-7} Torr, the speed of the pump is 115 l/sec, N_2 is pumped at a rate of ~2.5 × 10^{15} N_2 molecules/sec, and the rate of creation of free Ti atoms is twice this number. Therefore, though we have increased the pressure by ~×6.5, we have increased the rate of creation of free Ti by ~×11 merely because of I/P consideration of the N_2. A decrease in H_2 speed at yet higher N_2 pressures would suggest that the rate of H_2 liberation due to sputtering of the cathodes by N_2 is exceeding the net gain in pumping of H_2 due to the sputtering of free Ti.

As shown in the $H_2 \rightarrow Ar$ tests conducted by Audi, there will be a precipitous reduction in H_2 speed at certain pressures. The H_2 pressure at which this occurs is a balance between: 1) the concentration of H_1 *stored* in the implant laminate of the cathode; and 2) the sputtering rate of Ar^+ ions, both as it relates to being a source of free Ti and erosion of the cathodes to liberate implanted H_1. Arguments of a similar nature apply to the pumping of all gases.

2.9 High Pressure Operation

The problem of thermal *run-away* in sputter-ion pumps is well known.[156] It is particularly pronounced when pumping hydrogen, or when operating at high pressures of other gases, after the pump cathodes were previously loaded with hydrogen (*e.g.*, as a consequence of pumping significant amounts of water vapor).[49,102,121,157,171] In the late 1950's, the primary product of Varian continued to be microwave tubes. Many of these tubes had *oxide* cathodes; that is, nickel-matrix cathodes coated with a mixture of carbonates of Ba, Sr and Ca, suspended in a nitrocellulose binder. The major source of gas from these cathodes came from a process called *cathode conversion*, which occurred some time during the bakeout of the tubes. That is, cathode heater power was applied and the carbonate mixture decomposed to form oxides of Ba, Sr and Ca; thus, the term *oxide* cathode. This decomposition process resulted in large throughputs of CO and CO_2. Because of heating effects, sputter-ion pumps could not handle the gas loads during cathode conversion. This in part led Jepsen to develop the first water-cooled sputter-ion pump. He applied for patent on such a pump in 1961,[172] and a modified version of this pump, with the cathodes water-cooled, in 1965.[173]

When Jepsen developed the first water-cooled pump, I was a *maverick* engineer at Raytheon. I was building equipment to process tubes - some having the dreaded oxide cathodes. They were *dreaded* by people using ion pumps to process their tubes, as the CO and CO_2 given off during conversion often *swamped* the pumps. A prototype of the earlier version, water-cooled pump was lent to me by a Varian salesman. The problem then was that I didn't have the vaguest notion what throughput meant. I was building hundreds of thousands of dollars of tube processing equipment, but didn't understand some of the fundamental vacuum concepts. Therefore, this water-cooled, sputter-ion pump languished in a storage cabinet - I was afraid to even use it - until it was retrieved by the salesman. It is ironic that decades later, I would develop the next generation of water-cooled pumps at Varian,[138] and what became known as the Hi-Q® pump.[45]

These water-cooled pumps helped somewhat in handling high gas loads. But, even this approach had its limits; it is primarily a question of power consumption. Throughput, Q, as defined by (1.11.1), is simply $Q = S \times P$. Assume that the sensitivity, s, of each cell in a sputter-ion pump is 10 A/Torr and that the speed of each cell is ~ 1.0 $\mathcal{L}$/sec. Assume the exponent n ~ 1.0 in (2.2.4b). If we were to build a pump with a speed of 500 $\mathcal{L}$/sec, we would need ~ 500 cells in the pump, each having a sensitivity of 10 A/Torr. Therefore, we would have a total *pump sensitivity* of 5000 A/Torr. This pump would have a pumping efficiency of $\eta = 0.1$ Torr-$\mathcal{L}$/coul., for N_2. (The speed rating for sputter-ion pumps is usually published for N_2 or air). If we were to operate the pump at 5×10^{-4} Torr, current drawn by the pump, at this pressure, would be 2.5 A. If the power supply maintained a constant 7.5 kV at this high pressure, the power dissipated in the pump would be ~ 19 kW. In this example, the throughput of the pump would be 0.25 Torr-$\mathcal{L}$/sec. A 15 cm ϕ turbo-molecular pump, cryopump or diffusion pump would be able to pump greater than $\times 10$ this throughput, with only $\sim 10\%$ of the power consumption of an ion pump.

This is obviously a contrived example, used to make the point that sputter-ion pumps are not high throughput devices. To minimize current drawn in Penning gauges, ballast resistors are put in series with the gauge power supply.[23] Similarly, because of changes in the impedance of the pump discharge with pressure, in the early days of pump development, ballast resistors were used to limit the pump current. Some resorted to the use of lightbulbs for these ballast resistors.[78] Of course, the maximum power dissipated in the pump occurs when the impedance of the pump is exactly that of the resistance of the ballast resistors. This implies that only half of the voltage is across the elements of the pump.

Some form of power-limiting circuit is required on even the very large sputter-ion pumps.[158] As noted, large pumps may be constructed with numerous elements inserted into individual *pockets*. Using this technique, sputter-ion pumps have been constructed having speeds $\gtrsim 5000$ $\mathcal{L}$/sec.[83] In order to operate these pumps at pressures of $\sim 5 \times 10^{-5}$ Torr, the DC power supply must be somewhat robust. But, this can lead to problems. For example, the diode sputter-ion pumps used on the LINAC at BNL have rated speeds of 1500 $\mathcal{L}$/sec. For years these pumps were plagued with cathode warping problems, under what was thought to be low throughput conditions. The pumps are constructed with 14 pockets, each of which contain two pump elements. The total pump sensitivity is ~2 × 10^4 A/Torr (*i.e.*, η = 0.075 Torr-$\mathcal{L}$/coul., for N$_2$). Each of the three power supplies used to power the 45, 1500 $\mathcal{L}$/sec pumps distributed along the LINAC is capable of supplying 12 A at 5.5 kV. Separate ballast resistor banks are used to limit the current to each of the pumps. Each bank once comprised 60, 500 Ω, wire-wound resistors, providing a total ballast resistance for each pump of 1875 Ω. With this resistance, we calculate that the maximum power which can be dissipated in each pump is ~4.0 kW. The 2.5 mm thick cathode plates each measured 10 × 22 cm. We calculate that the maximum power density dissipated in each of the cathodes plates is ~0.3 W/cm^2. Assume that, because cathode plates are in poor contact with the walls of the pump, they must dissipate this heat through black-body radiation. The maximum temperature they would achieve, under these conditions, is ~200° C. Therefore, this didn't seem to be the cause of the cathode warping problem.

The *flaw* in the above calculation is that I have assumed that the power is uniformly dissipated in each of the pump elements. This proved not to be the case. We observed from time to time that, because of excessive heating, cathodes in pumps would buckle and short to the anodes. In some instances, melting of the Ti cathodes occurred in one or two spots having the same diameter as the anode cylinders. Surface tension kept the material in place, but it was evident the material had melted completely through the plate. Near the melting temperature, the vapor pressure of Ti is ~3 × 10^{-3} Torr. This suggested that a self-sustaining, localized thermal run-away process was occurring, and that power dissipation in spots was > 100 W/cm^2.

We sealed the input flange of one of these pumps with a viewing window, roughed the pump with a turbo-molecular pump to ~10^{-5} Torr, and started the pump. We were able to observe the discharge within most of the elements. Sure enough, though the discharge in all of the elements was confined, the discharge in one of the elements glowed a bright orange-red, where the others glowed with a quiescent blue. We maintained the power into the pump for several minutes, and on turning it off, noted that the cathode plates of the element which had the bright, orange-red discharge were now glowing *orange*, from heating. We repeated the process several times. Eventually, the orange-red discharge aged out of the first pump element. However, it was observed to move to a second and then third element. Before the glowing problem went away, we modified the ballast resistor bank, in series with the pump, to have a resistance of 7.5 kΩ. Thereafter, with *early* start-up (*i.e.*, P > 10^{-5} Torr), we were still able to initiate thermal run-away problems; the orange-red glow still *wandered* about the pump from element to element.

Lamont suggested that thermal run-away in pumps could be caused by hysteresis effects in the discharge.[27] He noted that with near-constant voltage and *decreasing* pressure, I/P tended to be somewhat constant at high pressures. But, with *increasing* pressure, because of this hysteresis phenomenon, the I/P of a discharge peaks to a value as much as ×5 that observed with the same but decreasing pressure. This effect is shown in Fig. 2.9.1. The very high value of I/P, shown in this figure at low pressures, stems from the very long anode used in this experiment (*i.e.*, d ~25 mm and ℓ ~51 mm).

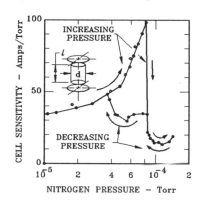

Figure 2.9.1. Hysteresis effect reported by Lamont, for d ~25.4 mm, ℓ ~50.8 mm, B = 0.01 T, and V_a = 3000 volts.[27]

If outgassing, due to heating, was sufficient it would cause all of the elements to share in the gas load, and uniformly share the power dissipation throughout the pump. However, if, after first pumping down and passing through the anomalous pressure region, the pressure increased locally in one of the elements, the current drawn to that element would increase to ×5 of that previously observed and cause the power dissipated in that element to correspondingly increase. A local increase in pressure could be due to metallic contaminants, local outgassing, or even the vapor pressure of the cathode material. For example, Tom reported on the effects of using high vapor pressure cathode materials in sputter-ion pumps.[175]

The *orange-red* discharge problem in the LINAC pumps stemmed from the materials we used when refurbishing these pumps. The process included disassembling the pump elements and *sandblasting* the cathodes and anodes with what we thought to be pure SiO_2 grit. We noted that more often than not, the orange-red glow discharge was confined to one of the gaps in a pumping element. On disassembling a problem element, we noted there were strange metallic streaks on the outer fringes of the cathode plate which had evidenced the red glow. Technicians commented: "Oh, yeah, we've seen that on lots of cathode plates." An Auger analysis of these streaks indicated the presence of Si and Ca.

Of course, it became evident that we had *shot ourselves in the foot* when we *cleaned* the cathodes by sandblasting. We discovered that the sandblasting material was in fact soda-lime glass beads, comprising only 73% SiO_2, and a remainder of 15% NaO, 7% CaO, 4% Mg, and 1% $A\ell_2O_3$. When sandblasting a cathode which had been extensively used, the fine grit was implanted in the microfissures created by the prior pumping of hydrogen (e.g., water vapor). A subsequent washing and 800° C vacuum firing process did not completely remove the grit. On reuse of the cathodes, ions desorbed the imbedded impurities, creating a local high pressure. As the cathodes heated, the vapor pressure of the volatile constituents, listed above, became operative. Local heating caused an increase in local pressure, which caused further heating, etc. A local thermal run-away condition was created in the pump by the presence of the contaminants introduced by sandblasting the pump cathodes. Of course, in retrospect, we know the orange-red glow was probably due to the Ca. The NaO component condensed on the cold wall of the vacuum furnace. We altered our pump refurbishment process thereafter.

Three lessons were learned by this: 1) how not to clean sputter-ion pumps; 2) the presence of contaminants on the surface of one element, in a pump containing many elements, can cause overheating in the one element, while the others go unscathed; 3) power delivered to a pump should be such that if it is all diverted to one element, the element will not self-destruct.

Advantages in Low Pressure Operation

On the brighter side, what is a disadvantage with sputter-ion pumps at high pressures is a major advantage in their use at low pressures. For example, you will recall that each of the SLAC vacuum sectors was pumped with four, 500 $\mathcal{L}$/sec pumps, and there were thirty such sectors, making a total of ~120 such pumps. Assume that for the gas N_2, η = 0.1 Torr-$\mathcal{L}$/coul. for each pump, and a pump voltage of 5.0 kV. Then, if the average pressure in each sector was 10^{-8} Torr, the *total* power drawn by the 120 pumps would be 30 W, or 0.25 W/pump! Where turbo-molecular, diffusion and cryopumps were more efficient at the higher pressures, of the order of 100 - 200 kW would be required to power ~120, of the latter, 500 $\mathcal{L}$/sec pumps. The low power consumption of sputter-ion pumps at low pressures is one of the primary reasons they have received such wide acceptance in the high-energy physics community.

2.10 Pumping Element Located in Antechambers or "Pockets"

Assume that a pump element is located in an antechamber or pocket situated adjacent to a much larger chamber, as shown in Fig. 2.10.1. This pocket may represent one of many pockets built into the body of a pump having numerous pump elements. It might also represent a DIP (distributed ion pump) located in the beam pipe of a storage ring. Though rarely cited, Jepsen was the first to publish the solution to this boundary value problem.[176] Referring to Fig.

2.10.1, Jepsen's results and derivations thereof are given below.[177]

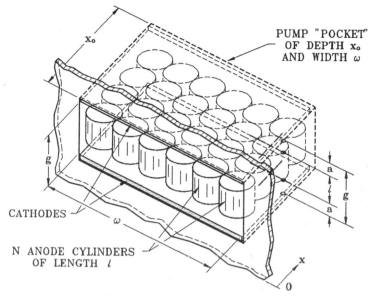

Figure 2.10.1. Pump "pocket" dimensions needed to calculate the effective speed at the opening, for a known intrinsic speed.

Assume there are N cells in the given element which is inserted into the pocket, each of which has an intrinsic speed of S_0. The intrinsic speed of the total element is simply NS_0. The I/P of a given cell is directly proportional to the length of the cell, ℓ. Therefore, assuming that the gap, g, between the two cathodes is fixed, the intrinsic speed of any one cell is given by:

$$S_0 = K\ell = K(g - 2a), \qquad (2.10.1)$$

where the constant K is a function of magnetic field, voltage, cathode material and cell aspect ratio, and a is the anode to cathode spacing. The intrinsic speed of the total element then becomes:

$$S = NS_0 = NK(g - 2a) = \omega N_1 K(g - 2a), \qquad (2.10.2)$$

where N_1 is the number of anode cells per unit width of the anode element (i.e., the length of the pocket). The effective speed of the pump, S', depends on the intrinsic speed of the total element, and the conductance C, leading into the antechamber. The approximation is made that this conductance, C, say for N_2, is equivalent to that of a slab projecting into the two gaps between anode and cathodes, in the antechamber where:[178]

$$C \quad = 2 \times 30.9 \, a^2 \, w/x_0 \, \mathcal{L}/\text{sec-cm}^2$$

$$\triangle \quad k_1 \, a^2 \, w/x_0 \, \mathcal{L}/\text{sec-cm}^2 \tag{2.10.3}$$

The ratio of the effective speed, S', delivered to the chamber, to the intrinsic speed of the element is found by solution of the following two differential equations:

$$dQ(x) = - P(x) \, (S/x_0) \, dx, \tag{2.10.4}$$

$$dP(x) = - Q(x) \, (Cx_0)^{-1} \, dx, \tag{2.10.5}$$

where x_0 is the depth of the pocket, and $Q(x)$ is the throughput at any plane x in the pocket. Equation (2.10.5) is less obvious than (2.10.4). Using a transmission line analogy, the expression $(Cx_0)^{-1}$ can be looked at as the impedance, Z, per unit depth in the pocket, where $(Cx_0)^{-1} = Z/x_0$. Using (2.10.4) and (2.10.5), and the boundary condition $dP(x)/dx = 0$, at $x = x_0$, the expression for the pressure, $P(x)$ is:

$$P(x) = c_1 \, [\exp(kx) + (\exp(2kx_0)) \times (\exp(-kx))], \tag{2.10.6}$$

where c_1 is some constant, and $k = (S/x_0^2 C)$. Substituting the value of $P(x)$ in the right hand side of (2.10.4) and integrating from $x = 0$ to x_0 yields the total rate at which gas is pumped throughout the pocket. By the same token, the product $S'P(x)$, evaluated at $x = 0$, represents the same throughput. Or,

$$S'P(0) = \int_0^{x_0} P(x)(S/x_0) \, dx \tag{2.10.7}$$

This results in the following expressions for the intrinsic speed, S', and the effective speed per unit width, w, of the pocket:

$$S'/S = (\tanh(S/C)^{\frac{1}{2}})/(S/C)^{\frac{1}{2}} \tag{2.10.8}$$

Equations (2.10.2) and (2.10.3) may now be used in (2.10.8) to solve for the optimum pump configuration. Note that k_1 in (2.10.3) varies depending on the type of gas and temperature of the chamber. The assumption in (2.10.3) is that the chamber is at RT and the gas is N_2. Assume that the gaps, a, are held constant, and the total intrinsic speed remains the same. The total intrinsic speed may be expressed by $S = x_0 \times w \times S_a$, where S_a is the pumping speed per unit area of the anodes. By making these substitutions, differentiating (2.10.8) with respect to w, and setting this value equal to zero, we find the maximum effective speed, S', delivered to the chamber is given by the following relationship:

$$x_0 \sim 0.125 \, a \, (k_1/S_a)^{\frac{1}{2}} \tag{2.10.9}$$

In that the pump comprises integer arrays of anode cells, we should not *over interpret* the significance of (2.10.9). However, assuming an anode-to-cathode gap of ~0.75 cm, and S_a ~0.25 $\mathcal{L}$/sec-cm^2, leads to the value x_0 ~1.7 cm. This suggests what was already intuitively obvious; that is, because of conductance limitations, for a constant value $x_0 \times \omega$, making $\omega \gg x_0$ will yield the most effective use of cathode material, and the highest S'/S ratio. Of course, in component pumps, magnetic circuit design considerations impose practical limits on the ratio ω/x_0.

Distributed Sputter-Ion Pumps (DIPs)

The development of electron-positron storage rings was ultimately responsible for the development of DIPs. G.K. O'Neill noted in 1963 that, in the Princeton-Stanford storage ring, there was "... a rather violent pressure rise due to synchrotron light.".[179] This ring was originally pumped with diffusion pumps. He speculated that "... soft gamma rays from the synchrotron light make photoelectrons which are in turn responsible for the cracking of pump oil." This effect is modeled as follows: 1) As the result of being bent in a circular orbit (*i.e.*, accelerated radially), the electron (or positron) beam emits a highly collimated beam of photons (*i.e.*, synchrotron radiation) tangent to the orbit of the electron beam; 2) The photon beam impacts on the walls of the beam chamber, dislodging photo-electrons; 3) The photoelectrons leave the surface, but are bent in a circular orbit by the magnetic field used to bend the primary beam; 4) On impact with the chamber walls, the returning photo-electrons desorb gas. The desorbed gas, generated the full length of the beam chamber, raises the average pressure in the machine, and results in decay in primary beam current due to scattering. Pumping on each end of the beam chamber is only a partial solution, due to conductance limitations of the beam chamber.

Fischer and Mack, using synchrotron radiation from the Cambridge Electron Synchrotron Accelerator, did the first quantitative measurements of gas desorption off of Cu surfaces resulting from the synchrotron radiation.[180] Their work was done under UHV conditions and led them to the conclusions that a "separate function system" would be needed to pump the gas desorbed by synchrotron radiation. By this they meant that there would be two pumping systems, one to handle the ambient gas load, and a second "distributed pump" to handle the gas desorbed by synchrotron radiation. Their proposed *distributed pump* comprised TSP (titanium sublimation pumps) filaments, strung along the length of the beam chamber. The filaments were located behind parallel-plate electrodes, located in the top and bottom of the beam chamber. These plates served to separate counterrotating electron and positron beams, and also served to shield the beam from sublimed Ti, which because of its high Z, would cause beam scattering.

Anashin, *et al.*, in 1968, were the first to report the use of distributed sputter-ion pumps in the VEPP-2 electron-positron storage ring, located at the Institute for Nuclear Physics in Novosibirsk, Russia.[181] These diode pump assemblies were distributed along the lower plane of the beam chamber, behind the lower beam separation electrode. They made use of the same magnetic field used to bend the electron-positron beam, to support the Penning discharge. Actually, as early as 1959, Jepsen suggested such a possibility, stating: "In special applications ... elements may be placed directly in the system ..., thus improving conductance ... In some cases incidental or stray magnetic fields associated with the system can be utilized".[24] This scheme, as in the case of the Fischer-Mack TSP approach, required a larger gap in the dipole bending magnets and resultant increase in cost.

Two years later, Cummings, *et al.*, described the proposed vacuum chamber configuration of the Stanford (SLAC) Electron-Positron storage ring. As shown in Fig. 2.10.2, it featured an extruded Aℓ beam chamber, with the DIP housed in an electrostatically shielded antechamber on the inboard side of the chamber radius of curvature.[133,182] This was the first major application of extruded Aℓ beam chambers in storage rings. The approach was met with some skepticism. Norman Dean, who managed construction of the vacuum system, related to me: "They said we were *crazy* to use extruded Aℓ chambers in SPEAR."

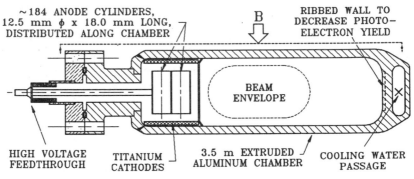

Figure 2.10.2. Distributed pump in SLAC e^-e^+ storage ring.[133]

The SLAC DIP, also making use of the magnetic field of the dipole magnets, was located in the same plane as the electron-positron beam, thus minimizing the gap of the dipole bending magnets. Since the original work at Novosibirsk, and then SLAC, the use of both diode and triode DIPs, of varying configurations, has found wide acceptance in storage rings (*e.g.*, 70,140, 184 - 194,233). Also, the use of extruded Aℓ beam chambers has found wide use in accelerator applications.

2.11 Pump Power Supplies

Little new has transpired in the development of pump power supplies in the last twenty-five years. Switching power supplies have the advantage of being light-weight and may in the future be used for sputter-ion pumps.[195] Varian Associates first exhibited a sputter-ion pump, switching power supply, at the 1989 IUVS Symposium, held in Köln, Germany. These types of power supplies are widely used in airborne radar and ECM systems. However, sputter-ion pumps tend to be inhospitable loads. They arc, sputter and frequently short out. As noted in Section 2.9, this requires that some form of current limiting provision exist in any power supply used on these pumps. Also, unlike the simple bridge rectifier supplies, with *soft transformers*, switching power supplies tend to be inherent noise generators. The high voltage cables connecting power supplies to sputter-ion pumps are essentially TEM waveguides. Therefore, provisions must exist for filtering out the noise from the supply to the pump or this noise might prove troublesome in the user's system. The reverse is also true: the TEM waveguide (*i.e.*, pump high voltage cable) will transmit noise generated in the pump back to the power supply.

In order to deal with thermal problems at high pressures, pump power supplies in the early days had ballast resistors. With a robust power supply, the power dissipated in these ballast resistors can be excessive. This prompted Hall to use conventional household light bulbs for these resistors.[4] Rutherford applied for patent on a power supply which *pulsed* the high voltage to the pump at a pulse repetition frequency (*i.e.*, duty cycle) which was varied in an inverse relationship to the peak power drawn by the pump.[49] In this manner, it was possible to limit the power drawn by the discharge at high pressures.

The industry *workhorse* presently comprises a high voltage transformer in series with a full-wave, bridge rectifier circuit. The high voltage transformer is loosely coupled, so as to become saturated as the impedance of the pump decreases (*i.e.*, with increasing currents at high pressures). This has the effect of limiting power drawn by the pump. This scheme, along with a voltage-doubler circuit, was patented by Quinn and Mandoli in 1967.[196,197] Because of the loose coupling of the transformer, it is called a *soft* transformer. This technology, along with a log-reading circuit (*i.e.*, a circuit to measure pump current over a broad dynamic range), patented by Mandoli,[198] comprises the only significant developments in pump power supplies in the last 30 years. Some manufacturers are now offering pump power supplies with RS 232 interfaces to facilitate remote monitoring and control through a computer interface.

If pump current drawn from the power supply is to be used as an inference of very low pressure, care must be taken to filter out power supply ripple. Capacitance of the pump and high voltage cable will cause AC currents which will obscure low pressure pump readings. For example, the capacitance of a 20 $\mathcal{L}$/sec diode pump is of the order of ~60 pF. The high voltage cable leading to the pump has a capacitance of ~90 pF/m. Assume that the power supply provides 5.0 kV to the pump, with 1.0% regulation. The ripple of a full wave, bridge rectifier circuit is at a frequency of 120 Hz. At *zero* pressure, AC currents in the system would be 12 μA and be equivalent to an operating pressure of ~4 × 10^{-8} Torr.

The problem is much more pronounced when very long runs of high voltage cable are used, as in accelerator applications. For example, the power supplies attending the Brookhaven AGS are located on the average ~200 m from the pumps. The pumps terminating the cables have capacitances of only ~200 pF, but the cables have capacitances of ~0.018 μF. Assuming similar regulation and operating voltages, AC currents (*i.e.*, *noise* in the system) would be equivalent to operating pressures of ~4 × 10^{-7} Torr; with 0.1% regulation, this would be ~4 × 10^{-8} Torr. If some sort of computer-aided current sampling system was used to measure pump current, signal noise would appear as *flutter* on top of the DC discharge currents.

2.12 Magnet Designs

In the early days of sputter-ion pump development in the United States, various forms A$\mathcal{L}$ NiFeCo alloy magnets were used. These alloys, identified by the trade-name Alnico®, were first developed in the mid 1930s. In 1940 a breakthrough in magnet design occurred with the development of Alnico-V, a domain oriented, permanent magnet alloy comprising the above elements and a *pinch* of copper.[199]

Casimir tells of the accidental discovery of the ceramic-type magnets which are extensively used in today's sputter-ion pumps.[200] Apparently, some time in 1950, a technician at Philips, when mixing up a chemical recipe for an experiment, made an error in the proportions of the mixture, and the result was a permanent magnet of unique properties. Of course a great deal of further development work followed (and continues) at Philips, but this error resulted in what came to be known as ferroxdure in Europe and magnedure in the United States; a sintered, ceramic compound of BaO · 6Fe$_2$O$_3$ known as barium hexaferrite. An excellent review paper was published on the properties of forms of this ferroxdure in 1977 by van den Broek and Stuijts of Philips.[201] They point out in this article the financial significance of the discovery, as magnets could now be produced using relatively inexpensive and readily available materials. A comprehensive treatment of ferromagnetic materials, including the Alnico alloys and all forms of the hexaferrite materials, is found in a recent work edited by Wohlfarth.[202]

Helmer published the first comprehensive theoretical treatment of the use of hexaferrite (ferrite, hereafter) magnets in pump applications, including multiple gap, periodic structures.[203] He filed for a patent in 1961 on, what I later *reinvented* and coined, a *confined magnetic field circuit*, using ferrite magnets.[138,204] Two years after Helmer's publication, Kearns, of the General Electric Company, published an article on some of the more practical considerations dealing with pump magnetic circuit configurations in which either the ferrite or Alnico magnets are used.[205] Kraus's chapter on ferromagnetic materials serves as an excellent *refresher* on the design of permanent magnet circuits.[206] The above references most adequately cover the design of magnetic circuits for sputter-ion pumps.

Sputter-ion pumps manufactured throughout the world today are almost exclusively provided with ferrite magnets with steel magnetic return circuits. Minor exceptions to this are the very small appendage pumps (*e.g.*, ~0.2 - 2.0 $\mathcal{L}$/sec); some of which are still offered with cast, Alnico-type magnets. These Alnico magnets are used in this case strictly for reasons of economics. It is more cost-effective to purchase these small, cast-alloy magnets from, for example, Indiana General or the Crucible Steel Company, rather than buy the ferrite magnets and build the required magnetic circuitry.

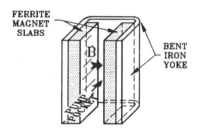

Figure 2.12.1. Iron yoke pole piece with ferrite magnets.

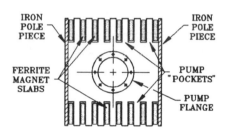

Figure 2.12.2. Closed loop magnetic circuit with iron pole pieces.

In ferrite magnet packages, slabs of the material are located on the inner faces of steel (iron) yoke assemblies such as shown in Fig. 2.12.1. These assemblies are then slipped over the pockets containing the pump elements. An example of a magnetic circuit which might be used in a very large, multi-pocketed, sputter-ion pump is shown in Fig. 2.12.2. Depending on the number of pump pockets, all magnetic circuit designs presently used are variations of these two circuits.

For over 15 years pumps similar to that shown in Fig. 2.4.9, and the smaller appendage pumps, were offered with horseshoe-type, cast Alnico-V or Alnico-VIII magnets. These magnets, besides being very heavy, had excessive stray magnetic fields. Because of concern for the effects of these stray fields near microwave tubes, Helmer developed a magnet package with a confined field.[204] That is, there was negligible stray field even at the surface of the magnetic circuit. It was specifically intended for use with pumps similar to that shown in Fig. 2.4.9, which appended large klystron tubes, having military applications. As a spin-off of work on the HiQ® pump, in 1975 I developed a compact "confined field" magnet package for use with this same pump, and to replace the Alnico-VIII, horseshoe magnets offered in the commercial market. This magnet package, shown in Fig. 2.12.3, had the advantage of weighing only 8.9 lbs., where the horseshoe magnet weighed ~18 lbs. It also had the advantage of very low fringing magnetic fields. It is noted herein to emphasize that it is possible, when needed, to construct magnetic circuits for pumps having very low fringing fields.

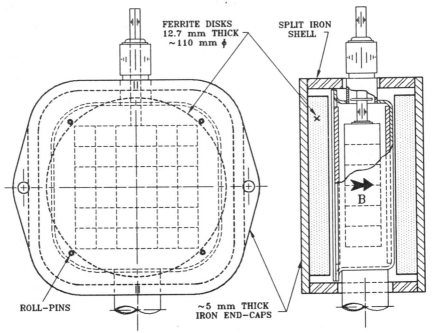

Figure 2.12.3, "Confined field" magnet package with ferrite magnets.

The magnetic circuit *shell* was made of annealed, 1010C steel. Plots of the field, in the magnet gap and proximate to the package, are given in Figures 2.12.4 and 2.12.5, respectively. There is an inherent efficiency associated with a *confined field* circuit. The normally *bugling* or fringing fields are wasted in the conventional horseshoe ferrite or alloy magnets. Because of the boundary conditions imposed by the steel circuitry, the fields for the greater part are confined to the interaction region. For reasons cited by Helmer, you will note that field uniformity within the gap is better with the alloy magnet than with the ferrite package, though in speed measurements this proved of no consequence. Helmer suggested that use of steel sheets on the ferrite faces would *flatten* out the field in the ferrite package.[2 0 3]

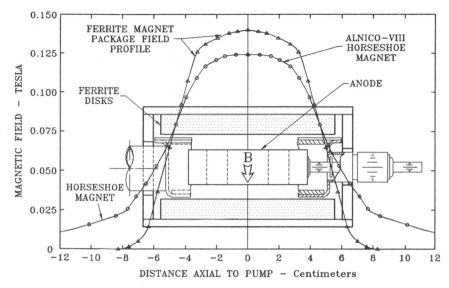

Figure 2.12.4. Comparison of both fringing and useable magnetic fields in "confined field" package and the conventional horseshoe magnets.

In special applications, requiring low fringing fields, such *confined field* packages could be fabricated for much larger sputter-ion pumps. However, because of the comparatively low energy product of the ferrite materials, magnetic discs of these materials (*i.e.*, see Fig. 2.12.3) would have to be too thick for use in smaller appendage pump applications. Because of this, about this same time, I designed a smaller confined field package for use with a smaller appendage pump (*e.g.*, ~2 ℒ/sec). I used SmCo magnet discs, stabilized to 300° C, in this application. The design, a scale-down of the package shown in Fig. 2.12.3, had SmCo discs measuring 9.5 mm thick × ~38 mm ϕ. Using a shell wall fabricated of 3.2 mm thick steel, TIG welded to end-caps ~4.8 mm thick, a field of ~0.13 T (1300 Gauss) was obtained in the 50.8 mm gap. This was slightly greater than the maximum 0.125 T observed with the alloy magnets. The design, though a technical "success" at the time, was not fiscally viable as the cost of SmCo magnets was prohibitive. The relative cost of SmCo magnets has not decreased in the last 15 years, though in some special applications, this design is still in use.

It is best to anneal the steel magnetic circuit after forming and welding, as these processes cause stresses which in turn decrease the effective permeability of the steel. Also, if magnetic fields of the order of 20-30 Gauss are measured near the surface of the steel circuit, the steel is near saturation and the return path not optimized. For example, the field at the center of the broad face of the magnet package, shown in Fig. 2.12.5, could be reduced from ~15 Gauss at contact, to a value approaching the magnitude of the earth's magnetic field merely by increasing the thickness of the 5 mm end-caps. The maximum magnetic flux density which must be supported by the end-caps is proportional to d/t, where d is the diameter of the ferrite disk, and t is the thickness of the end-caps. Therefore, to further reduce the fringing fields, we have the option of either using a material with higher permeability (e.g., silicone-bearing irons), or increasing the thickness of the end-caps to support the high flux.

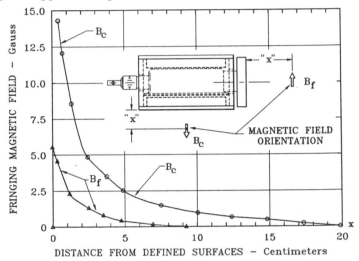

Figure 2.12.5. Fringing magnetic field of ferrite magnet package.

Pumps with ferrite magnets may be baked at temperatures of ~350° C without loss of field in the gap on returning to room temperature. For example, the magnetic circuit shown in Fig. 2.12.3 was baked at ~400° C with a total recovery in the field on returning to room temperature.

2.13 More on the Nature of Penning Discharges

2.13.1 Space Charge Distribution in Penning Cells

The development of sputter-ion pumps and various forms of Penning gauges prompted considerable study into the properties of the negative space charge stored in the cells, particularly in the LPPDs (i.e., low pressure Penning discharges at $P \leq 10^{-4}$ Torr). Since that time these discharges have been measured for oscillations and RF content, *poked* and probed with charged beams and Langmuir probes, optically studied, and otherwise investigated and modeled with attention rarely given a phenomenon. Just a few of these studies will be cited. In that the speed of sputter-ion pumps is directly proportional to the cell sensitivity, I/P, and I/P is known to be directly proportional to the stored space charge in the cells, knowledge of the characteristics of this discharge became of paramount importance.

First, it is important to note that there appear to be numerous *modes*, or SPD (*i.e.*, space charge distribution) configurations in a simple cell. The SPD will vary with V_a, B_z, d, ℓ, a, and P, the anode potential, axial magnetic field, cell diameter, cell length, anode-to-cathode gap and pressure, respectively. Because of this, findings on the SPD characteristics have at times led authors to what appear to be conflicting conclusions, as all of these parameters were not the same from one study to the next. This does not mean that the findings were not scientifically accurate. Most probably they were accurate, and many of these experiments showed a scientific ingenuity and technical understanding which I to this day hold in awe. However, the findings in most cases were not sufficient to suggest a particular unified SPD theory.

Discharge Modes

Hooper qualitatively describes the various discharge modes as a function of V_a, P and B_z, as shown in Fig. 2.13.1.[37] This figure was a modification of a mode diagram first proposed by Shuurman in 1966.[207]

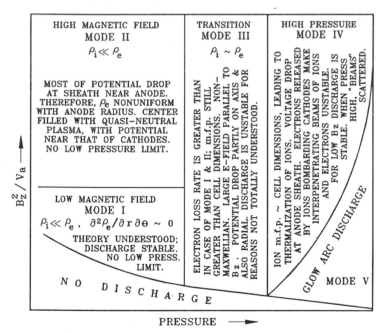

Figure 2.13.1. Hooper's diagram of discharge modes.[37]

This figure may not be used as a rigorous interpretation of all existing modes. For example, when making noise measurements with magnetic loops, located within a sputter-ion pump operating in the LPPD region, I observed four distinct modes when traversing Hooper's modes I and II. They were evidenced by abrupt changes in noise, with associated changes in I/P. However, Fig. 2.13.1 is qualitatively a reasonable model.

The distinctive feature of all modes of LPPD is that negative space charge is trapped within the cell, and depresses the cell potential on axis to a value approaching that of the cathodes. This results in radial electric fields within the discharge (e.g., see Fig. 2.2.2). As the pressure is increased to $\gtrsim 10^{-5}$ Torr, space charge neutralization effects occur, and a neutral column develops at the cell center, growing as the pressure is further increased.[143]

The Stark Effect

Knauer and Lutz[209] made use of the Stark effect[210] to measure the degree of depression in a Penning cell with the geometry (1.2, 1.6 cm) at an H_2 pressure 2×10^{-4} Torr. This notation "(1.2, 1.6 cm)" is used hereafter to denote a cell aspect ratio (i.e., ℓ/d) of 1.2 and cell diameter of 1.6 cm. The Stark effect is the splitting of the spectral lines of, for example, ionized hydrogen, on recombination, as a consequence of the presence of an electric field. The magnetic field was 0.15 T and anode voltage 3.0 kV. This classic paper has been described by others and the results reproduced in numerous subsequent publications. They introduced H_2 into the cell, and by measuring the splitting of the H_β line, they deduced the magnitude of the electric field within the cell. From this, they could calculate both the space charge density consistent with the field and the potential within the anode. The SPD was configured as a hollow sheath. The inside diameter of the sheath was ~0.25 the diameter of the anode, peaking at a density of ~1.6×10^{10} electrons/cm^3 at ~0.44 the diameter of the anode, and gradually falling to zero at the anode wall. Using the implied space charge density and electric fields, their calculations of electron current reaching the anode were within 30% of that predicted using classical cross-field mobility calculations.

At the above pressure, a film exposure time of ~$\frac{1}{2}$ hour was required. Therefore, using this technique for determining space charge within the cell at significantly lower pressures was impractical. At higher pressures and magnetic fields, excessive broadening of line-structure occurred. This effect was attributed to the presence of fluctuating electric fields, thereby deduced as having amplitudes of the order of 4 kV/cm. This departure (i.e., increase) in I/P with increasing pressure is the effect noted by Lamont (2.0, 2.5 cm) and shown in Fig. 2.9.1.[27] Lamont suggests that the mode observed in this experiment corresponds to mode II of Fig. 2.13.1.

Frequency Shift in RF Cavity

In another classic experiment, Lange took an entirely different approach in an attempt to define and quantify the SCD of an LPPD.[211] He measured the frequency shift in a cell (1.2, 2.5 cm) also serving as an RF cavity, resonant in the $TM_{0\,1\,0}$ mode. From this frequency shift he was able to determine the magnitude of electron space charge stored in the cell. At this resonant mode, it was possible for the cathode plates to be electrically isolated from the anodes. They were inserted within the anode cylinder. Conn and Daglish indicated that cell configurations of this type (i.e., cathodes inserted within the anode cylinder) were very unstable at pressures $\gtrsim 10^{-5}$ Torr.[2 3] This did not appear to be the case over the majority of the pressure range of Lange's experiments (i.e., $10^{-1\,1}$ to 10^{-4} Torr, Ar). Lange calculated expected frequency shifts from 18 possible space charge distributions, with variations both in r, the radius of the anode, and axial distance, z. The SPD possibilities entertained at most, would result in errors $< \times 10$ of the absolute space charge density implied by frequency shift measurements. He then arbitrarily normalized all his data to a particular electron space charge distribution, ρ_e (ρ_e uniform in r, and linearly decreasing with z). It is interesting to evaluate Rutherford's data of Fig. 2.2.7, for the different cell sizes and a magnetic field of 0.1 T. If we assume the space charge is uniform and configured as right-angle cones, the bases of which meet at the center of each cell (i.e., one of the distributions evaluated by Lange), we find very close correspondence between I/P findings and cell sizes in Rutherford's data. This also serves as further verification that $I/P \propto \ell$, the cell length.

Lange's findings pointed out three very important features of LPPD: 1) depending on anode voltage, ρ_e is nearly constant over 5-6 orders of magnitude; 2) at low pressures, ρ_e is directly proportional to anode voltage; and, 3) there is close correlation between Lange's observed variation in ρ_e as a function of pressure and Lamont's findings[2 7] of I/P vs. pressure. This last point emphasized the correspondence between I/P and ρ_e which was pointed out in Section 2.2 (i.e., $\sum q \triangleq \iiint \rho_e \, dv$). The correspondence between I/P and ρ_e, B_z and V_a, in LPPD was verified by this experiment. The absolute value of ρ_e, assuming the above SCD and pressures $\lesssim 10^{-6}$ Torr, varied from ~0.5 to 6.0×10^9 electrons/cm^3, depending on V_a and B_z. This SCD density converged to values approaching Knauer's findings at the higher pressures.

Similar frequency-shift measurements were conducted by Agdur and Ternström with an rf cavity operating in the $TM_{0\,1\,0}$ mode.[2 1 2] However, these measurements, as others referenced therein, were conducted at much higher pressures. Lange's work represents the only work done of this sort at very low pressures.

Ion Sputtering Patterns

Prior to reporting on the Stark Effect measurement, Knauer reported on an experiment where he used the sputtering pattern appearing on the cathode to deduce the characteristics of the space charge distribution, $\rho_e(r)$, in a

Penning cell (1.2, 1.9 cm).[36] He made a study of the sputtering patterns for several different gases in the pressure interval 10^{-6} to 10^{-4} Torr. In this case he used a magnetic field of 0.25 T and varied the anode voltage up to 4.0 kV. He noted that there was a mode shift when reducing the magnetic field to < 0.12 T, but at fields > 0.12 T, there was "... little change in behavior (of the discharge)." He observed that a plateau of cathode material was preserved on the cell center line during sputtering. He noted that the diameter of this plateau varied depending on the magnetic field, anode voltage and species of gas. He made calculations of the probable origin of ions initiated in the discharge based on the cathode sputtering patterns. Such sputtering patterns were explained by Knauer by assuming that all of the sputtering ions originated in a sheath close to the inner anode surface. Using Fig. 2.13.2a, where r_c is the ion cyclotron radius and r_p the plateau radius, he calculated that:

$$r_p = (r_a^2 + r_c^2)^{\frac{1}{2}} - r_c. \tag{2.13.1}$$

Given that $r_c = m_i v_i / q_i B_z$, where m_i, v_i and q_i are the mass, velocity and charge of the ion, respectively, and $v_i = (2q_i V_a / m_i)^{\frac{1}{2}}$, then (2.13.2) is:

$$r_p = (r_a^2 + 2m_i V_a / q_i B_z^2)^{\frac{1}{2}} - (2m_i V_a / q_i B_z^2)^{\frac{1}{2}}$$

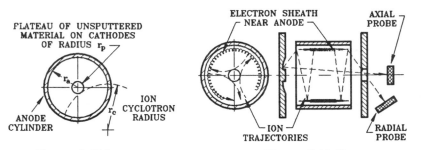

Figure 2.13.2a.   Figure 2.13.2b.

Figure 2.13.2.   Knauer's observations of cathode sputtering patterns and ion probe measurements leading to an anode "sheath" model.[36]

Equation (2.13.2) was found to be in excellent agreement with experimental results, totally explaining the plateau effect. Hirsch also reported on this effect two years later.[213] However, this seems to be the only other instance it has been observed. This may be due to the fact that slight anode cylinder asymmetries (e.g., 0.05-0.1 mm) would "wash out" such a plateau. Note that the streaks of Si and Ca observed on the cathodes of the BNL LINAC pump elements (i.e., see Section 2.9) were all oriented in a direction suggesting curvature of these ions by the pump's magnetic field.

This same year, Knauer applied for a patent on a sputter-ion pump comprising solid anodes.[214,227] The claim of this patent was that a sheath of electrons would be *stored* on the outer surface of the solid cylinders. This configuration was not subsequently exploited, as I suspect that it was probably more applicable at higher pressures and magnetic fields.

Ion Currents and Retarding Potential Probes

Using two retarding potential probes, one on axis with the cell and the second radially off center and *hidden* behind the cathode plate (*i.e.*, see Fig. 2.13.2b), Knauer measured the energy distribution of ions created in the discharge.[36] From these results and subsequent calculations, he determined that even a 5% potential depression along the axis of the cell would cause the axial velocity component of even the most energetic ions (*i.e.*, in terms of the radial component) to miss the radial probe. From this, Knauer concluded that essentially all of the potential depression occurs near the anode. He then cut an 8 mm ϕ hole in the center of the cathode plate opposite that having the probes. An 80% reduction of axial ion probe current was then observed, while current to the radial probe remained the same. From this he concluded that most of the ion current initially measured with the axial probe was due to particle reflections from the opposite cathode.

Earlier experiments of a similar nature, conducted by Helmer and Jepsen,[35] led them to different conclusions. A Faraday cup was used to measure the axial ion current from a cell (1.7, 1.2 cm). They determined the following: 1) ion energy, measured at the Faraday cup, is almost monoenergetic, having an energy spread of only 10% of the average; 2) the beam is highly collimated; 3) there is a direct relationship between ion energy and anode voltage, becoming linear above a certain value; 4) there is a decrease in average ion energy with increasing pressure; and, 5) ion current to the collector represented $\times 0.2$ of the total current to the respective cathode, whereas the aperture represented only $\times 0.063$ of the total cathode area. They concluded from these measurements that the mean ion energy was a measure of the potential depression at the center of the cathode, but that this potential depression was not as significant as that suggested by Knauer's experiments. Note, however, that essentially the same results were obtained by Knauer, and Helmer and Jepsen when their experiments overlapped (*i.e.*, the same B_z and pressure).

Other investigators have made similar use of ion current measurements to deduce something about SCD in an LPPD. For example, Young used a Faraday cup to measure total ion current passing through a 3 mm ϕ hole on the axis of a cell (1.0, 1.2 cm), over the pressure range of $\sim 10^{-10}$ to 10^{-4} Torr.[58] He found that at 10^{-10} Torr, 60% of the ion current was collected on the central 6% of the cathodes and at $\sim 10^{-4}$ Torr this number reduced to 20% of the total discharge current. That same year, Smirnitskaya and Nguyen-Khiu-Tee made observations similar to Young.[60] Results of experiments published by Rudnitskii a year earlier showed that all the above experimental results were probably consistent and correct.[57] He graphically illustrated how cathode current densities shifted between two mode patterns with changes in V_a and B_z. There also appeared to be some mode mixing. With low magnetic fields and voltages, such as used by Young (0.097 T, 2.0 kV), Rudnitskii's data suggests that most of the current would be confined to the central part of the cathodes. This also proves to be the case when comparing Rudnitskii's and Smirnitskaya's data for the case of low B_z (0.13 T) and high V_a (7.0 kV). With higher magnetic fields and lower voltages, as used by Knauer (0.15 - 0.25 T, 3.0 kV), Rudnitskii's data is in very close agreement with Knauer's results. That is, a cylindrical sheath exists as described by Knauer, in the above section discussing use of the Stark Effect.

Hirsch used a radial probe, running the full length of the anode cylinder, to intercept azimuthal currents in a Penning cell.[213] He measured ion current as a function of probe depth in the cell, for various pressures. This and other similar tests give questionable results, as the discharge often will *readjust* itself to the new boundary conditions created by the presence of the probe, and give misleading results.

Electron Beam Probes

Dow used a transverse *ribbon*, electron beam as a Penning discharge probe.[31] He hoped to use the measure of azimuthal drift of the beam, on transversing the discharge, to deduce the average radial field therein, and therefore, the SCD. He was not able to make self-consistent, quantitative determinations of these fields. However, he did make qualitative determinations of the radial fields in a small cell (2, 1.2 cm) as a function of pressure, V_a and B_z. The ribbon beam, on passing through the cell, appeared on a phosphorescent screen. Though forming a straight line when launched into one end of the cell, it was distorted by the electric fields in the cell on passing through the discharge. The *squiggles* observed indicated distinct variations in discharge modes of the SCD, as noted above. Others have reported use of electron beams to probe Penning cells, at low pressures, including Knauer (without a discharge present),[36] and Hirsch.[215]

Jepsen's Smooth-bore Magnetron Model

As noted earlier, Jepsen recognized that the space charge potential depression on axis of an LPPD resulted in a *crossed-field* type configuration similar to that of a magnetron. Though others had commented on the similarity between these cells and magnetrons, Jepsen expanded on this point in modeling the discharge in a 1961 article.[30] This, and the subsequent work of Schuurman,[216] served as the basis from which subsequent semi-empirical models of the discharge were formulated, including both normal and *inverted* (magnetron) discharges.[217] The discharge mode modeled by

Jepsen is represented as mode II in Fig. 2.13.1, predicting behavior of the discharges at and above the transition from mode I to mode II (*i.e.*, decreasing I/P with increasing B_z), where Schuurman's model dealt with mode I, and as pointed out by Redhead, did not agree with experimental findings in mode II.[5] Jepsen's model was a smooth-bore, coaxial magnetron, with a filamentary cathode radius. It showed a linear relationship between I/P and ℓ, anode length. Values of I/P predicted by the model were a factor of 2-5 higher than values empirically observed with these discharges.[108]

In the years that followed, many excellent articles appeared dealing with optimizing the speed (I/P) of various pump geometries. This includes works by Malev and Traktenberg,[188] Hartwig and Kouptsidis,[218] and the more theoretical work of Wutz.[219] The working models which were subsequently developed became semi-empirical modifications of Jepsen's and Schuurman's work.

2.13.2 More on Sputtering Patterns on Pump Cathodes

When disassembling a used, multi-celled, diode sputter-ion pump, I have invariably observed distinct star-like, sputter-errosion patterns on the cathode plates. These observations have been made on hundreds of sputter-ion pump cathodes. The *stars* will be observed regardless of the cathode material, or gases pumped. If the pump has been operated exclusively at very low pressures, the width-to-depth ratio of the sputtered *stars* will be small; perhaps 1:5. This ratio increases with higher operating pressures, and may be of the order of 20:1 in an extensively used pump. Nevertheless, the stars are always present. The stars will take shapes similar to that depicted in Fig. 2.13.3. For a cell size of the order of 20 mm, and B_z of ≤ 0.18 T - the maximum field in any commercial pump which I have studied - the *stars* will have four points or *spokes*. If the cells are nested in a rectangular array (as opposed to close packed), additional stars will appear in the parasitic cell regions created between the cells. Stars in these pseudo-cells will usually have eight spokes.

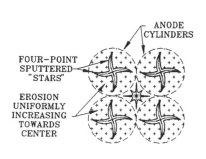

FOUR-POINT SPUTTERED "STARS"

EROSION UNIFORMLY INCREASING TOWARDS CENTER

ANODE CYLINDERS

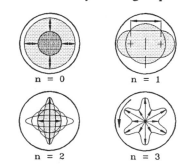

$n = 0$ $n = 1$

$n = 2$ $n = 3$

Figure 2.13.3. Sputtering patterns invariably observed on cathodes.

Figure 2.13.4. Nodes of oscillation in a smooth-bore magnetron.[220]

Each spoke of a star is always curved, and the curvature of the spokes of all the stars will be oriented in the same clockwise or counter-clockwise direction. Star patterns of facing cathode plates are mirror images. The orientation of star spokes in close-packed cell arrays differs from that observed in rectangular arrays, such as shown in Fig. 2.13.3. Lastly, the spokes of adjacent stars are always oriented in a preferred direction, suggesting a mutual coupling mechanism exists between cells. I have never seen an exception to the above observations in a cathode of a multi-celled, diode pump element. The *stars* would not be discernible in triode pumps.

Though these erosion *star* patterns are prevalent, sputtering still occurs over the entire face of the cathode, and is greatest on center-line with the cell. However, existence of the stars suggests some nonuniform SPD exist in all multi-celled sputter-ion pumps. A half-century ago, Brillouin described various possible modes of oscillation in the SPD of a cylindrical, crossed-field diode (*i.e.*, a magnetron).[220] These are reproduced in Fig. 2.13.4. The $n = 0$ mode, shown in the upper left corner of the figure, corresponds to the SPD uniformly expanding and contracting radially within the cell (magnetron). The $n = 1$ mode is a dipole-type of oscillation. The $n = 2$ mode corresponds to an oscillation of the space charge about two fixed axes. The SPD at one extreme takes the shape of an ellipse of distorted space charge along the y-axis, where at the other extreme of the oscillation it is distorted into an ellipse having a major axis along the x-axis. I suggest that the above noted stars are the result of $n = 2$ space charge oscillations in the sputter-ion pumps. The stars in the parasitic cells, having eight spokes, correspond to the $n = 4$ mode of oscillations. Helmer and Jepsen first reported on the existence of these stars, noting eight-point stars in cells having square cross section, and four-pointed stars in cells with slightly elliptical, but near-round shape.[35] A photograph of the discharge in a ~8 mm square cell, shown in that paper, evidenced eight distinct stars (*i.e.*, $n = 4$) at a pressure of ~10^{-4} Torr. They attributed these stars to space charge oscillations. Short of the modes suggested by Brillouin, I know of no one to date having modeled the dynamics of the SCD which would create these stars. The magnetron analogy persists.

Noise in Sputter-Ion Pumps

Conn and Daglish first reported measuring low frequency oscillations in these discharges.[22] Using a Lecher wire system, they observed oscillations in LPPDs in the frequency range of 10 - 26 MHz. Helmer and Jepsen also reported on noise and oscillations in LPPDs.[35] They reported noise in these discharges stemming from plasma oscillations, occurring at ~1.0 GHz; electron cyclotron frequency oscillations, f_c, occurring at ~3.3 GHz (depending on the magnetic field), and, low frequency, *rotational* oscillations at ~150 MHz. Helmer and Jepsen used a split-ring magnetron configuration to measure some of the low frequency oscillations. Redhead has done the most extensive work on these LPPD instabilities and oscillation at very low pressures.[34] A review paper by Hooper cites much of the early work done in this area.[37]

In a recent review paper, Redhead cites additional important work done in this area, to the present.[5] He suggests that the decrease in I/P at very low pressures is probably caused by the loss of energetic electrons from the discharge as a consequence of inherent instabilities in the discharge, including diocotron oscillations. Diocotron oscillations occur at higher magnetic fields, and are proportional to the amount of space charge stored in the discharge. Reporting on work by Malmberg and Driscoll,[221] he noted how at a pressure less than $\leq 10^{-8}$ Torr, the electron containment time in a discharge became constant, suggesting an anomalous electron transport phenomenon. By improving uniformities in the electric fields within cells, and the magnetic field, containment times could be increased. Oscillations in SCD, suggested by the sputtering patterns in cathodes, probably stem from space charge instabilities caused by slight asymmetries in each anode cell. Where on the one hand the ordered nature of the sputtered patterns from cell to cell may be due to electromagnetic coupling of the discharges between each of the cells, on the other hand, it may be due to cell asymmetries introduced by fabrication techniques used in building the anode structure (*e.g.*, spot welding).

In work at SLAC, using very sensitive superheterodyne receivers with bandwidths of ~25 MHz and MDSs of −90 dbm, I made measurements of noise emanating from sputter-ion pumps of various configurations in the frequency range of 1.9 - 8.0 GHz. The pumps were coupled to, in one case, by inserting the pump element into an S-band waveguide. In the other cases, magnetic loops were inserted into the pump housing. Of course, magnetic loops often have inherent resonances. Therefore, data taken with one loop configuration could never be considered conclusive. The following observations were made: 1) Noise was manifest in several forms, including broadband Gaussian-type noise, increasing with pressure, V_a and B_z. 2) There was a pressure dependent, low frequency amplitude modulation of the Gaussian-type noise. 3) There were impulsive-type, narrow-band bursts of noise within the Gaussian noise. 4) Depending on the pressure, V_a and B_z, there was at times a striking similarity in the plots of noise amplitude as a function of frequency and that reported by Glass and his colleagues for magnetrons near cut-off.[222] 5) Noise thresholds existed at f_c as a function of voltage, where the amplitude of the noise would increase three orders in magnitude with slight changes in V_a. 6) No noise was evident at frequencies $> f_c$, to a level of −90 dbm.

2.13.3 Striking Characteristics

Penning discharges do not instantaneously ignite with the application of high voltage (*e.g.*, see (223)). The delay in ignition of the discharge depends on pressure, cell geometry, applied voltage and magnetic field. This delay is sometimes called the *striking characteristics* of the cell. There have been numerous articles published on this topic, and the theory is fairly well understood. Redhead was the first to publish a comprehensive theoretical treatment of the striking characteristics of an inverted magnetron Penning cell vs. pressure,[224] and he and Hobson the I^+/P characteristics of such cells at extremely low pressures,[217] and Lassiter the striking characteristics specifically for He at very low pressures.[226]

The striking characteristics of an ensemble of Penning cells, of different configurations, was reported on by Laurent.[140] The delay in the striking of the discharge, particularly at very low pressures, was reported on by de Chernatony and Craig.[223]. This delay-effect prompted the use of hot filament, electron sources in some single cell devices. For example, Westendorp's patent on single-cell pumps included configurations with hot filaments,[19] as did a multi-cell pump patented by Mark and Henderson several years later.[228] The General Electric "trigger discharge" cold cathode gauge makes use of such a hot electron source to initiate the discharge at low pressures.[229,230] The use of hot electron sources, radioactive beta emitters and sharp points (field emitters) to initiate a Townsend (i.e., starting) discharge was suggested by Conn and Daglish in 1953.[22] Komiya used a ^{63}Ni beta source to initiate a discharge in a multi-cell pump at a pressure of 3.5×10^{-11} Torr.[122]

There is no *quick fix* for this ignition problem, short of increasing both B_z and V_a to practical maximums. The problem plagues microwave tube engineers who, after baking both tube and appendage pump (without magnet) thereafter, when high voltage testing the tubes, have problems initiating discharges in the pumps. The appendage pumps used on small microwave tubes such as TWTs are usually the *mini-pumps*, and have very small cells, further exacerbating starting problems. I have often wondered if it might be possible to integrate a Tesla coil in a pump power supply, and by intermittent use of this coil, facilitate greater ease in starting of these mini-pumps. In the larger, multi-celled pumps, once discharge starts in one cell, it quickly propagates throughout the rest of the cells. I observed this effect in noise measurements of pumps at very low pressures, noting *quantum* increases in the noise level associated with the sequential *ignition* of cells.

2.13.4 Use of Very High Magnetic Fields

The use of very high magnetic fields to support LPPDs is often a requirement in DIP (distributed ion pump) applications.[185,190] These pumps often exploit use of the fringing magnetic field of magnets used to bend particle beams in storage rings and accelerators (e.g., see Section 2.10). This was first proposed by Jepsen in 1959.[24,231] He did much of the early work relating to the use of very high magnetic fields (*i.e.*, up to ~1.0 T) in sputter-ion pumps which here-to-fore has not been referenced.[232] He showed, in a 1959 patent application, that for a given cell diameter, d_c, there was a critical magnetic field below which the discharge extinguished. For a fixed voltage, this field was defined by:

$$d_c = k_1 / B_Z, \tag{2.13.3}$$

where k_1 is a constant. This relationship has its origin from the *cutoff parabola* of magnetrons.[155] Based on empirical measurements, he noted that for cells of the same length, ℓ, and $\ell/d > 1$, to a first approximation the speed of the cell, S_0 is given by:

$$S_0 = k_2 (d - d_c). \tag{2.13.4}$$

The constant k_2 is a function of V_a, the cathode material and the gas being pumped. The number of cells which can be arrayed per unit area is given by:

$$n_a = k_3 / d^2, \tag{2.13.5}$$

where the constant k_3 depends on the packing geometry of the array. For *close packed* and rectangular arrays, k_3 has values of $(4/3)^{1/2}$ and 1.0 respectively. The product of (2.13.4) and (2.13.5) yields the speed of the array assembly. On differentiating this product with respect to d, and setting this value equal to zero, we find the optimum cell diameter for the maximum speed at a given magnetic field; this is found to be when $d = 2d_c$. Jepsen provided empirical data in the above reference for fields ranging from 0.04 to 1.0 T. He suggested an optimum value of $k_1 = d_c B_Z \sim 5.6 \times 10^{-2}$ T-cm.

Jepsen's patent application was made four years prior to Rutherford's paper, showing that the speed of pumps mysteriously decreased as a function of pressure, for the same magnetic fields (*e.g.*, see Fig. 2.2.7).[26] Nevertheless, this early work by Jepsen is still applicable and was used by others over a decade later in the design of DIPs using high magnetic fields.[188,234] However, at low pressures, the onset of moding or oscillations at high magnetic fields may cause pumping instabilities. As noted, these instabilities have not been adequately modeled. Therefore, empirical work is required when tailoring DIPs for use in high magnetic fields.

2.14 Other Considerations

Maintenance and Trouble Shooting

These pumps are actually very simple devices to trouble shoot. When experiencing difficulties, we must first make sure that it is truly a pump problem, rather than a system problem (see Section 1.17). Assuming it is a pump problem, there are only a few items which need be checked. If the pump current is the only measure of system pressure, we must determine if we truly have a pressure problem, or if the indicated pressure is high as a consequence of leakage current. Note that a sputter-ion pump at air behaves the same as a sputter-ion pump at very low pressures. This is often the cause of problems in a system. Short of the existence of some other form of gauging, there is no way to be completely sure that your pump is not at air.

In the case of triode pumps, field emission leakage current may be eliminated by *hi-potting*, while under vacuum. Prior to this, we might want to determine if there is leakage current in the anode insulators or high voltage feedthrough. This may be checked by: 1) turning off the power supply; 2) disconnecting or locking out the primary power source; 3) switching the meter on the high voltage supply to the voltage setting, to make sure the voltage has drained to zero; 4) removing the high voltage cable and measuring the resistance across the pump electrodes. Note that the resistance should be measured using both polarities, as the equivalent of a thin-film diode may exist on an insulator, and yield a high impedance in one direction and a much lower impedance with reversed polarity. Because of thin film effects, insulator leakage may also disappear after venting the pump to atmosphere. If a low resistance is noted, this may stem from leakage across an anode insulator or across the high voltage feedthrough. In the case of the latter, if it is due to internal sputtering, there is no simple remedy. A replacement must be purchased from the OEM (original equipment manufacturer). Because of high electric fields, in the presence of high humidities and moderate temperatures, an accelerated corrosion of the exterior parts of the high voltage feedthrough may take place. The exterior fields may be reduced by the careful design of the high voltage connector.

When leakage currents exist as the result of sputtered material on the anode insulators, the pump element may be disassembled, and the insulators completely refurbished by sandblasting (preferably with Al_2O_3 grit), washing in an industrial dishwasher, and then firing them in air at $1000°$ C.

Are the magnets weak or improperly installed? Irreversible changes in the magnetic field will occur if the magnets are exposed to either very high, or, in the case of ferrite magnets, very low temperatures. If the magnets are improperly installed so that the fields are *bucking*, though a discharge may exist in the cells, it will be comparatively weak. If the magnets were cracked because of some prior thermal or mechanical shock, this too will significantly reduce the field. Lastly, you should make sure that some extraneous magnetic member is not shunting the field in the gap.

If high voltage is being applied to the elements (*i.e.*, there is not an internal or external open circuit) and the magnetic field is adequate, then the pump has the same speed it had when new. The problem then is that the *effective* speed has been diminished because of sources of gas. We saw in Section 1.15, that the speed of a pump will be diminished at low pressures depending on the base pressure of the pump. This is not to be confused with I/P effects. If there is a source of gas leaking into the pump, or originating from within the pump, though the speed of the pump may be the same as when new, the effective speed at low pressures is diminished as given by (1.15.1). Remember, certain types of sputter-ion pumps effectively pump He. Therefore, to get the highest sensitivity when leak checking the pump, the pump should first be turned off. Also, in Section 1.18, we made calculations showing how an atmosphere of He will break down under high voltage much more readily than an atmosphere of air. Because of this, use caution when leak checking around a pump which is energized.

If there is no leak into the pump, then the source of gas must either be due to contamination inadvertently (or deliberately) introduced, or outgassing from saturated cathodes. Outgassing from saturated cathodes is usually the last probable cause for poor performance of a pump, as with the exception of H_2 bearing gases, most of the air gases form very stable compounds with the cathode materials. End of life in this case happens when the cathodes are sputtered through. For reasons cited in Section 2.5, there is a rate limitation in the pumping of H_2-bearing gases. In this case, the cathodes of diode pumps can be refurbished by sandblasting them with $A\ell_2O_3$ grit, washing them in an industrial dishwasher and then firing them at ~800° C, for a few hours, in vacuum. This same process is also used to refurbish the anodes. The element insulators may also be refurbished at the same time, using the previously noted procedures. Lastly, cathodes of diode pumps which have been eroded through are easily replaced with fabricated parts. Alternate methods of cleaning stn. stl. parts, such as the pump body, are reported on by Sasaki, at the American Vacuum Society 37th Annual Symposium. Transactions of this symposium had not yet been published at the time of this writing.

2.15 A Summary of Advantages and Disadvantages

Operating Efficiency: At low pressures, the operating efficiency of these pumps, in terms of Watts/ℓ/sec., exceeds all other pumps. However, we have seen that at very high pressures this is not the case. These pumps have very poor throughput handling capability.

Capacity: In that these are capture pumps, they have a finite capacity for all gases. End of life may be manifest as a gradual decrease in the base pressure of the pump, and resultant decrease in speed at the lower pressures for a given gas. A more dramatic example of end of life might be when, in the case of diode pumps, holes are sputtered through the walls of the pump body. This does not have catastrophic results on the system, as the *drilled* holes are usually initially very small and are evidenced as a minor system leak. When either the throughput or total sorptive capacity of the pumps is exceeded, pumps will become difficult to start, or evidence thermal run-away problems. These pumps can also be rendered useless when pumping forms of hydrocarbons.[235,236] For reasons noted, their capacity for pumping the noble gases is limited, particularly in the case of conventional diode pumps.

Source of System Contaminants: In 1963, it was shown by Litchman[237] Riviere,[125] and Reich[238] that these pumps synthesize methane. Carbon, which is always present in the Ti cathodes, combines with H_2 therein to create this gas. This effect was also pointed out by Jepsen in 1963, when he noted it was prevalent when pumping H_2 and H_2O.[239] This can be troublesome when using these pumps at very low pressures.[46] However, components in the vacuum system, other than the sputter-ion pumps, often prove to be a far greater source of contamination.[240]

These pumps are also sources of charged particles which can cause damage to instrumentation and cause system *noise* problems. In fact, when starting these pumps, the glow discharge originating in the pump can spread throughout the entire system to which it is appended. Because of this, screens are used at the pump inlet flange to shield the system from these charged particles.[241,242] The spread of the discharge is much more prevalent in diode pumps, as with triode pumps the walls of the pump are at the anode potential. Note, however, that use of these screens will not shield the system from neutrals and sputtered material generated within the pump.[243] These neutrals can also cause system contamination problems (*e.g.*, the coating of optical surfaces).

Pump Starting: The starting of these pumps can be more difficult with pump age, or if they are contaminated by hydrocarbons or water vapor from the roughing pumps or some other source. Usually, the older the pump (*i.e.*, the greater its use), the harder it is to start. This is primarily because of thermal desorption effects resulting from ion bombardment of the cathodes. Bell and his colleagues showed that pumps in good condition would very effectively pump down systems with large water vapor loads.[244] With diode pumps, on starting, the walls of the pump are bombarded with low energy ions which desorb gas. This compounds the difficulty in starting. This is not the case with triode pumps.[156,235] On the other hand, it has been shown where triode pumps can present starting problems as the result of delayed, thermal run-away. Because of wall and pump element gas desorption considerations, the use of an isolation valve between the pump and system has obvious benefits. Use of isolation valves is fiscally impractical in very large accelerators, having numerous pumps. The SLAC two-mile accelerator made use of isolation valves at each of the pumps distributed along the accelerator (*e.g.*, see Fig. 2.4.13). The valves eventually were no longer used. However, dry N_2 was always used when venting the sectors and they were always sorption roughed.[245] These precautions effectively minimized both the hydrocarbon contamination and water vapor levels in the system.[170,174]

Triode pumps must be *hi-potted* after starting to eliminate field emission problems (*i.e.*, assuming current at low pressures is used to indicate pressure). Thermal run-away problems can also occur when starting all sputter-ion pumps. Because of this, when starting these pumps the high voltage should be cycled on and off according to some *recipe* which depends on the characteristics of the power supply, size and surface area of the pump, the surface area and aspect ratio of the system, and the roughing provisions. The fact that the roughing pump might be capable of achieving low pressures, even $< 10^{-5}$ Torr, is no assurance that starting problems will be avoided. For the above reasons, the pump starting recipe must be empirically established for the total system. Roughing pumps of various forms can be the source of hydrocarbon contamination. Sorption pumps are used for roughing in instances when absolutely no risk of system hydrocarbon contamination can be tolerated.

Magnetic Fields and High Voltages: The possible existence of high fringing magnetic fields can cause problems, as noted in Section 2.12. Pump magnets and the requirement of the use of high voltages present operation problems and represent potential safety hazards. Therefore, when designing a vacuum system in which sputter-ion pumps are to be used, the potential hazards of both the pump magnets and high voltage must be given as much consideration as the vacuum performance. Also, in UHV applications, these pumps are often baked, when operating, to temperatures as high as 350° C. Care must be taken that the appropriate type of high voltage cables are used in such applications.

Advantages of Sputter-Ion Pumps: They afford clean, *dry* pumping. Excluding methane, these pumps neither synthesize nor become the origin of hydrocarbon contaminants. They are bakeable, immune to very high radiation fields, can be operated in any orientation and are not a source of system vibration. When used in conjunction with TSP pumps, and baked at temperatures of ~250° C, pressures $< 10^{-11}$ Torr may be achieved. This combination of TSP and sputter-ion pumping is the least expensive and most reliable means of achieving very low pressures. Sputter-ion pumps can be constructed to suit the application, including built-in, distributed pumps, and the appendage and large component pumps. At low pressures they are by far the most efficient method of pumping. When designing, building, or purchasing a sputter-ion pump, one key to success is in first defining the pressure at which you need a given speed.

APPENDIX 2A
ELECTROSTATIC GETTER-ION PUMPS

Herb, R.G. and Davis, R.H., "Evapor-Ion Pump", Phys. Rev. 89, 897 (1953).

Divatia, A.S. and Davis, R.H., "Construction and Performance of Evapor-Ion Pumps", Proc. 1st Nat. AVS Symp., 1954 (W.M. Welch Manufacturing Company, 1955), p. 40.

Alexeff, I. and Peterson, E.C., "Evapor-Ion Pump Performance with Noble Gases", Proc. 2nd Nat. AVS Symp., 1955 (Committee on Vacuum Techniques, Inc., 1956), p. 87.

Swartz, J.C., "Evapor-Ion Pump Characteristics", Proc. 2nd Nat. AVS Symp., 1955 (Committee on Vacuum Techniques, Inc., 1956), p. 83.

Reich, G. and Nö ller, H.G., "Production of Very Low Pressures with Getter-Ion Pumps", Proc. 4th Nat. AVS Symp., 1957 (Pergamon Press, Inc., New York, 1958) p. 97.

Herb, R.G., "Evapor-Ion Pump Development at the University of Wisconsin", Proc. 1st Int. Vac. Cong., June, 1958, Vacuum 9, 97 (1959).

Holland, L., "Theory and Design of Getter-Ion Pumps", J. Sci. Instrum. 36(3), 105 (1959).

Klopfer, A., Ermrich, W., "Properties of a Small Titanium-ion Pump", Proc. 6th Nat. AVS Symp., 1959 (Pergamon Press, Inc., New York, 1960) p. 297.

Holland, L., and Laurenson, L., "Pumping Characteristics of a Titanium Droplet Getter-Ion Pump", Brit. J. Appl. Phys. 2(9), 401 (1960).

Kumagai, H., et al., "Characteristics of Titanium Evapor-Ion Pump", J. Vac. Sci. Technol. 1, 433 (1960).

Warnecke, M. and Moulou, P.C., "On a Miniature Titanium Pump", Le Vide 85, 41 (1960).

Adam, H. and Bachler, W., "Operational Procedures and Experiences with a High-Speed Ion Getter Pump", Proc. 2nd Int. Vac. Cong., 1961 (Pergamon Press, Inc., New York, 1962) p. 347.

Gould, C.L. and Dryden, R.A., "One Year of Operating Experience with Getter-Ion Pumps", Proc. 2nd Int. Vac. Cong., 1961 (Pergamon Press, Inc., New York, 1962) p. 369.

Gould, C.L. and Mandel, P., "A Sublimation Pump", Proc. 9th Nat. AVS Symp., 1962 (The Macmilian Company, New York, 1963) p. 360.

Hooverman, R.H., "Charged Particle Orbits in a Logarithmic Potential", J. Appl. Phys. 34(12), 3505 (1963).

Herb, R.G., Pauly, T., Welton, R.D., Fisher, K.J., "Sublimation and Ion Pumping in Getter-Ion Pumps", Rev. Sci. Inst. 35(5), 573 (1964).

Maliakal, J.C., Limon, P.J., Arden, E.E., Herb, R.G., "Orbitron Pump of 30-cm Diameter", J. Vac. Sci. Technol. 1(2), 54 (1964).

Mourad, W.G., Pauly, T., Herb, R.G., "Orbitron Vacuum Gauge", Rev. Sci. Instrum. 35(6), 661 (1964).

APPENDIX 2A, CONTINUED
ELECTROSTATIC GETTER-ION PUMPS

Nazarov, A.S., Ivanovskii, G.F., Men'shikov, M.I., "Getter-Ion Pump with Filamentary Titanium and Chromium Evaporators", Instr. & Exper. Tech., No. 6, 934 (1964).

Douglas, R.A., Zabritski, J., Herb, R.G., "Orbitron Vacuum Pump", Rev. Sci. Instrum. 36(1), 1 (1965).

Andrew, D., "The Performance Assessment of Sputter Ion Pumps", Vacuum 16(12), 653 (1966).

Gretz, R.D., "A High Vacuum Titanium Getter Pump", Vacuum 16(10), 537 (1966).

Holland, L., Laurenson, L., Fulker, M.J., "The Vacuum Performance of a Combined Radial Electric Field Pump and Penning Pump", Vacuum 16(12), 663 (1966).

Bills, D.G., "Electrostatic Getter-Ion Pump Design", J. Vac. Sci. Technol. 4(4), 149 (1967).

Denison, D.R., "The Effect of Gas Quantity Pumped on the Speed of Getter-Ion and Sputter-Ion Pumps", Proc. 14th Nat. AVS Symp., 1967 (Herbrick and Held Printing Company, Pittsburgh, 1968), p. 87.

Denison, D.R., "Performance of a New Electrostatic Getter-Ion Pump", J. Vac. Sci. Technol. 4(4), 156 (1967).

Maliakal, J.C., "Residual Gas Analysis in an Orbitron Pump System", Proc. 14th Nat. AVS Symp., 1967 (Herbrick and Held Printing Company, Pittsburgh, 1968) p. 89.

Della Porta, P. and Ferrario, B., "Magnetless Gauge Appendage Pump Utilizing Non-Evaporable Getter Material", Proc. 4th Int. Vac. Cong., 1968 (The Institute of Physics and The Physical Society, London, 1969) p. 369.

Denison, D.R., "Getter Properties of Yttrium as Used in a Getter-Ion Pump", Proc. 4th Int. Vac. Cong., 1968 (The Institute of Physics and The Physical Society, London, 1969) p. 377.

Maliakal, J.C., "Residual-Gas Analysis in an Orb-Ion Pump System", Proc. 4th Int. Vac. Cong., 1968 (The Institute of Physics and The Physical Society, London, 1969) p. 361.

Naik, P.K. and Herb, R.G., "Glass Orbitron Pump of 5-cm Diameter", J. Vac. Sci. Technol. 5(2), 42 (1968).

Kuznetsov, M.V. and Ivanovsky, G.F., "New Developments in Getter-Ion Pumps in the U.S.S.R.", J. Vac. Sci. Technol. 6(1), 34 (1969).

Bills, D.G., "Electrostatic Getter-Ion Pump Performance", J. Vac. Sci. Technol. 10, 65 (1973).

Singleton, J.H., "The Performance Characteristics of Modern Vacuum Pumps", J. Phy. E6, 685 (1973).

APPENDIX 2A, CONTINUED
ELECTROSTATIC GETTER-ION PUMPS

Naik, P.K. and Verma, S.L., "Performance of the Modified Orbitron Pump", J. Vac. Sci. Technol. 14, 734 (1977).

Feidt, M.L. and Petit, B., "Measurement of Ion Energy Distribution in an Orbitron Device", J. Vac. Sci. Technol. 18(3), 987 (1981).

Feidt, M.L. and Paulmier, D.F., "A Model for Optimum Ionizing Characteristics of an Orbitron Device", Vacuum 32(8), 491 (1982).

Problem Set

1) Assume that the speed of a sputter-ion pump is 100 $\mathcal{l}$/sec at a pressure of 10^{-5} Torr, the magnetic field is 0.10 T, and the anode cells have a diameter of 12.7 mm. What would be the speed of the pump at a pressure of 10^{-8} Torr and for the magnetic field settings of 0.1, 0.15 and 0.2 T?

2) Assuming the sputter-yield of 1.05 keV ions, normal to a target, is 1.94 for A$\mathcal{l}$ and 1.13 for Ti, what would the sputter-yield be for these two materials, at this same argon ion energy, if the ions were incident at an angle of 70° ?

3) Assume that the sensitivity of a cell of "zero" length is proportional to the diameter of the cell. Assume that the diameters of anode rings of Fig. 2.2.1 are each ~16 mm ϕ, and the I/P of the cell is 1.8 A/Torr at a pressure of 10^{-5} Torr. Neglecting, $\mathcal{l}$, the length of the cells in Fig. 2.2.7, what would be the I/P of each of the anode arrays given, for a magnetic setting of 0.1 T?

4) Assume that the I/P of a cell is directly proportional to the space charge stored in the cell. Assume that the space charge is of uniform density, and the shape of the discharge forms two right-angle cones, the bases of which join at the mid-point of each cell. Based on these assumptions, what would be the total I/P for the three configurations given in Fig. 2.2.7, for a magnetic field of 0.1 T and at a pressure of 10^{-5} Torr?

5) Assume that each cell of the pump shown in Fig. 2.4.9 has an I/P of 5 A/Torr at pressures $\gtrsim 10^{-6}$ Torr, and that the pump is connected to a robust power supply capable of maintaining a voltage of 5.0 kV even at very high pressures. What would be the total power drawn by this pump at pressures of 10^{-6}, 10^{-5} and 10^{-4} Torr? What would the average power density dissipated in the cathodes be at these three pressures?

6) When operating at high pressures, why don't the anodes get as hot as the cathodes?

7) Assume that there are two pumps attached to a vacuum system, with all metal valves. One pump has ten cells, each with an I/P of 10 A/Torr and the second pump has 250 cells, each with an I/P of 10 A/Torr. Assume that gas is let into the system through a controlled leak. Prove that for the

same leak rate, the current drawn by each pump, when individually pumping on the same leak, would be the same.

8) Assume that each of the cells in the pump elements shown in Fig. 2.4.10 have the same I/P as in problem 5. What would be the power drawn by the pump for the same voltage and pressures of problem 5?

9) You have built a very large accelerator to which 100 pumps, comparable in size to that shown in Fig. 2.4.10, are appended. Assume that each pump cell has an I/P which decreases according to the data given in Fig. 2.2.7, for the 12.7 mm ϕ cells and a field of 0.15 T. What is the total power drawn by the pumps at a pressure of 10^{-9} Torr?

10) Calculate the results of Fig. 2.4.14, using the given assumptions.

11) You suspect your pump has a field emission problem, and with a 6.0 kV setting, the current drawn by the pump is 100 μA. How can you differentiate between current due to field emission, leakage currents and real I/P effects?

12) What is the maximum possible scattering angle of Kr ions off of Aℓ and Ti cathodes?

13) Assume that the density of TiH_2 is ~86.5% that of pure Ti. Prove that the dimensions of a cube of Ti pure would have to increase by ~6.4% to accommodate sufficient H_2 to form TiH_2.

14) Using (2.5.2) and (2.5.6), show that for large times, t, the total amount of gas which has diffused through a semi-infinite slab from time 0 - t is given by $\int Qdt = (DC/d)(t - d^2/6D)$, and is called the "breakthrough" equation.

15) What is the ratio of the speeds of pumps when the ratio $\omega:x_0$, of Fig. 2.10.1, is 2:1, vs. 1:1 for the same anode area and cathode-to-anode gap.

References

1. Herb, R.G. and Davis, R., "Evapor-Ion Pump", Phys. Rev. 89, 897 (1953).
2. Herb, R.G., Pauly, T., Welton, R.D., Fisher, K.J., "Sublimation and Ion Pumping in Getter-Ion Pumps", Rev. Sci. Instrum., 35(5), 573 (1964).
3. Adam, H. and Bachler, W., "Operational Procedures and Experiences with a High-Speed Ion Getter Pump", Proc. 2nd Int. Vac. Cong., 1961 (Pergamon Press, Inc., New York, 1962), p. 374.
4. Hall, L.D., Helmer, J.C., Jepsen, R.L., U.S. Patent No. 2,993,638, "Electrical Vacuum Pump Apparatus and Method", filed 7/24/57, awarded 7/25/61.
5. Redhead, P.A., "Instabilities in Crossed-Field Discharges at Low Pressures", Vacuum 38(8-10), 901 (1988).
6. Gaede, V.W., "Tiefdruckmessungen" (vacuum gauge review paper), Zeitschr. f. Techn. Physik 12, 664 (1934).
7. Guthrie, A. and Wakerling, R.K., Vacuum Equipment and Techniques (McGraw-Hill Book Co., Inc., New York, 1949), p. 115.
8. Philips, C.E.S., Proc. Roy. Soc. (London), A64, 172 (1898).
9. Cobine, J.D., Proc. Nat. Acad. Sci. U.S., 2, 683 (1916).
10. Penning, F.M., U.S. Patent, "Coating by Cathode Disintegration", filed 11/7/36, awarded 2/7/39.
11. Penning, F.M., "Die Glimmentladung Bei Niedrigem Druck Zwischen Koaxialen Zylindern in Einem Axialen Magnetfeld" (sputtering apparatus paper), Physica 3(9), 873 (1936).
12. Guentherschulze, A., "Cathodic Sputtering, an Analysis of the Physical Processes ", Vacuum 3(4) (1953).
13. Gill, W.D. and Kay, E., "Efficient Low Pressure Sputtering in a Large Inverted Magnetron Suitable for Film Synthesis", Rev. Sci. Instrum. 36(3), 277 (1965).
14. Wasa, K. and Hayakawa, S., "Low Pressure Sputtering System of the Magnetron Type", Rev. Sci. Instrum. 40(5), 693 (1969).
15. Welch, K.M., "New Materials and Technology for Suppressing Multipactor in High Power Microwave Windows", Stanford Linear Accelerator Pub. No. SLAC-174, 1974.
16. Penning, F.M., "Ein Neues Manometer fur Niedrige Gasdrucke, Insbesondere Zwischen 10^{-3} und 10^{-5} mm" (paper on various configurations of cold cathode gauges), Physica 4(2), 71 (1937).
17. Penning, F.M. and Nienhuis, K., "Construction and Application of a New Design of the Philips Vacuum Gauge", Philips Tech. Rev. 11(4), 116 (1949).

References

18. Soddy, F., Proc. Roy. Soc. (London), A80, 92 (1908).
19. Westendorp, W.F., and Gurewitsch, A.M., U.S. Patent No. 2,755,014, "Ionic Vacuum Pump Device", filed 4/24/53, awarded 7/17/56.
20. Gurewitsch, A.M. and Westendorp, W.F., "Ionic Pump", Rev. Sci. Instrum. 25(4), 389 (1954).
21. Gale, A.J., "Cold Sealed Getter/Ion Pumped Supervoltage X-ray Tubes", Proc. 2nd Nat. AVS Symp., 1955 (Committee on Vacuum Techniques, Inc., 1956), p. 12.
22. Conn, G.K.T. and Daglish, H.N., "Cold Cathode Ionization Gauges for the Measurement of Low Pressure", Vacuum 3(1), 24 (1953).
23. Conn, G.K.T. and Daglish, H.N., "The Influence of Electrode Geometry on Cold-Cathode Vacuum Gauges", Vacuum 3(2), 136 (1954).
24. Jepsen, R.L., "Important Characteristics of a New Type Getter-Ion Pump", LeVide 80, 80 (1959).
25. Rutherford, S.L., U.S. Patent No. 3,258,194, "Magnetically Confined Glow Discharge Apparatus", filed 7/29/63, awarded 6/28/66.
26. Rutherford, S.L., "Sputter-Ion Pump for Low Pressure Operation", Proc. 10th Nat. AVS Symp., 1963 (The Macmillan Company, New York, 1964), p. 185.
27. Lamont, L.T., "Physical Processes in the Penning Discharge from 10^{-8} to 10^{-3} Torr", Proc. 14th Nat. AVS Symp., 1967 (Herbrick and Held Printing Co., Pittsburgh, 1968), p. 149.
28. Pierini, M., "Use of Discharge Intensity for the Evaluation of Pumping Characteristics of a Sputter Ion Pump", J. Vac. Sci. Technol. A2(2), 195 (1984).
29. Tom, T. and James, B.D., U.S. Patent No. 3,398,879, "Asymmetric Ion Pump Method", filed 10/7/66, awarded 8/27/68.
30. Jepsen, R.L., "Magnetically Confined Cold-Cathode Gas Discharges at Low Pressures", J. Appl. Phys. 32(12), 2619 (1961).
31. Dow, D.G., "Electron-Beam Probing of a Penning Discharge", J. Appl. Phys. 34(8), 2395 (1963).
32. Lange, W.J., "Microwave Measurement of Electron Density in a Penning Discharge", J. Vac. Sci. Technol. 7(1), 228 (1970).
33. Chen, F.F., Introduction to Plasma Physics (Plenum Press, New York, 1974), p. 154.
34. Redhead, P.A., "Oscillations in Magnetically Confined Cold-Cathode Discharges at Very Low Pressures", Canadian J. Phys. 43, 1001 (1965).
35. Helmer, J.C. and Jepsen, R.L., "Electrical Characteristics of a Penning Discharge", Proc. I.R.E. 49, 1920 (1961).
36. Knauer, W., "Mechanism of the Penning Discharge at Low Pressures", J. Appl. Phys. 33(6), 2093 (1962).

References

37. Hooper, E.G., "A Review of Reflex and Penning Discharges", Advances in Electronics and Electron Physics, 27, 295 (1969).
38. Blodgett, K.B. and Vanderslice, T.A., "Mechanism of Inert Gas Cleanup in a Gaseous Discharge", J. Appl. Phys. 31(6), 1017 (1960).
39. Dallos, A., "The Pressure Dependence of the Pumping Speed of Sputter Ion Pumps", Vacuum 19(2), 79 (1969).
40. Peacock, N.T. and Peacock, R.N., "Some Characteristics of an Inverted Magnetron Cold Cathode Ionization Gauge with Dual Feedthroughs", J. Vac. Sci. Technol. A6(3), 1141 (1988).
41. Ono, S. and Yokoo, K., "The Multipactor Ion Pump for Ultra-High Vacuum", Proc. 8th Int. Vac. Cong., 1980 (Supplé ment à la Revue, "Le Vide, les Couches Minces, No. 201", 1981), p. 299.
42. Yokoo, K. and Ono, S., "High-Speed Ion Pumps with a Multipactor Cathode--the Multipactor Ion Pump", Vacuum 32(5), 265 (1982).
43. Welch, K.M., "Diode Sputter Ion Pump Speed for H_2 and N_2 as a Function of Axial and Transverse Magnetic Fields up to 1.2 Tesla", Stanford Linear Accelerator Technical Note No. TN-72-10, July 1972.
44. Jepsen, R.L., "The Physics of Sputter-Ion Pumps", Proc. 4th Int. Vac. Cong., 1968 (The Institute of Physics and the Physical Society, London, 1969), p. 317.
45. Welch, K.M., "New Developments in Sputter-Ion Pump Configurations", J. Vac. Sci. Technol. 13, 498 (1976).
46. Chou, T.S. and McCafferty, D., "Pumping Behavior of Sputter-Ion Pumps", J. Vac. Sci. Technol. 18(3), 1148 (1981).
47. Hamilton, A.R., "Some Experimental Data on Parameter Variation with Triode Getter-Ion Pumps", Proc. 2nd Int. Vac. Cong. and 8th Nat. AVS Symp., 1961 (Pergamon Press, Inc., New York, 1962), p. 388.
48. Konjevic, R., Grant, W.A., Carter, G., "Ion Pumping Effects in Inert Gas Mixtures in an Ionization Tube", Vacuum 18(9), 511 (1968).
49. Rutherford, S.L., U.S. Patent No. 3,159,332, "Methods and Apparatus for Enhanced Sputter-Ion Pump Operation", filed 8/14/61, awarded 12/1/64.
50. Audi, M., "Pumping Speed of Sputter Ion Pumps", Vacuum 38(8-10), 669 (1988).
51. Berman, A., "Methods of Pumping Speed and Gas Release Measurement in Ionization Gauge Heads--A Review", Vacuum 32(8), 497 (1982).
52. Bowden, P. and Brandon, D.G., "The Generation of Dislocations in Metals by Low Energy Ion Bombardment", Phil. Mag. 8, 935 (1963).
53. Oechsner, H., "Sputtering--A Review of Some Recent Experimental and Theoretical Aspects", Appl. Phys. 8, 185 (1975).

References

54. Maissel, L.I., Physics of Thin Films, Vol. 3 (Academic Press, New York, G. Hass and R.E. Thun, Eds.), Chapt. 2, "The Deposition of Thin Films by Cathode Sputtering", p. 61.

55. Andersen, H.H. and Bay, H.L., "Sputter-Yield Studies on Silicon and Silver Target", Radiation Effects 19, 139 (1973).

56. Brubaker, W.M., "A Method for Greatly Enhancing the Pumping Action of a Penning Discharge", Proc. 6th Nat. AVS Symp., 1959 (Pergamon Press, London, 1960), p. 302.

57. Rudnitskii, E.M., "Ion Current Density Distribution at the Cathode of a Penning Discharge", Soviet Phy.-Tech. Phys. 12(5), 666 (1967).

58. Young, J.R., "Pressure Dependence of the Axial Ion Current in a Penning Discharge", J. Vac. Sci. Technol. 5(4), 102 (1968).

59. Komiya, S. and Yagi, N., "Enhancement of Noble Gas Pumping for a Sputter-Ion Pump", J. Vac. Sci. Technol. 6(1), 54 (1969).

60. Smirtnitskaya, G.V. and Tee, N.K., "On the Ion Current Distribution on the Cathode in an Ion Pump Discharge at Low Pressures", Proc. 4th Int. Vac. Cong., 1968 (The Institute of Physics and The Physical Society, London, 1969), p. 356.

61. Ho, W., Wang, R.K., Keng, P.T., "Calculation of Sputtering Ion Pump Speed", J. Vac. Sci. Technol. 20(4), 1010 (1982).

62. Grant, W.A., Carter, G., "Ion Trapping and Gas Release Phenomena", Vacuum 15(10), 477 (1965).

63. Reed, D.J., Harris, F.T., Armour, D.G., Carter, G., "Thermal Evolution Spectrometry of Low Energy Helium Ions Injected into Stainless Steel and Nickel Targets", Vacuum 24(4), 179 (1974).

64. Redhead, P.A., Hobson, J.P., Kornelsen, E.V., The Physical Basis of Ultrahigh Vacuum (Chapman and Hall, Ltd., London, 1968).

65. Winters, H.F., Horne, D.E., Donaldson, E.E., "Absorption of Gases Activated by Electron Impact", J. Chem. Phy. 41(9), 2766 (1964).

66. Bond, G.C., Catalysis by Metals (Academic Press, Inc., New York, 1962) p. 69.

67. Wagener, S., "Efficiency and Mechanism of Barium Getters at Low Pressure", Brit. J. Appl. Phys. 2, 132 (1951).

68. Trapnell, B.M.W., "The Activities of Evaporated Metal Films in Gas Chemisorption", Proc. Roy. Soc. A218, 566 (1953).

69. Reikhrudel, E.M., Smirnitskaya, G.V., Burnisenico, A.I., "Ion Pump with Cold Electrodes and its Characteristics", Radiotekh. Electron 2, 253 (1956).

70. Momose, T., Kanazawa, K., Hisamatsu, H., Ishimaru, H., "Inversely Operated Aluminum Alloy Distributed Ion Pump", J. Vac. Sci. Technol. A7(5), 3092, (1989).

References

71. McCracken, G.M. and Maple, J.H.C., "The Trapping of Hydrogen Ions in Molybdenum, Titanium, Tantalum, and Zirconium", Brit. J. Appl. Phys. 18, 919 (1967).

72. Liu, Y.C., Lin, C.C., Lee, S.F., "Pumping Mechanism for N_2 Gas in a Triode Ion Pump with A1100 Aluminum Cathode", J. Vac. Sci. Technol. A6(1), 139 (1988).

73. Welch, K.M., "Diode Sputter-Ion Pump Speed with Al Cathodes and Ti Anodes, for H_2, N_2, CO and Air", submitted to the Journal of Vacuum Science and Technology, June 1991.

74. Okano, T., Ohsaki, A., Tuzi, Y., "A Zr-Al Composite-Cathode Sputter-Ion Pump", J. Vac. Sci. Technol. A2(2), 191 (1984).

75. Ishimaru, H., "All-Aluminum-Alloy Ultra High Vacuum System for a Large-Scale Electron-Positron Collider", J. Vac. Sci. Technol. A2(2), 1170 (1984).

76. Lu, M. and Fang, P.W., "New Cathode for Sputter Ion Pumps: Aluminum Mixed Rare-Earth Alloy", J. Vac. Sci. Technol. A5 (4), 2591 (1987).

77. Holland, L., "Theory and Design of Getter-Ion Pumps", J. Sci. Instrum. 36, 105 (1959).

78. Hall, L.D., "Electronic Ultra-High Vacuum Pump", Rev. Sci. Instrum. 29(5), 367 (1958).

79. Wheeler, W.R. and Carlson, M., "Ultra-High Vacuum Flanges", Proc. 2nd Int. Vac. Cong. and 8th Nat. AVS Symp., 1961 (Pergamon Press, Inc., New York, 1962), p. 1309.

80. Hall, L.D., "Properties and Behavior of Electronic Ultra-High Vacuum Pumps", Proc. 5th Nat. AVS Symp., 1958 (Pergamon Press, Inc., London, 1958), p. 158.

81. Halliday, D., Resnick, R., Physics for Students of Science and Engineering, Part I (John Wiley and Sons, Inc., New York, 1963), p. 191.

82. Lloyd, W.A. and Huffman, G.A., U.S. Patent No. 2,983,433, "Getter Ion Vacuum Pump Apparatus", filed 8/1/58, awarded 5/9/61.

83. Zaphiropoulos, R. and Lloyd, W.A., "Design Considerations for High Speed Getter-Ion Pumps", Proc. 6th Nat. AVS Symp., 1959 (Pergamon Press, Inc., New York, 1960), p. 307.

84. Andrew, D., "The Development of Sputter-Ion Pumps", Proc. 4th Int. Vac. Cong., 1968 (The Institute of Physics and The Physical Society, London, 1969), p. 325.

85. Jepsen, R.L., Francis, A.B., Rutherford, S.L., Kietzmann, B.E., "Stabilized Air Pumping with Diode Type Getter-Ion Pumps", Proc. 7th Nat. AVS Symp., 1960 (Pergamon Press, Inc., New York, 1961), p. 45.

86. Malter, L. and Mandoli, H., "Electron Tube Processing with Getter-Ion Pumps", Vacuum 10, 121 (1960).

References

87. Neal, R.B., "The Stanford Two-Mile Linear Electron Accelerator", J. Vac. Sci. Technol. 2(3), 149 (1965).
88. Rutherford, S.L., Mercer, S.L., Jepsen, R.L., "On Pumping Mechanisms in Getter-Ion Pumps Employing Cold-Cathode Gas Discharges", Proc. 7th Nat. AVS Symp., 1960 (Pergamon Press, Inc., New York, 1961), p. 380.
89. Andrew, D., Sethna, D.R., Weston, G.F., "Inert-Gas Pumping in a Magnetron Pump", Proc. 4th Int. Vac. Cong., 1968 (The Institute of Physics and the Physical Society, London, 1969), p. 337.
90. Jepsen, R.L., U.S. Patent No. 3,070,719, "Cathodes for Magnetically-Confined Glow Discharge Apparatus", filed 10/11/60, awarded 12/25/62.
91. Vaumoron, J., "The Study of the Influence of Certain Parameters of a Penning Pump on its Stability in Argon", LeVide 25(145), 26 (1970).
92. Vanderslice, T.A., U.S. Patent No. 3,080,104, "Ionic Pump", filed 9/25/58, awarded 3/5/63.
93. Clarke, P.J. and Roman, N., U.S. Patent No. 3,325,086, "Triode Ion Pump", filed 9/20/65, awarded 6/13/67.
94. Brubaker, W.M. and Berry, C.E., U.S. Patent No. 3,535,055, "Cold-Cathode Discharge Ion Pump", filed 5/25/59, awarded 10/20/70.
95. Lafferty, J.M. and Vanderslice, T.A., "The Interplay of Electronics and Vacuum Technology", Proc. I.R.E. 49, 1136 (1961).
96. Kawasaki, K., Sugita, T., Kohno, I., Watanabe, J., "Study of Pumping Action of a Diode Type Getter Ion Pump for Inert Gases by the Tracer Technique", Japan J. Appl. Phys. 2(2), 315 (1963).
97. Willis, R.D., Allman, S.L., Chen, C.H., Alton, G.D., Hurst, G.S., "Pumping Inert Gases by Electron-Impact Ionization Sources and Associated Memory Effects", J. Vac. Sci. Technol. A2(1), 57 (1984).
98. Pierini, M. and Dolcino, L., "A New Sputter-Ion Pump Element", J. Vac. Sci. Technol. A1(2), 140 (1983).
99. Audi, M. and Pierini, M., "Surface Structure and Composition Profile of Sputter-Ion Pump Cathode and Anode", J. Vac. Sci., Technol. A4(3), 303 (1986).
100. Audi, M., Dolcino, L., Doni, F., Ferrario, B., "A New Ultrahigh Vacuum Combination Pump", J. Vac. Sci. Technol. A5(4), 2587 (1987).
101. Audi, M. and DeSimon, M., "The Influence of Heavier Gases in Pumping Helium and Hydrogen in an Ion Pump", J. Vac. Sci. Technol. A6(3), 1205 (1988).
102. Singleton, J.H., "Hydrogen Pumping by Sputter-Ion Pumps and Getter Pumps", J. Vac. Sci. Technol. 8(1), 275 (1971).
103. Dallos, A. and Steinrisser, F., "Pumping Speeds of Getter-Ion Pumps at Low Pressures", J. Vac. Sci. Technol. 4(1), 6 (1967).

References

104. Wear, K.B., "Electrical Characteristics of Sputter-Ion Pumps", J. Appl. Phys. 38(4), 1936 (1967).
105. Bloomer, R.N. and Cox, B.M., "Some Effects of Gases Upon Vacuum Breakdown Initiated by Field Emission of Electrons", Vacuum 18(7), 379 (1968).
106. Tom, T. and James, B.D., "Inert Gas Ion Pumping Using Differential Sputter-Yield Cathodes", Proc. 13th Nat. AVS Symp., 1966 (Herbick and Held Printing Company, Pittsburgh, Pennsylvania, 1967), p. 21.
107. Tom, T. and James, B.D., "Inert Gas Ion Pumping Using Differential Sputter Yield Cathodes", J. Vac. Sci. Technol. 6(2), 304 (1969).
108. Jepsen, R.L., "The Physics of Sputter-Ion Pumps", Proc. 4th Int. Vac. Cong., 1968 (The Institute of Physics and The Physical Society, London, 1969), p. 317.
109. Brodie, I., Lamont, L.T, Jepsen, R.L., "Production of High-Energy Neutral Atoms by Scattering of Ions at Solid Surfaces and Its Relation to Sputtering", Phys. Rev. Lett. 21(17), 1224 (1968).
110. Kornelsen, W., "The Ionic Entrapment and Thermal Desorption of Inert Gases in Tungsten for Kinetic Energies of 40 eV to 5 keV", Can. J. Phys. 42, 364 (1964).
111. Snoek, C. and Kistemaker, J., "Fast Ion Scattering Against Metal Surfaces", Advances in Electronics and Electron Physics, Vol. 21 (Academic Press, New York, 1965, L. Marton and C. Marton, Eds.), p.67.
112. Vaumoron, J.A., "Influence of Certain Parameters of a Penning Pump on its Stability in Argon", LeVide 145, 26 (1970).
113. Harra, D.J., Abstract: "Pumping Mechanisms of Sputter-Ion Pumps", J. Vac. Sci. Technol. 11, 331 (1974).
114. Hall, L.D., "Multicomponent Getter Films for Getter-Ion Pumps", J. Vac. Sci. Technol. 6(1), 44 (1969).
115. Vaumoron, J.A. and DeBiasio, M.P., "Argon and Rare Gas Instability With Heavy Metal Cathode Pumps", Vacuum 20(3), 109 (1970).
116. Baker, P.N. and Laurenson, L., "Pumping Mechanisms for the Inert Gases in Diode Penning Pumps", J. Vac. Sci. Technol. 9(1), 375 (1972).
117. Rutherford, S.L., Jepsen, R.L., Frances, A.B., "Study Program on the Utilization of Vacion® Pump Elements for DCX", Final Report, August, 1960, prepared for Oak Ridge National Laboratory, Varian Engineering Report No. 267-1F.
118. Vissers, D.R., Holmes, J.T., Nelson, P.A., Bartholme, L.G., U.S. Patent No. 3,683,272, "Method and Apparatus for Determining Hydrogen Concentration in Liquid Sodium Utilizing an Ion Pump to Ionize the Hydrogen", filed 11/24/70, awarded 8/8/72.

References

119. Welch, K.M., U.S. Patent No. 4,047,102, "Dual Chamber, High Sensitivity Gas Sensor and Pump", filed 5/3/76, awarded 9/6/77.
120. Singleton, J.H., "Hydrogen Pumping Speed of Sputter-Ion Pumps", J. Vac. Sci. Technol. 6(2), 316 (1969).
121. Singleton, J.H., "The Performance Characteristics of Modern Vacuum Pumps", J. Phys. E6, 685 (1973).
122. Komiya, S., Sato, S., Hayashi, C., "Triggering Sputter Ion Pump in Extreme High Vacuum", Proc. 13th Nat. AVS Symp., 1966 (Herbick and Held Printing Company, Pittsburgh, Pennsylvania, 1967), p. 19.
123. Dushman,S., Scientific Foundations of Vacuum Technique (John Wiley and Sons, Inc., New York, 1949), p. 581.
124. Zäidel, A.N., Spectra-Isotopic Method for Determination of H2 in Metals (Butterworths, London, 1961).
125. Riviere, J.C. and Allison, J.D., "Gas Evolution During Baking of Sputter-Ion Pumps", Supplemento Al Nuovo Cimento, 1(2), 520 (1963); also, Vacuum 14(3), 97 (1964).
126. Fast, J.D., "Foreign Atoms in Metals", Philips Tech. Rev. 16(12), 341 (1955).
127. Klaichko, I.A., Kunin, L.L., Fedorov, S.P., Larionov, I.N., "Studies of Interaction Between Gases and Metals", UCRL Translation #1375, from Russian Akademiia Nauk SSSR Komissiia po Analiticheskoi Khimii, Trudy 10, 49 (1960).
128. Carter, G., Nobes, M.J., Armour, D.G., "The Erosion Energy Efficiency of Sputtering", Vacuum 32(8), 509 (1982).
129. Colligon, J.S., "Ion Bombardment of Metal Surfaces", Vacuum 11, 272 (1961).
130. KenKnight, C.E. and Wehner, G.K., "Sputtering of Metals by Hydrogen Ions", J. Appl. Phys. 35(2), 322 (1964).
131. Rutherford, S.L. and Jepsen, R.L., "Enhanced H2 Pumping With Sputter-Ion Pumps", Rev. Sci. Instrum. 32, 1144 (1961).
132. Dean, N., "Helium Treatment of Vac-Ion Pumps to Remove Deposited Argon", Rev. Sci. Instrum. 36, 1266 (1965).
133. Cummings, U., Dean, N., Johnson, F., Jurow, J., Voss, J., "Vacuum System for the Stanford Storage Ring, SPEAR", J. Vac. Sci. Technol. 8(1), 348 (1971).
134. Jepsen, R.L., U.S. Patent No. 3,147,910, "Vacuum Pump Apparatus", filed 8/30/61, awarded 9/8/64.
135. Hill, E.F., U.S. Patent No. 4,097,195, "High Vacuum Pump", filed 10/10/75, awarded 6/27/78.

References

136. Lamont, L.T., U.S. Patent No. 3,460,745, "Magnetically Confined Electrical Discharge Getter Ion Vacuum Pump Having a Cathode Projection Extending into the Anode Cell", filed 8/23/67, awarded 8/12/69.
137. Lamont, L.T., "A Novel Diode Sputter-Ion Pump", J. Vac. Sci. Technol. $\underline{6}$(1), 47 (1969).
138. Welch, K.M., U.S. Patent No. 3,994,625, "Sputter-Ion Pump Having Improved Cooling and Improved Magnet Circuitry", filed 2/18/75, awarded 11/30/76.
139. Brothers, C.F., Tom, T., Munro, D.F., "Design and Performance of a 50,000 ℓ/sec Pump Combining Cold Cathode Ion Pumping and Active Film Gettering", Proc. 10th Nat. AVS Symp., 1963 (The Macmillan Company, New York, 1964), p. 202.
140. Laurent, J.M., "Study of Ion Pumps for Operation in the Low Magnetic Bending Field of a Large Electron Storage Ring", Proc. 8th Int. Vac. Cong., 1980 (Supplément à la Revue, "LeVide, les Couches Minces, No. 201", 1981), p. 164.
141. Zeng-Rui, L. and Ming, L., "Pumping Mechanism for H_2 and Cathode Material in a Sputter Ion Pump", op. cit., p. 295.
142. Donachie, M.J., Ed., Titanium and Titanium Alloys (American Society for Metals, Metals Park, Ohio, 1982).
143. Lamont, L.T., "A Model for the Penning Discharge in the Neighbourhood of the Transition Point", Proc. 4th Int. Vac. Cong., 1968 (The Institute of Physics and The Physical Society, London, 1969), p. 345.
144. Beanland, D.G., Ion Implantation and Beam Processing (Academic Press, New York, 1984, J.S. Williams and J.M. Poate, Eds.), Chapt. 8, "High-Dose Implantation", p. 297.
145. Young, J.R., "Penetration of Electrons and Ions in Aluminum", J. Appl. Phys. $\underline{27}$(1), 1 (1956).
146. Paton, N.E. and Williams, J.C., Titanium and Titanium Alloys (M.J. Donachie, Ed., American Society for Metals, Metals Park, Ohio, 1982), Section IV, "Effect of Hydrogen on Titanium and Its Alloys", p. 185.
147. Barret, C.R., Nix, W.D., Tetelman, A.S., The Principles of Engineering Materials (Prentice-Hall, Inc., Englewood Cliffs, New Jersey, 1973), p. 152.
148. Wanhill, R.J.H., "Aqueous Stress Corrosion in Titanium Alloys", Brit. Corrosion J. $\underline{10}$(2), 69 (1975).
149. Philibert, J., "Diffusion in Solids", Sci. & Technol., August, 1968, p. 47.
150. Crank, J., The Mathematics of Diffusion (Oxford University Press, Oxford, 1975), p. 254.
151. Ibid, p. 50.

References

152. Kubaschewski, O., Titanium: Physico-Chemical Properties of Its Alloys, Special Issue No. 9 (International Atomic Energy Agency, Vienna, 1983, K.L. Komarek, Ed.), Chapt. IV, "Diffusion", p. 441.

153. Crank, J., The Mathematics of Diffusion (Oxford University Press, Oxford, 1975), p. 21.

154. Ibid, p. 61.

155. Spagenberg, K.R., Vacuum Tubes (McGraw-Hill Book Company, Inc., New York, 1948), p. 125.

156. Snouse, T., "Starting Mode Differences in Triode and Diode Sputter-Ion Pumps", J. Vac. Sci. Technol. $\underline{8}$(1), 283 (1971).

157. Baker, L.C. and Bance, R.D., "The Pumping of Deuterium by Triode Ion Pump in the 10^{-5} Torr Region on a Van de Graaff Accelerator", Vacuum $\underline{20}$(6), 251 (1970).

158. Turner, C.M., Marinuzzi, J.G., Bennett, G.W., "Improvement of Hydrogen Pumping of Penning Discharge Getter-Ion Pumps", IEEE Trans. Nucl. Sci., $\underline{NS-14}$, 831 (1967).

159. Denison, D.R., "Comparison of Diode and Triode Sputter-Ion Pumps", J. Vac. Sci. Technol. $\underline{14}$(1), 633 (1977).

160. Hatch, J.E., Editor, Aluminum, Properties and Physical Metallurgy (American Society of Metals, Metals Park, Ohio, 1984), p. 18.

161. Dushman, S., Scientific Foundations of Vacuum Technique (John Wiley and Sons, New York, 1962), p. 570.

162. Celik, M.C. and Bennett, G.H.J., "Effects of Hydrogen Inclusions on Blistering in High-Purity Aluminum Sheet and Foil on a Laboratory Scale", Metals Technol. $\underline{6}$(4), 138 (1979).

163. Barnes, R.S. and Mazey, D.J., "The Migration and Coalescence of Inert Gas Bubbles in Metals", Proc. Roy. Soc. (London), $\underline{A275}$, 47 (1963).

164. Kleuh, R.L. and Mullins, W.W., "Some Observations on Hydrogen Embrittlement of Silver", Trans. Metallurgical Soc. AIME, $\underline{242}$, 237 (1968).

165. Norton, F.J., "Gas Permeation Through the Vacuum Envelope", Proc. 8th Nat. AVS Symp. and 2nd Int. Vac. Cong., 1961 (Pergamon Press, New York, 1962, L.E. Preuss, Editor), Vol. I, p. 8.

166. Eschbach, H.L., Gross, F., Schulien, S., "Permeability Measurements with Gaseous Hydrogen for Various Steels", Vacuum $\underline{13}$, 543 (1963).

167. McCracken, G.M. and Maple, J.H.C., "Reflection of Hydrogen Ions From Metal Surfaces in the Energy Range 5-30 keV", Proc. 7th Int. Conf. on Phenomena in Ionized Gases, Beograd, 1965 (Gradevinska Knjiga Publishing House, Beograd, 1966, B. Perovic and D. Tosic, Editors), Vol. I., p. 137.

References

168. PNEUROP, Editorial Body of Twelve European Countries, Vacuum Technology Sub-committee, "Acceptance Specifications, Part VI, Acceptance Rules for Sputter Ion Pumps" (British Compressed Air Society, PNEUROP Editorial Body, 1976).

169. Rapp, D. and Englander-Golden, P., "Total Cross Sections for Ionization and Attachment in Gases by Electron Impact. I. Positive Ionization", J. Chem. Phys. 43(5), 1464 (1965).

170. Jepsen, R.L., King, L.T., Rogers, R.M., U.S. Patent No. 3,116,764, "High Vacuum Method and Apparatus", filed 3/30/59, awarded 1/7/64.

171. Souchet, R., "Study of the Degassing Spectra of Materials for the Construction of the Vacuum Chamber of an e^+e^- Colliding Beam at 1.8 GeV", Institut National de Physique Nucleaire et de Physique des Particules, Laboratoire de L'Accelerateur Lineaire, Report No. 4-72, June, 1972.

172. Jepsen, R.L., U.S. Patent No. 3,149,774, "Getter Ion Pump Method and Apparatus", filed 1/27/61, awarded 9/22/64.

173. Jepsen, R.L., U.S. Patent No. 3,331,975, "Cooling Apparatus for Cathode Getter Pumps", filed 2/19/65, awarded 7/18/67.

174. Johnson, J.W., Atkins, W.H., Dowling, D.T., McConnell, J.W., Milner, W.T., Olsen, D.K., "Heavy Ion Storage Ring for Atomic Physics Vacuum Test Stand for Pressures of $10^{-1\,2}$ Torr", J. Vac. Sci. Technol. A7(3), 2430 (1989).

175. Tom, T., "Use of a Metallic Ion Source in Cold Cathode Sputter Ion Pumps", Proc. 5th Int. Vac. Cong., 1971, J. Vac. Sci. Technol. 9(1), 383 (1972).

176. Jepsen, R.L., U.S. Patent No. 3,094,639, "Glow Discharge Method and Apparatus", filed 10/6/60, awarded 6/18/63.

177. Helmer, J.C., private communication.

178. Guthrie, A. and Wakerling, R.K., Vacuum Equipment Techniques (McGraw-Hill Book Company, Inc., New York, 1949), p. 39.

179. O'Neill, G.K, "Super-High-Energy Summer Study No. 8, Storage Ring Work at Stanford", Brookhaven National Laboratory, Internal Report AADD-8 (1963).

180. Fischer, G.E. and Mack, R.A., "Vacuum Design Problems of High Current Electron Storage Rings", Cambridge Electron Accelerator Report CEAL-1013 (1964).

181. Anashin, V.V., Auslender, V.L., Bender, E.D., Blinov, G.A., Malev, M.D., Osipov, B.N., Popov, A.T., Trakhtenberg, E.M., "The Ultrahigh-Vacuum Pumping System of the VEPP-2 Storage Ring", Proc. All-Union Conf. on Particle Accelerators, Moscow, 1968, Vol. 2, 560 (1970).

References

182. Cummings, U., Dean, N., Johnson, F., Jurow, J., Voss, J., "Vacuum System for Stanford Storage Ring, SPEAR", Stanford Linear Accelerator Center, SLAC-PUB-797 (1970).

183. Zeilinger, A., and Pochman, W.A., "New Methods for the Measurement of Hydrogen Diffusion in Metals", J. Appl. Phys. $\underline{47}$(12), 5478 (1976).

184. Falland, Chr., Hartwig, H., Kouptsidis, J., Kueppershaus, R., Schumann, G., Schwartz, M., Wedekind, H.-P., "First Operational Experience with the 2.3 km Long UHV System of the Electron Storage Ring PETRA", Proc. 8th Int. Vac. Cong., 1980 (Supplé ment à la Revue, "LeVide, les Couches Minces, No. 201", 1981), p. 126.

185. Blechschmidt, D., Bojon, J.-P., Chapman, G., Jensen, D., Monnier, B., Nordstrom, H., "Linear Sputter-Ion Pump for Integration in High Magnetic Fields", Proc. 8th Int. Vac. Cong., 1980 (Supplé ment à la Revue, "LeVide, les Couches Minces, No. 201", 1981), p. 159.

186. Hartwig, H. and Kouptsidis, J.S., "Design Performance of Integrated Sputter-Ion Pumps for Particle Accelerators", Proc. 7th Int. Vac. Cong. and 3rd Int. Conf. on Solid Surfaces, 1977 (R. Dobrozemsky, F. Rüdenauer, F.P. Viehböck, A. Breth, Postfach 300, A-1082 Vienna, Austria, 1977), p. 93.

187. Momose, T., Chen, J.R., Kanazawa, K., Hisamatsu, H., Ishimaru, H., "A New Distributed Ion Pump Made of Aluminum Alloy with an Aluminum or a Titanium Cathode in the Transposable Ring Intersecting Storage Accelerators in Nippon e$^+$e$^-$ Colliding Ring", J. Vac. Sci. Technol. $\underline{A7}$(5), 3098 (1989).

188. Malev, M.D. and Trachtenberg, E.M., "Built-in Getter-Ion Pumps", Vacuum $\underline{23}$(11), 403 (1973).

189. Ishimaru, H., Nakanishi, H., Horikoshi, G., "Distributed Sputter Ion Pump and Titanium Getter Ion Pump Combination", Proc. 8th Int. Vac. Cong., 1980 (Supplé ment à la Revue, "Le Vide, les Couches Minces, No. 201", 1981), p. 331.

190. Pingel, H. and Schulz, L., "A High-Field Integrated Sputter Ion Pump for BESSY - The Berlin Electron Storage Ring for Synchrotron Radiation", Proc. 8th Int. Vac. Cong., 1980 (Supplé ment à la Revue, "LeVide, les Couches Minces, No. 201", 1981), p. 147.

191. Pingel, H. and Schulz, L., "The Ultra High Vacuum System for BESSY - The Berlin Electron Storage Ring for Synchrotron Radiation", Proc. 8th Int. Vac. Cong., 1980 (Supplé ment à la Revue, "LeVide, les Couches Minces, No. 201", 1981), p. 119.

192. Ryabov, V.V. and Saksagansky, G.L., "Influence of Asymmetry of Magnetic and Electric Fields on the Parameters of Sputter-Ion Pumps", Vacuum $\underline{22}$(5), 191 (1972).

References

193. Trickett, B.A., "Vacuum Systems for the Daresbury Synchrotron Radiation Source", Proc. 7th Int. Vac. Cong. and 3rd Int. Conf. on Solid Surfaces, 1977 (R. Dobrozemsky, F. Rüdenauer, F.P. Viehböck, A. Breth, Postfach 300, A-1082 Vienna, Austria, 1977), p. 347.
194. Tsujikawa, H., Chida, K., Mizobuchi, T., Miyahara, A., "Vacuum System for the Test Accumulation Ring for Numatron Project (TARN)", Proc. 8th Int. Vac. Cong. 1980 (Supplé ment à la Revue, "Le Vide, les Couches Minces, No. 201", 1981), p. 151.
195. Pressman, A.I., Switching and Linear Power Supply, Power Converter Design (Hayden Book Company, New Jersey, 1977).
196. Quinn, D.L., U.S. Patent No. 3,412,310, "Power Supply for Glow Discharge Type Vacuum Pumps with Voltage-Doubler Bridge-Rectifier and a Soft Transformer", filed 3/6/67, awarded 11/19/68.
197. Mandoli, H.A., U.S. Patent No. 3,118,103, "Voltage Doubling Power Supply", filed 6/1/59, issued 1/14/64.
198. Mandoli, H.A. and Rorden, J.R., U.S. Patent No. 3,267,377, "Measuring Circuit for Providing an Output either Linearly or Logarithmically Related to an Input", filed 6/1/59, awarded 8/16/66.
199. Underhill, E.M., Ed., Permanent Magnet Design (Crucible Steel Co. of America, Pittsburgh, PA, 1957).
200. Casimir, H.B.G., Haphazard Reality (Harper and Row, New York, 1983), p. 281.
201. van den Broek, C.A.M. and Stuijts, A.L., "Ferroxdure", Philips Tech. Rev. 37(7), 157 (1977).
202. Wohlfarth, E.P., Editor, Ferromagnetic Materials (North Holland Publishing Company, Inc., 1982).
203. Helmer, J.C., "Magnetic Circuits Employing Ceramic Magnets", Proc. I.R.E., 1528 (1961).
204. Helmer, J.C., U.S. Patent No. 3,159,333, "Permanent Magnet Design for Pumps", filed 8/21/61, awarded 12/1/64.
205. Kearns, W.J., "A High Efficiency Magnetic Field Design for Large Ion Pumps", Proc. 10th Nat. AVS Symp., 1963 (The Macmillan Company, New York, 1964), 180.
206. Kraus, J.D., Electromagnetics (McGraw-Hall Book Company, Inc., New York), p. 206.
207. Schuurman, W., "Investigation of a Low Pressure Penning Discharge", Physica 36, 136 (1967).
208. Calder, R. and Lewin, G., "Reduction of Stainless-Steel Outgassing in Ultra-High Vacuum", Brit. J. Appl. Phys. 18, 1459 (1967).

References

209. Knauer, W. and Lutz, M.A., "Measurement of the Radial Field Distribution in a Penning Discharge by Means of the Start Effect", Appl. Phys. Lett. 2(6), 109 (1963).
210. Herzberg, G., Atomic Spectra and Atomic Structure (Dover Publications, New York, 1944).
211. Lange, W.J., "Microwave Measurements of Electron Density in a Penning Discharge", J. Vac. Sci. Technol. 7(1), 228 (1969).
212. Agdur, B. and Ternström, U., "Instabilities in Penning Discharges", Phys. Rev. Lett. 13(1), 5 (1964).
213. Hirsch, E.H., "On the Mechanism of the Penning Discharge", Brit. J. Appl. Phys. 15, 1535 (1964).
214. Knauer, W., U.S. Patent No. 3,216,652, "Ionic Vacuum Pump", filed 9/10/62, awarded 11/9/65.
215. Hirsch, E.H., "Excess Energy Electrons", Brit J. Appl. Phys. 15, 909 (1964).
216. Schuurman, W., "Investigation of a Low-Pressure Penning Discharge", Ph.D. Thesis, Rijksuniversiteit Te Utrecht, March 21, 1966.
217. Hobson, J.P. and Redhead, P.A., "Operation of an Inverted-Magnetron Gauge in the Pressure Range 10^{-3} to 10^{-12} mmHg", Can J. Phys. 36, 271 (1958).
218. Hartwig, H. and Kouptsidis, J.S., "A New Approach to Evaluate Sputter-Ion Pump Characteristics", J. Vac. Sci. Technol. 11(6), 1154 (1974).
219. Wutz, M., "Getter-Ion Pumps of the Magnetron Type and an Attempted Interpretation of the Discharge Mechanism", Vacuum 19(1), 1 (1969).
220. Brillouin, L., "Practical Results from Theoretical Studies of Magnetrons", Proc. I.R.E., 216 (1944).
221. Malmberg, J.H. and Driscoll, C.F., "Long-Time Containment of a Pure Electron Plasma", Phys. Rev. Lett. 44(10), 654 (1980).
222. Glass, R.C., Sims, G.D., Stainsby, A.G., "Noise in Cut-Off Magnetrons", Proc. Inst. Elec. Eng. (London), B102, 81 (1955).
223. de Chernatony, L. and Craig, R.D., "The Discharge Mechanism in a Magnetic Ion Pump", Vacuum 19(9), 393 (1969).
224. Redhead, P.A., "The Townsend Discharge in a Coaxial Diode with Axial Magnetic Field", Can. J. Phys. 36(8), 255 (1958).
225. Andersen, H.H. and Ziegler, J.F., Hydrogen Stopping Powers and Ranges in All Elements, Vol. 3 of Stopping and Ranges of Ions in Matter (Pergamon Press, New York, 1977).
226. Lassiter, W.S., "Extension of Measurements of the Striking Characteristics of the Magnetron Gage into Ultrahigh Vacuum", Proc. 14th Nat. AVS Symp., 1967 (Herbrick and Held Printing Company, Pittsburgh, 1968), p. 45.

References

227. Knauer, W. and Stack, E.R., "Alternative Ion Pump Configuration - Penning Discharge", Proc. 10th Nat. AVS Symp., 1963 (The Macmillan Company, New York, 1964), p. 180.
228. Henderson,W.G. and Mark, J.T., U.S. Patent No. 3,540,812, "Sputter Ion Pump", filed 4/12/68, awarded 11/17/70.
229. Bryant, P.J., "On the Use of a Penning Cell for Pressure Measurements", Proc. 14th Nat. AVS Symp., 1967 (Herbrick and Held Printing Company, Pittsburgh, 1968), p. 55.
230. Bryant, P.J. and Gosselin, C.M., "Response of the Trigger Discharge Gauge", J. Vac. Sci. Technol. 3(6), 350 (1966).
231. Jepsen, R.L., U.S. Patent No. 3,174,069, "Magnetically Confined Glow Discharge Apparatus", filed 11/29/61, awarded 3/16/65.
232. Jepsen, R.L., U.S. Patent 3,028,071, "Glow Discharge Apparatus", filed 3/6/59, awarded 4/3/62.
233. Reid, R.J. and Trickett, B.A., "Optimization of Distributed Ion Pumps for the Daresbury Synchrotron Radiation Source", Proc. 7th Int. Vac. Cong. and 3rd Int. Conf. on Solid Surfaces, 1977 (R. Dobrozemsky, F. Rüdenauer, F.P. Viehböck, A. Breth, Postfach 300, A-1082 Vienna, Austria, 1977), p. 89.
234. Hartwig, H. and Kouptsidis, J.S., "A New Approach for Computing Diode Sputter-Ion Pump Characteristics", J. Vac. Sci. Technol. 11(6), 1154 (1974).
235. Bance, U.R. and Craig, R.D., "Some Characteristics of Triode Ion Pumps", Vacuum 16(12), 647 (1966).
236. Kelly, J.E. and Vanderslice, T.A., "Pumping of Hydrocarbons by Ion Pumps", Vacuum 11(4), 205 (1961).
237. Lichtman, D., "Hydrocarbon Formation in Ion Pumps", J. Vac. Sci. Technol. 1(1), 23 (1964).
238. Reich, G., "Investigation of Titanium Sheets for Sputter-Ion Pumps", Supplemento Al Nuovo Cimento 1(2), 487 (1963).
239. Jepsen, R.L. and Francis, A.B., "Interactions Between Ionizing Discharges and Getter Films", Supplemento Al Nuovo Cimento 1(2), 694 (1963).
240. Sheraton, W.A., "Contamination in Sputter Ion Pumped Systems", Vacuum 15(12), 577 (1965).
241. Lloyd, W.A. and Zaphiropoulos, R., U.S. Patent No. 3,042,824, "Improved Vacuum Pumps", filed 6/22/60, issued 7/3/62.
242. Staph, H.E., "Protection of Electrical Components from Damage by Arcing or Plasma Discharges in Vacuum Equipment", Vacuum 15(11), 545 (1965).

References

243. Wear, K., "By-Products Emitted from Sputter-Ion Pumps", R&D Magazine, February, 1967.
244. Bell, P.R., Moore, E.C., Wyrick, A.J., "Starting Sputter-Ion Pumps and the Outgassing of Wet Metal Surfaces", Vacuum 17(2), 87 (1967).
245. Neal, R.B., "Adsorption Pumping of the Two-Mile Accelerator, Stanford Linear Accelerator Center, Stanford, California, Report No. SLAC TN-67-77, 1962.
246. Bance, U.R., "Development of a diode/triode (D/T) ion pump", Vacuum 40(5), 457(1990).

CHAPTER 3

TITANIUM SUBLIMATION PUMPING

3.0 Introduction

Titanium sublimation pumping is called in the vernacular *TSP pumping*, or titanium sublimation pumping pumping (sic). In the same vein, titanium sublimation pumps are referred to as TSP pumps. Oh well, I won't attempt to change tradition. TSP pumping is used extensively in UHV applications. Such pumping is a form of chemical getter pumping, or chemisorption pumping. Titanium is sublimed or evaporated onto a surface - the walls of the pump - where it getters gas. To *getter* a gas means to remove the gas from a system by chemisorption, or the formation of a stable chemical compound of the gas with some chemically active metal. The terms *evaporation* and *sublimation* are used synonymously in the literature, when referring to the operation of these pumps.

The gettering of gases by various metals has been put to practical applications for over a century.[1] A survey of some of the earlier work on the properties of getters, including his own substantial work, is given by Wagener.[2] Stout conducted one of the early investigations of the sorption of various gases by heated, nonsubliming, Ti filaments.[1] A more recent review of the subject of gettering was done by Giorgi, Ferrario and Storey.[3] Also, papers have been published on the gettering properties of large-scale, sublimed materials other than titanium. For example, Hayward[4] and Okano[5] have reported on the pumping of sublimed Ta films, Prevot and Sledziewski[6] on the effectiveness of sublimed TiO films in pumping gases, including Ar(!), Haque[7] the pumping effectiveness of simultaneously sublimed films of Ni and Ti, and Nazarov's[8] pumping comparison studies between sublimed Cr and Ti films. All evidence to date suggests that, when sublimation pumping in large systems (*i.e.*, *vs.* gettering in sealed-off devices, such as microwave tubes), the most practical and effective material proves to be Ti. For reasons other than chemisorption properties, alloys of Ti and refractory materials are often used.[9,10] However, Ti is the important, chemically active component. Therefore, this chapter will deal exclusively with the properties and applications of TSP pumping. There have been numerous clever schemes used to evaporate or sublime Ti onto surfaces. I will describe some of these later. Note that many of the original electrostatic ion pumps (*e.g.*, see Appendix 2A) were primarily clever methods of dispensing Ti onto pumping surfaces. Granted, ionic pumping plays an important role in these devices. However, the high speed observed for the chemically active gases stems primarily from the Ti sublimed, or otherwise spewed onto the walls of the pump. Because of this, Herb and his colleagues might be viewed as the founders of TSP pumping.[11] To understand TSP pumping, we must first have an appreciation for the concept of the *sticking coefficient* of a gas.

3.1 Sticking Coefficients

The *sticking coefficient*, α, of a gas molecule is merely a measure of the probability that the molecule, when landing on some surface, will permanently stick. Therefore, it is the measure of the effectiveness with which the pump (or pumping surface) removes gas from the system. The concept is simple; but, as noted in numerous publications, including the definitive work of Hayward and Taylor[12] and the earlier work of Giorgi and Pisani,[13] the measurement of sticking coefficient values is difficult, and fraught with potential error. Fortunately, most of us need not make such measurements. However, we do need to understand the concept and to interpret the data at our disposal.

In Chap. 1, I noted that the rate at which gas molecules impinge on a surface is given by:

$$\nu \quad = k_1 \, P \text{ molecules/sec-cm}^2, \tag{3.1.1}$$

where, k_1 $= 3.51 \times 10^{22}/(MT)^{1/2},$
  P $=$ the pressure in Torr,
  M $=$ the molecular weight of the gas,
and T $=$ the temperature of the gas in $^\circ$K.

Further, in Chap, 1, I defined the concept of conductance as a convenient mathematical model for describing the behavior of gas in a linear (*i.e.*, *molecular flow*) system. For example, the conductance of an aperture was discussed in Section 1.12.1. We noted in Fig. 1.12.1b, that the rate at which gas was conducted from Manifold B to Manifold A, Q_B, was directly proportional to the pressure in Manifold B and the area of the aperture separating the two manifolds. Or,

$$Q_B \quad = C_a P_B \tag{3.1.2}$$

where, C_a $=$ the conductance of the aperture in $\mathcal{L}$/sec,
and P_B $=$ the pressure in Manifold B.

If the pressure in Manifold A is maintained at a *perfect* vacuum, no gas molecules will pass from Manifold A into Manifold B. Therefore, in this case, every molecule which impinges on the area defined by the aperture will be permanently removed from the vacuum system of Manifold B.

Now, imagine that the aperture of Fig. 1.12.1b is covered over with a metal plate of the exact same area. Assume that the metal plate is somehow cooled to a temperature near absolute zero. If, for example, the gas in Manifold B is N_2, every gas molecule which impinges on the cooled surface will permanently stick to the plate. This corresponds to a unity sticking coefficient for the gas and it is permanently removed from the system. There is no difference in the pressure in Manifold B over the case when there was a perfect vacuum in Manifold A and gas could pass through the aperture. In the former case, the rate at which gas was removed from the system depended on the conductance (*i.e.*, area) of the aperture; in the latter case, it depends on the area of the cooled surface. Where the aperture is said to have a conductance, the cold plate is said to have a *surface conductance*. If all of the gas which impinges on it sticks, making the plate even colder will have no benefit in terms of the further pumping of gas. Therefore, the speed of the surface pump is limited only by the area of the cold plate. If we wished to increase the speed of the surface pump, we would have to increase its area. Therefore, the speed of this surface pump is said to be *surface conductance limited*. We must keep in mind that the conductance of an aperture or pumping surface differs with the temperature and species of gas (*i.e.*, see (1.12.7)).

A similar situation exists when gas is chemisorption pumped on surfaces. We noted in Section 2.4.1, that when a gas molecule, such as N_2, lands on a surface on which resides a metal with which it is chemically active, the N_2 molecule will perhaps stick, dissociate and form a stable chemical compound with that metal. It is reasonable to assume that: 1) there is a relationship between the number of active metal *energy sites* per unit surface area (defined in Section 2.4.1) and the probability that a molecule, when landing on the surface, will stick and form a chemical compound; and, 2) the sticking probability (coefficient) will diminish with time if the sites are not in some way replenished, but the flux of gas impinging on the surface persists. Both assumptions are correct. For example, for one *gas-metal system* - this term is often used in the literature when noting the reaction of a specific gas with a specific metal - the sticking coefficient, α, might vary from an initial value of unity and, with additional gas exposure, decay to zero. Therefore, though the material sublimed on the surface may initially have very effectively removed gas from the system, in time the sticking coefficient of the gas on the pumping surface will decay to zero. Understanding of the dynamics of maintaining a balance between the incident flux of gas and the availability of chemisorption sites is what TSP pumping is all about.

Two experimental methods were proposed by Clausing for determining the sticking coefficients of various gases on Ti films.[10] The approach used exploits the anisotropic state created in a vacuum system when some or all of the surfaces in the system have nonzero sticking coefficients for a gas (*e.g.*, see the work of Giorgi and Pisani[13]). An *anisotroptic state* in a vacuum system merely means that the flux of gas in the system is not equal and uniform in all directions. Clausing's experimental technique, in somewhat refined form, was used by Harra some time later to determine the precise sticking coefficients of N_2 on Ti films.[14] The method, as reported by Clausing and Harra, is given below.

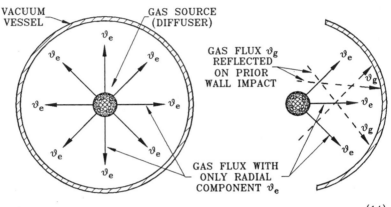

Figure 3.1.1.   Apparatus for achieving, ϑ_e, a radial flux of gas.[14]

Figure 3.1.1 represents a spherical vacuum vessel. Gas is introduced into the vessel through a diffuser located at the center of the chamber. The purpose of the diffuser is to ensure that all gas entering the chamber is dispersed in a radially uniform fashion to the walls of the chamber, as though originating from a point source located at the center of the chamber. The rate at which gas enters the chamber is measured with some form of throughput, or *Q-meter*. The flux density of entering gas molecules impinging on the walls of the *pump* (*i.e.*, chamber), ν_e, is proportional to Q/A_c, or,

$$\nu_e \quad \propto \quad Q/A_c \text{ Torr-}\mathcal{L}/\text{sec-cm}^2, \tag{3.1.3}$$

Q = the gas throughput entering the chamber, Torr-$\mathcal{L}$/sec,
A_c = the area of the chamber in cm^2.

Knowing the value of Q, or rate at which gas enters the vessel, we can solve for ν_e by using (1.13.3), or

$$\nu_e \quad = \quad k_2\, Q \text{ molecules/sec-cm}^2, \tag{3.1.4}$$

where, k_2 = $N_0/\Re TA_c$,
            N_0 = 6.02×10^{23} molecules per mole of gas,
            $\Re$ = 62.36 Torr-$\mathcal{L}$/mole °K,
            T = the temperature in °K.

Assume that we have sublimed a fresh film of Ti on the inner surface of the chamber. On introducing a gas, such as N_2, through the diffuser, some of the molecules of the gas flux, ν_e, will stick to the chamber on their first encounter with the wall. However, this is not a *perfect* pump, so some of the gas is reflected back off the walls. This gas is reflected away from the walls according to the *cosine law* (*i.e.*, see (2.4.3a)). In the second and subsequent encounters this reflected gas has with the walls, the angle of incidence on impact can vary from 0 to $\pi/2$ radians. Therefore, the total flux density of gas incident on the walls at any time (*i.e.*, the total number of molecules striking the wall per sec-cm^2), ν_t, is:

$$\nu_t = \nu_e + \nu_g, \tag{3.1.5}$$

where ν_e = $k_2\,Q$ molecules/sec-cm^2, which impinge $\perp$ to the walls,
ν_g = the subsequent impingement rate of molecules, not sorbed on the first wall collision, per sec-cm^2.

From the definition of sticking coefficient, α, given at the beginning of Section 3.1, clearly:

$$\alpha = \nu_e/\nu_t. \tag{3.1.6}$$

The key to determining the value of α is to be able to measure any two of the three components of (3.1.5). For example, if we measure ν_t and ν_g, we can solve for ν_e using (3.1.5) and express the sticking coefficient with

$$\alpha = (\nu_t - \nu_g)/\nu_t = 1 - \nu_g/\nu_t. \tag{3.1.7}$$

On the other hand, if we were to able to measure only ν_e and ν_g, we could solve for the value of the sticking coefficient by substituting for ν_t, found with (3.1.5), into (3.1.6), and we find:

$$\alpha = \nu_e/(\nu_e + \nu_g)$$
$$= (1 + \nu_g/\nu_e)^{-1}. \tag{3.1.8}$$

The above analysis involves only simple algebra and the *counting of molecules* which we learned how to do in Chap. 1. The difficulty arises when attempting to experimentally differentiate or separately identify two of the three flux components. Gas beaming effects can lead to large experimental errors (*e.g.*, see the work of Kornelsen and Domeij[15]). Clausing was the first to propose methods for the separation of the components.[10] Harra and Hayward described an apparatus of similar nature, but with added features.[14] This latter apparatus is represented in Fig. 3.1.2. The RGA (residual gas analyzer) shown was used to measure the rate at which Ti vapor is deposited on the chamber wall. It is calibrated by measuring ion current (*i.e.*, Ti$^+$, or amu 48), and by subsequently normalizing changes in ion current, with time, to the subsequently weighed Ti source (*e.g.*, a filament). Baffle "A" serves to shield the BAG1 (Bayard-Alpert Gauge #1) from the line-of-sight gas flux from the diffuser (*i.e.*, ν_e) and Ti vapors from the source. Therefore, BAG1 actually affords an indication of gas stemming from ν_g, as defined by (3.1.5). Baffle "B" shields BAG2 from Ti vapors from the source, but not from gas emitted from the diffuser. Therefore, measurements made with BAG2 are an indication of both components ν_e and ν_g (*i.e.*, ν_t).

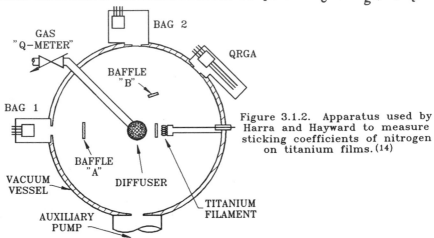

Figure 3.1.2. Apparatus used by Harra and Hayward to measure sticking coefficients of nitrogen on titanium films. (14)

Assume that both gauges are calibrated to read absolute pressure. Also, assume that they have negligible pumping speed for the test gas. This latter assumption implies that every gas molecule passing through the apertures and into the antechambers housing the gauges, later leaves these antechambers. Assume that the apertures leading into the gauge antechambers both have areas A_a cm^2. With this apparatus, we have two methods at our disposal for determining the values of α in time. First, we can use the *Q-meter* to directly establish ν_e at any instant in time. Then, using the indicated pressure, P_1 of BAG1, we can determine the value of ν_g with (3.1.1). Then, using (3.1.8) we find:

$$\alpha \quad = (1 + k_1 \, P_1 \, A_a / k_2 \, Q)^{-1}. \qquad (3.1.9)$$

On the other hand, we can measure implied values of ν_g and ν_t by the insertion of the reading of BAG1 and BAG2 respectively, into (3.1.1), and determine the value of α using (3.1.7). This reduces to:

$$\alpha \quad = (1 - P_1 / P_2). \qquad (3.1.10)$$

Using methods similar to those described above, the sticking coefficients for various gases on room temperature and LN2 (liquid nitrogen) cooled surfaces have been studied by a number of researchers. Harra has published an excellent review paper on this subject.[16] In a recent recapping of published data, Grigorov proposes slightly different interpretations of Harra's summary findings.[17] However, prior to discussing some of these specific results, I will first discuss some of the general results and how they are applicable to building and using a TSP pump.

3.2 Pump Speed vs. Sticking Coefficient

The components of a pump are quite simple and include only three elements, as depicted in Fig 3.2.1: 1) a *source* from which titanium is sublimed or evaporated through heating; 2) a *power supply* to heat the titanium; and 3) a *surface* onto which the Ti is sublimed and is accessible to the arriving, chemically active gas. Liquid nitrogen (LN$_2$) is sometimes used to cool the pumping surfaces. However, for the moment, let us neglect the possible beneficial effects of such cooling.

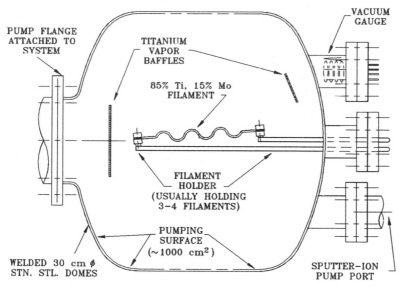

Figure 3.2.1. Example of a titanium sublimation pumping system.

There are two procedures used to deposit Ti when operating a TSP pump. Procedure #1: Ti is periodically sublimed onto the pumping surface, and the film is allowed to saturate with gas over a period of time. This is called *batch sublimation*. If the Q into the system is constant, as the film saturates (*i.e.*, α diminishes), the pressure will rise. Therefore, when the pressure approaches a value which we do not wish to exceed, we will have to sublime additional material onto the surface to restore the original sticking coefficient, etc. The time between *batch sublimations* will depend on the pressure, the thickness and temperature of the film, and, of course, the gas species. For example, at a starting pressure of 10^{-11} Torr, it might take days, or even weeks before a film is saturated to the point where the pressure rises above the desired level. Procedure #2: Ti is continuously sublimed onto the surface, throughout the process, at a rate which for a given system throughput, Q, the pressure remains constant. That is, we replenish the pumping sites at a rate that is in perfect stoichiometric balance with the rate at which they are being occupied. Continuous deposition is usually used in high throughput applications. In either case the speed of the *surface*, S_s, for a gas species "i" is simply:

$$S_s \quad = \alpha_i \, C_i A_c \quad \mathcal{L}/sec, \quad\quad\quad (3.2.1)$$

where, α_i = the sticking coefficient for gas species "i",
C_i = the aperture conductance of 1 cm^2 area for species "i",
A_c = the total pumping area of the chamber, cm^2.

In Procedure #1, the batch deposition method, α_i will change in time as the film becomes saturated. Such an effect is shown in the two curves given in Fig. 3.2.2, where the change in sticking coefficient, α_i, is given for N$_2$, on a thick deposit of Ti.[14] Curve "A" gives values of α for a batch deposited film of 8.0×10^{14} Ti atoms/cm^2 deposited onto a previously exposed base Ti film of $\sim 1.8 \times 10^{17}$ atoms/cm^2. Curve "B" shows values of α for a batch film of 8.3×10^{14} Ti atoms/cm^2 deposited on a previously exposed base of Ti atoms $\sim 10^{18}$ Ti atoms/cm^2. Rather than show the increase in pressure as a consequence of film saturation, it is customary to show the decrease in sticking coefficient as a function of the amount of gas pumped. We see from this figure that major changes in α occur with the amount of gas pumped, as well as the apparent thickness of the Ti substrate onto which the film is deposited.

For any given amount of gas pumped, the speed of the film for N$_2$ is found by substituting the value of α for the amount of gas sorbed into (3.2.1). For example, assume that the area of the pump, A_c, is 10^3 cm^2, and we have thus far pumped 10^{14} N$_2$ molecules per cm^2. From Curve "A" of Fig. 3.3.2, we can determine that the value of α has decayed to ~ 0.18. If the chamber is at RT (room temperature), C_{N_2} is ~ 11.7 $\mathcal{L}$/sec-cm^2. Therefore, the resultant instantaneous speed of the pump for N$_2$ is ~ 2110 $\mathcal{L}$/sec. We will later establish some approximate values of α for the different gases. However, for now the troubling aspect of the result of Fig. 3.2.2 is that in order to predict the behavior of the pump, we must know something about the Ti film exposure up to that point in time.

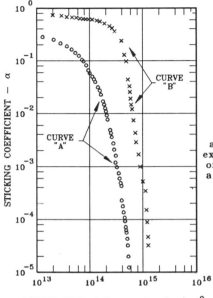

Figure 3.2.2. Harra and Hayward's sticking coefficient, α, data for N_2 on Ti. Curve "A" gives values of α for a "batch" layer of 8.0×10^{14} Ti atoms/cm² deposited on a previously exposed base of $\sim 1.7 \times 10^{17}$ atoms/cm² of Ti. Curve "B" gives values of α for a fresh layer of 8.3×10^{14} atoms/cm² of Ti deposited on an exposed base of $\sim 10^{18}$ Ti atoms/cm². [14]

For high throughput, continuous deposition applications, data are usually presented in a form equivalent to that shown in Fig. 3.2.3.[10,14,18] In this case, the sticking coefficient of N_2 on the Ti film is given as a function of the ratio of the rate at which gas molecules are sorbed to the rate of Ti deposition per cm². This method of presenting data, as in the above batch case, also normalizes out pressure data in the presentation of results. Data of Curve "C" of Fig. 3.2.3 were taken with the sublimation rate held constant, where Curve "D" data were taken with the throughput held constant. Harra noted that with either a constant sublimation rate, R, with varying pressure, or constant pressure with varying sublimation rate, R, the sticking coefficient could be approximated by:

$$\alpha = \alpha_m (1 - k_2 \, k_3 \, Q / R),　\qquad (3.2.2)$$

where α_m = the maximum observed sticking coefficient,
 R = the Ti sublimation rate in atoms/sec-cm²,
 k_2 = $N_0 / \Re \, A_c \, T$,
and k_3 = a data "fitting" constant.

Fitting the data of Fig. 3.2.3, Harra found values of $\alpha_m \sim 0.49$, $k_3 \sim 1.8$ for the Curve "C", the constant Ti sublimation rate data, where $\alpha_m \sim 0.87$, $k_3 \sim 1.6$ for the Curve "D", the constant leak rate, Q, data. Clausing reported on continuous sublimation data for a number of gases (e.g., H_2, D_2, N_2, CO and CO_2) on both RT and LN_2 surfaces.[10] We will return to these and other results, and eventually establish *engineering* values of k_3 for the different cases (*i.e.*, see Table 3.2.1).

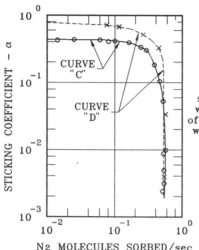

Figure 3.2.3. Harra and Hayward's N_2 sticking coefficient, α, data. Curve "C" was taken with a constant sublimation of Ti, with increasing pressure. Curve "D" was taken with constant N_2 throughput and varying Ti sublimation rate.[14]

N_2 MOLECULES SORBED/sec
per Ti ATOM DEPOSITED/sec

Some investigators have used measurements of the total pumping speed of a Ti film, sublimed onto a surface, to determine the sticking coefficients of gases.[19] In this case, speed per unit area is usually given as a function of time, and for the given gas. Assuming no measurement errors due to beaming effects, etc.,[12] then the total amount of gas pumped per unit area, up to time t, is simply $\int S(t)P(t)dt$ up to this time. When the speed, or sticking coefficient, decreases to zero, the film is said to have reached its *capacity*. Note that though the speed or sticking coefficient for a particular gas may be very high at the start, the rate at which it decreases with gas sorption is of equal importance. Therefore, in describing the effectiveness of a film in pumping a gas, we must quantify both the capacity of the film and the sticking coefficient or speed as a function of the amount of gas pumped. These are two distinct concepts. That is, we are dealing with both the *rate* of pumping as a function of the amount of gas pumped and the *capacity* of the film.

Titanium and Conductance Limited Operation

Let us recast (3.2.1) into a more intuitively useable form. First, we know that the throughput into the chamber is $Q = S_s P_c$, where S_s is given by (3.2.1) and P_c is the pressure in the pump chamber. If we substitute this value for Q in (3.2.2) and the resultant value of α into (3.2.1) and solve for S_s, we find the general equation for the pumping speed for the gas species "i", as a function of pressure, sticking coefficient, and sublimation rate, is:

$$S_{si} = (\alpha_{mi} C_i A_c) / (1 + (k_2 \; k_3 \; \alpha_{mi} \; C_i A_c P_{ci} / R)), \qquad (3.2.3)$$

where, the various terms are given in (3.2.1) and (3.2.2). This expression, though having a lot of terms, is actually quite simple and reduces in the limits to two important expressions. In the first case, when $1 \gg (k_2 \; k_3 \; \alpha_{mi} \; C_i A_c P_{ci} / R)$, or the rate of Ti sublimation is very high compared to the pressure in the pump, then,

$$S_{si} \approx (\alpha_{mi} C_i A_c). \qquad (3.2.4)$$

In this case, the speed of the pump is independent of the pressure, and depends solely on the area of the pumping surface and, of course, the maximum sticking coefficient for the gas species "i". This mode of operation is called *surface* or *conductance limited* pumping. Remember, this is a *surface conductance*, having to do with the product $C_i \times A_c$ of (3.2.1), rather than the conductance of some manifold or aperture separating the pump from the chamber.

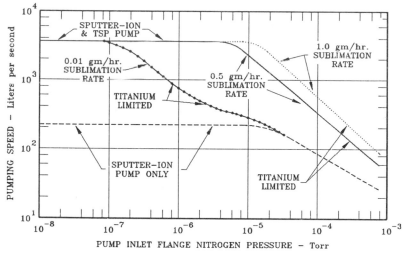

Figure 3.2.4. Nitrogen "combination" pumping speed for both titanium and conductance limited conditions.

When $1 \ll (k_2 \; k_3 \; \alpha_{mi} \; C_i A_c P_{ci} / R)$, or when the pump pressure is high compared to the rate of Ti sublimation, then,

$$S_{si} \approx R / k_2 \; k_3 \; P_{ci}. \qquad (3.2.5)$$

In this case, the speed of the pump is directly proportional to the sublimation rate and inversely proportional to the pressure. The implication here is that there will be a decrease in pump speed at the higher pressures, in this case limited by the rate of sublimation of Ti on the surface of the pump, This mode of operation is called *titanium limited* pumping. This difference in conductance vs. titanium limited pumping is illustrated in Fig. 3.2.4. Nitrogen speed data, on RT surfaces, are shown for three different Ti sublimation rates. These data are averages of unreported results, using a special pumping configuration reported on by Harra.[20]

Dependency of α on Gas, Film Thickness and Temperature

We see from the findings of just Figures 3.2.2 and 3.2.3 that this *business* of sticking coefficients is quite involved, even for the gas N_2, on Ti, sublimed on room temperature surfaces. To date, there is much disagreement in the literature regarding absolute values of initial sticking coefficients, whether or not one gas *poisons* the film in the pumping of another, the preferential displacement of one gas by another, etc. I will discuss some of these findings, and eventually establish some recommended *engineering* values of sticking coefficients and film capacities. The apparent sticking coefficient of a gas and the capacity of the film are known to markedly vary with the

1) thickness of the film at the time of exposure to the gas,
2) ratio of the rate gas is pumped to the rate Ti is sublimed,
3) surface temperature at the time of film sublimation,
4) surface temperature at the time of gas sorption,
5) method of film deposition (*i.e.*, batch *vs.* continuous),
6) gas species,
7) Ti deposition rate, independent of the presence of gas,
8) gas desorption and synthesis at the source (*e.g.*, H_2, CH_4),
9) partial pressures of gases at the time of sublimation,
10) contamination of the film for one gas by another gas,
11) dissociation and liberation of gases at the film surface,
12) effects of film *annealing*, and
13) time dependent effects (*e.g.*, surface and bulk diffusion processes).

Returning to Harra's data for N_2 on Ti, the effects of film thickness, implied in the results of Fig. 3.2.2, are graphically illustrated in Fig. 3.2.5, where a summary of the initial, maximum sticking coefficient, α_m, is given throughout the sequence of twelve experiments reported by Harra.[14] Data shown here are for both batch and continuous sublimation runs.

The implication of this figure is that the sticking coefficient for N_2 appears to increase with film thickness. This effect has been reported elsewhere for N_2 and other gases.[18,19,21] This effect is believed to stem from increases in the surface roughness or effective area vs. the projected surface area of the film. Of course, for the gases H_2 and D_2, capacity of the films would vary with film thickness due to simple diffusion related considerations.[10,19,21-24] Such increases in effective area increase the number of pumping sites per projected surface area and thereby increase the effective value of α.

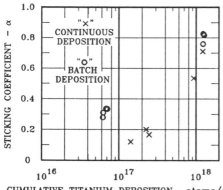

Figure 3.2.5. Harra and Hayward's data for the variation in α, the sticking coefficient of nitrogen on titanium, as a function of the cumulative thickness of titanium. [14]

Results of Elsworth, *et al.* seem to disagree to some extent with Harra's findings, regarding the effect of RT pumping of N_2 *vs.* film thickness. They indicate that after the first of many batch films were exposed to N_2, both α_m and film capacity remained the same thereafter. [19] The implication in their discussions was that the first batch exposure gave depressed values of α_m because of possible experimental errors stemming from outgassing from their Ti source. Others have observed this effect. [21] Both source outgassing and gas beaming effects will lead to errors in such measurements.

Clausing observed the effects of film texture (*i.e.*, high values of sticking coefficients) of films deposited under batch conditions, and in a high partial pressure of He. [10] He noted that when the film was sublimed under high partial pressures of He (*e.g.*, 2×10^{-3} Torr, on a surface at $10°$ C), the films deposited in this manner "... have a velvety black appearance and are very poorly crystalized." Such films had higher sticking coefficients for the gases H_2, N_2 and CO than those films batch deposited under high vacuum conditions. This same effect - that is, an apparent improvement in sticking coefficients of films deposited in high partial pressures of He - was also reported on by Clausing for films deposited on LN_2 cooled surfaces. [10] This enhancement of sticking coefficients on LN_2 surfaces was also reported on the same year (*i.e.*, 1961) by Sweetman. [41] This was markedly apparent for the gas H_2, where Clausing reported the difference in α_m was $\times 4$ higher for Ti films deposited under a high He pressure. In most applications, sublimation of films under high partial pressures of He is not practical. However, results of this experiment emphasized the importance of surface texture or roughness as it relates to values of α. In this same reference, Clausing reported significant increases in sticking coefficients and capacities of Ti films, deposited in high vacuum at near LN_2 temperatures, over those deposited on RT surfaces, for all gases tested. This effect was also reported the same year (*i.e.*, 1961) by Sweetman. [41] This effect has been widely noted in the literature (*e.g.*, see 9,17,25,26,41), though Gupta and Leck report it is significant only for the gases H_2 and N_2.[21]

Another outstanding early work on TSP pumping on surfaces at temperatures of $-190°$ C and $20°$ C was reported by Elsworth, Holland and Laurenson.[19] They tested the sorptive capacity of Ti films for the gases H_2, D_2 and N_2, at the two temperatures, and for varying film thickness. They expanded this work even further by measuring the sorptive capacities of Ti films at temperature T_m, which were evaporated into surfaces at temperature T_e, where T_m was either $-190°$ C or $20°$ C and T_e was either $-190°$ C or $20°$ C. Lastly, they measured the sorptive capacity of films at $-190°$ C, which were deposited at $-190°$ C and subsequently *degassed* at a temperature of $20°$ C, and they then remeasured the pumping speed of the film at $-190°$ C.

Using their notation "T_e/T_m" they determined for N_2: 1) under $20°$ C/$20°$ C conditions, virtually irreversible N_2 chemisorption occurred, even if the film was later baked to $\sim170°$ C; 2) under $-190°$ C/$-190°$ C conditions, the equivalent of ~100 monolayers of N_2 were sorbed at full capacity (one monolayer of N_2 corresponds to $\sim6 \times 10^{14}$ molecules/cm^2); 3) under $20°$ C/$-190°$ C conditions, the equivalent of ~3 monolayers were sorbed at capacity; 4) under $-190°$ C/$-190°$ conditions, and the film was subsequently *baked* at $20°$ C, ~3 monolayers of N_2 were sorbed on reaching the capacity of the film.

This and other work of Elsworth, *et al.*, verified the findings of Clausing, wherein he noted it appeared the beneficial effect of films deposited at LN_2 temperatures "... anneals out on warming to room temperature."[10] Clausing also noted that even slight traces of O_2 or N_2 "... drastically reduced the value (of α) obtained when hydrogen is sorbed onto a new deposit." Because of this latter effect, he submitted that what appeared to be a film *annealing* effect might be due to gas poisoning. However, Elsworth, *et al.*, verified Clausing's findings of an actual film annealing effect. Also, Gupta and Leck later verified that film *poisoning* by other gases was also a real effect which could lead to experimental error when making sticking coefficient measurements.[21]

The conclusions of Elsworth, *et al.*, regarding the pumping of N_2 on $-190°$ C surfaces, included: 1) once the capacity of fresh film, deposited on a $-190°$ C surface, is reached, the value of α_m was completely reversible on warming the surface to $20°$ C and then returning to $-190°$ C; 2) the capacity of films once saturated at $-190°$ C, when *baked* at $20°$ C under UHV conditions, and again cooled, was ~0.03 of that of fresh films deposited at $-190°$ C. However, this latter capacity was still $\times300$ greater than would be expected for simple physisorption on a cold metal surface. This led investigators to later speculate of the existence of an *intermediate* mechanism in the sorption on N_2 on cooled Ti surfaces.[18,25] That is, there is some process short of $N_2 + 2Ti \rightarrow 2TiN$ (*i.e.*, complete chemisorption), but stronger than physisorption.

The findings of Elsworth, *et al.*, regarding the pumping of H_2, were as follows: 1) for $20°\,C/20°\,C$ conditions, $\alpha_m \sim 0.01$ (compared with $\alpha_m \sim 0.1$ for N_2); 2) for $-190°\,C/-190°\,C$ conditions, $\alpha_m \sim 0.4$; 3) the capacity of a 350Å film under $20°\,C/20°\,C$ conditions was ξ; under $20°\,C/-190°\,C$ conditions it was $\sim 0.11\xi$; and, under $-190°\,C/-190°\,C$ conditions it was 7.4ξ, where $\xi = 7.8 \times 10^{15}$ H_2 molecules/cm^2.

Steinberg and Alger noted a very important effect relating to the *poisoning* of Ti films by one gas, and the subsequent pumping of H_2 and D_2.[23] They noted that the bulk Ti of a thickly deposited film, which has been once contaminated on the surface by some other gas, can be made to effectively pump H_2 and D_2 if a thin overlay of fresh Ti is deposited on the contaminated layer. We noted in Chapter 2 that H_2, on dissociation into atomic hydrogen, diffuses readily in many materials. Therefore, it is reasonable to assume that the fresh overlay of Ti produces *energy sites* which dissociate the H_2. On dissociation, the atomic hydrogen diffuses through the *poisoning* film barrier of oxide or nitride and on into the bulk Ti deposit. This is precisely the pumping model proposed by Steinberg and Alger.

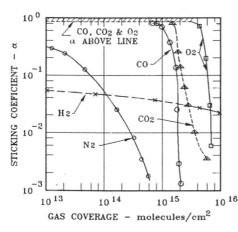

Figure 3.2.6. Sticking coefficients, α, found by Gupta and Leck, for titanium RT films of $\sim 10^{16}$ atoms/cm^2; each batch was deposited in ~ 100 sec. [21]

Gupta and Leck considerably expanded on the work of Clausing and Elsworth, *et al.*[21] Their work was done under rigorously controlled UHV conditions and after outgassing and characterizing their Ti sources, including the filament and holder. They studied the Ti film sorption of a wide variety of gases on both RT and LN_2 (*i.e.*, $77°$ K) surfaces. They classified the gases under study into three groups: Group #1: the *inert* gases CH_4 and Ar; Group #2: the intermediately chemically active gases H_2, D_2 and N_2; and Group #3: the very chemically active gases O_2, C_2H_2 (acetylene), CO, CO_2 and H_2O. Results of sorption of most of these gases, under $300°$ K/$300°$ K conditions, are given in Fig. 3.2.6.

In each case, the deposited Ti film had a thickness equivalent to $\sim 10^{16}$

Ti atoms/cm^2, each batch being deposited in a 100 second time-span.
Group #1: Argon was not sorbed on Ti films, even at 77° K. For methane
(*i.e.*, CH$_4$), sorption at RT was barely detectable, with values of α_m ~4 ×
10^{-4} and α ~10^{-4} with a coverage of ~10^{-1} ^{3} molecules/cm^2. Values of the
respective α's increased by ~×10 in conditions of 300° K/77° K. The CH$_4$
pumped on 77° K Ti films, even if covered over with additional Ti prior to
warming to ~150° K, resulted in "copious" desorption of the CH$_4$ at this
temperature.
Group #2: Nitrogen results were similar to those reported by others. That is:
1) the values of α_m can vary as much as ×2 depending on film batch deposition rate. Note that results for N$_2$, in Fig. 3.2.5, seem to be *bracketed* by
Harra's results shown in Fig. 3.2.2. 2) The values of α increased by ~×5 on Ti
films on 77° K surfaces, over that of RT surfaces. However, it was not clear if
the conditions of these experiments were 300° K/77° K or 77° K/77° K.
Group #3: 1) The values of α_m for all of these gases ranged from 0.9 to 0.99
when batch deposited at RT. 2) Substrate temperature had only a second
order effect on values of α_m, the value for CO varying the greatest, with α's
of 0.8 and 0.98 on 373° K and 77° K surfaces, respectively.

Engineering Values for Film Speeds and Capacities

Harra, in two review papers, summarized sticking coefficient data for several
gases, pumped under both batch and continuous conditions, and on surfaces
at 77, and 300° K.[16,36] This review paper comprised results and considerations of some 25 references reported in the literature, including his own
work. Results of this summary are given in Table 3.2.1.

Table 3.2.1. "Engineering" values for maximum sticking coefficients and speed for
various gases on Ti films at 77 °C and ~300 °C as summarized by Harra.(16,36)

TEST GAS	MAX. STICKING COEFFICIENT-α_m		MAX. SPEED[a] Liters/sec-cm^2		MAX. CAPACITY OF FILM − x 10^{15} molecules/cm^2 [b]		VALUES OF CONSTANT k3 [d]	
	300°K	77°K	300°K	77°K	300°K	77°K	283°K	77°K
H2	0.06	0.4	2.6	17	8–230[c]	7–70	1.1	0.7
D2	0.1	0.2	3.1	6.2	6–11[c]	–	–	–
H2O	0.5	–	7.3	14.6	30	–	–	–
CO	0.7	0.95	8.2	11	5–23	50–160	1.0	–
N2	0.3	0.7	3.5	8.2	0.3–12	3–60	1.9	1.8
O2	0.8	1.0	8.7	11	24	–	1.0 (300°K)	
CO2	0.5	–	4.7	9.3	4–12	–	1.0	–

NOTES: a) Speed calculated based on gas being at RT.
         b) Wide variations due in part to film roughness.
         c) Wide variations due in part to bulk diffusion into film.
         d) Constant used with (3.2.2); i.e., for continuous deposition.

These are very reasonable working numbers for use in the design of TSP
pumped systems. The value of film speed per unit area, S_s, given in the above
table, was calculated using (3.2.1), and while assuming that the temperature of

the gas was 293° K. In *counting molecules*, there is a limit to the number of gas molecules which can be pumped by a Ti film. Harra reported that for each atom of Ti, one molecule of H_2, CO, O_2 or CO_2 can be pumped; for each molecule of either H_2O or N_2, two atoms of Ti are required.[16]

3.3 Synthesis, Displacement and Dissociation of Gases

In the early days of development of electrostatic ion pumps, and in the use of TSP pumps in combination with sputter-ion pumps, it was believed that the presence of an electrical discharge in some way *activated* sublimed films so as to create a synergistic effect of the combined pumping modes. That is, the apparent combined speed of the sputter-ion and TSP pumps was greater than that observed for each, when independently operated. In these measurements, total system pressure was used as the indication of speed. For example, assume a fixed N_2 throughput, Q_1, is introduced into a system and that the sputter-ion pump has an N_2 speed of S_i and the TSP pump a speed of S_s, yielding pressures of P_i and P_s, respectively, when individually operated. If the combined speeds were the sum of the individual speed we calculate that:

$$P_{si} = P_s P_i / (P_s + P_i) \tag{3.3.1}$$

where P_{si} is the pressure observed with both pumps simultaneously operating. The problem is that in every case it was observed that combined speed $S_{si} > S_i + S_s$. Therefore, it was concluded that some form of "activation" was occurring to the sublimed film as the result of the discharge of the sputter-ion pump. The *flaw* in (3.3.1) was in assuming that the throughput of gas was constant under all three conditions. Indeed, the throughput of N_2 or other experimental gas might have been constant, but the total throughput of gas into the system was not the same when operating the TSP pump. This pump, and other forms of pumping involving the spewing of Ti onto the walls of the pump, creates methane (CH_4). The background of CH_4 only increases when using the TSP pump. The synthesized CH_4 increases the indicated total pressure so as to lead to lower indicated speeds for the test gas. When the sputter-ion pump is operated simultaneously with the TSP pump, the sputter-ion pump pumps the methane generated at the TSP. Therefore, this indeed does prove to be a synergistic effect, but not for the reasons assumed. This was verified in a classic experiment by Francis and Jepsen in 1963.[27]

This synthesis of CH_4 was observed by Klopfer and Ermrich early in the development of these pumps.[28] Wagener, at the same time, reported observing high CH_4 backgrounds above several types of getter films deposited from hot sources.[2] He also noted that though there was observable sorption of CH_4 on Ti films, it was slight, and the film had negligible capacity for this gas. Holland showed that the production of methane was related to the content of carbon in the Ti source.[55] He noted an enrichment of carbon in a molten droplet, replenished by wire feed, as the material sublimed. Yet, he concluded that the CH_4 and CH_3 synthesized during the experiments was manufactured at the deposited film. Kuz'min reached this same conclusion in a later work.[9] Gupta and Leck were the first to prove that the CH_4 was synthesized on the hot Ti source, rather than on the surface of the film.[21] They accomplished this by introducing D_2 into the system subsequent to the batch sublimation of the film. They then noted a displacement of CH_4 by the sorbed D_2, but no CD_4 (or, presumably CD_xH_y, where integers $x + y$ = 3 or 4). Conversely, they noted the production of CD_4, and CD_xH_y, $x + y$ = 4, when introducing D_2 into the system with the filament turned on.

Halama showed that CH_4 and H_2 were the limiting components in achieving pressures $< 10^{-10}$ Torr in proton colliders.[29] Because of the poor pumping of CH_4 by Ti films, Halama noted that sputter-ion pumps must be used in combination with TSP pumps in UHV applications. Edwards devised a recipe for accelerating the outgassing of CH_4 from sublimed films, for applications at pressures $< 10^{-10}$ Torr.[30] It essentially involved a post-sublimation, 100° C bake of the sublimed film. He hastened to advise us, in a paper submitted three months after the first, that the same benefit would not be achieved by subliming the Ti with the pump walls at elevated temperatures (i.e., during bakeout).[31] Chou and McCaferty expanded on Halama's work in characterizing the pumping of CH_4 by sputter-ion pumps at very low pressures (i.e., see ref. 46, Chap. 2).

Gupta and Leck reported on the displacement of one chemisorbed gas by a second.[21] By displacement, it is meant that one gas, previously chemisorbed on the Ti film, is subsequently replaced by a second and the first is liberated into the vacuum system. These results are somewhat controversial. Harra sites some conflicting publications regarding this displacement phenomenon.[16] Nevertheless, Gupta and Leck submit that experimental evidence indicates that a displacement process preferentially occurs as shown in Table 3.3.1. In this table, O_2 will displace all other gases noted; CO will displace all the gases but O_2; H_2 all gases but O_2 and CO, etc. I find it difficult to conceive that a displacement process of the nature "$O_2 + 2TiN \rightarrow 2TiO + N_2 (gas)$", or "$2O_2 + 3TiN \rightarrow 2TiO_2 + TiN + N_2 (gas)$" occurs. However, it is possible that gas molecules either physisorbed or sorbed in some *intermediate* phase on the Ti surface, are subsequently displaced by another gas. Gupta and Leck *coined* this gas displacement phenomenon as a pump "memory" effect (similar to that noted with sputter-ion pumps, in Section 2.4.3). They did not report on the liberation of H_2 on the pumping of H_2O on RT, Ti films, though they listed this as one of the test gases.

PUMPED GAS	DISPLACED GAS				
	CH4	N2	H2	CO	O2
CH4		no	no	no	no
N2	yes		no	no	no
H2	yes	yes		no	no
CO	yes	yes	yes		no
O2	yes	yes	yes	yes	
α_m	$<10^{-3}$	0.3	0.05	0.85	0.95

Table 3.3.1. Gupta and Leck's gas displacement and sticking coefficient data for several gases on RT titanium films. For example, N2 displaces CH4, H2 displaces N2 and CH4, CO displaces CH4, N2 and H2, etc. [21]

In the process of chemisorption, most gases are first dissociated. For example, I noted that this is the method by which H_2 and N_2 are pumped. This also proves to be the case with water vapor. In this case, however, H_2O + Ti → TiO + H_2 (gas). Of course this is not observed at LN_2 temperatures, as the water vapor has unity sticking coefficient on surfaces at this temperature. However, Harra (and others whom he cited) observed that on RT surfaces the chemisorption of H_2O on continuously deposited films results in the liberation of copious quantities of H_2.[20] As long as the sublimation rate, R, is sufficient (*i.e.*, the condition (3.2.4) prevails), the partial pressure of H_2 remains low. However, when conditions of (3.2.5) begin to prevail, the partial pressure of H_2 starts to rise, as the O in H_2O *wins* in the competition for the limited amount of fresh Ti. The poor pumping of Ti films for H_2O, reported on earlier by Wagener, was probably due to film poisoning effects (*i.e.*, this self-poisoning mechanism).[2]

3.4 Sublimation Sources

There are three primary sources used at this time for subliming Ti onto pumping surfaces: 1) *filamentary* sources; 2) *radiantly heated* sources; and 3) *E-gun* sources, where the Ti is heated as the result of bombardment with high energy electrons. These will be discussed in the order given.

Filamentary Sources

Titanium has poor hot strength at temperatures needed for reasonable sublimation rates. Because of this, wires of this material had to be *bundled* together with a second refractory material which possessed good hot strength and supported the Ti at or near its melting temperature. Several material combinations have been used to accomplish this goal. Some of these schemes are illustrated in Fig. 3.4.1. Kornelsen used a twisted pair of 0.076 mm ϕ W and 0.25 mm ϕ Ti wires wound over a 0.25 mm ϕ W mandrel.[33] Clausing used a scheme where he tightly overwound a 25 cm long, 4.3 mm ϕ Ta rod with one layer 0.76 mm ϕ Nb wire followed by windings of two layers of 0.89 mm ϕ Ti wire.[10] The Ti alloyed with the Nb wire forming an alloy with a higher melting temperature. With this scheme, he was able to achieve Ti sublimation rates as high as ~3.0 gm/hr for up to four hours and with a power input of ~1.0 - 1.5 kW. McCracken and Pashley mention an early Varian product consisting of a *trifilar* overwinding of a 0.76 mm ϕ W core with two 0.76 mm ϕ Ti wires and a third 0.38 mm ϕ Mo wire.[34] In a review paper, Herb sites these and some of the other early variations of sublimators, including other filamentary sources.[35]

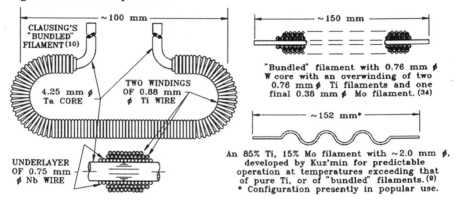

Figure 3.4.1. Varieties of "bundled" and alloy filaments as reported by Clausing,[10] McCracken,[34] and Kuz'min.[9]

Reliability problems and inconsistent sublimation rates are pointed out in many papers dealing with these bundled sources. The primary problem was that if the proper input power was exceeded in the wire bundles, *blobs* of molten Ti would form, fall from the bundle, and hot spots would develop in the assembly. This and the unpredictability of the thermal contact of the Ti with the refractory metal conductor, led to wide variations in sublimation rates for the same input power.

The technical *breakthrough* in these filamentary sources was made by Kuz'min, who developed and patented (in 1961) a sublimation filament comprising an alloy of Ti and Mo, with a 15 to 20% Mo content, by weight.[9] This Ti/Mo alloy had superior hot strength, and enabled operation of the filament at temperatures far exceeding the melting temperature of pure Ti. Using a 25 cm long, 2.0 mm ϕ wire of this material, he was able to achieve a sublimation rate of ~0.3 gm per hour with a current of ~44 A and total power of ~300 W. He stated that "For maintaining a constant evaporation rate within certain (unspecified) limits, it is sufficient to maintain a constant current through the evaporator." Publications in the following years took issue with the idea of constant current operation (*i.e.*, *vs.* constant voltage or

constant power operation). The objective is to be able to predict the sublimation rate at all times during the life of the filament. This is prompted by a
desire to get the best utilization of the available Ti, neither wasting Ti when
sublimating, nor *starving* the pumping surface by operating Ti limited. As the
Ti becomes depleted in the Ti/Mo alloy mixture, changes occur in the
physical and electrical characteristics of the wire making correlation of
sublimation rate *vs.* filament power, in time, difficult.

Constant Current Operation

McCracken and Pashley published the first study of sublimation rate characteristics of the 85% Ti, 15% Mo alloy filaments with constant current
usage.[34] The filaments, manufactured by Imperial Metal Industries, Ltd.,
were 25 cm long, with a diameter of 2.1 mm. With constant current operation, they observed sublimation rates as shown in Fig. 3.4.2 for operation at a
constant current of 50, 55 and 57 amperes.

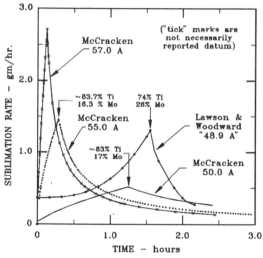

Figure 3.4.2. Sublimation rate
data found by differentiating
McCracken's filament weight
loss data,[34] and compared
with normalized Lawson and
Woodward constant current
sublimation rate data.[24]
Where Lawson and Woodward
maintained constant current
in their filament, McCracken
had to cycle his filaments
from RT to the "constant"
current settings in order to
measure changes in weight.

Data of Fig. 3.4.2 were obtained by differentiating their published weight loss
data with respect to time. Changes in sublimation rates observed in this figure
were comparable, with exceptions noted below, to rates noted by Lawson and
Woodward when operating a smaller diameter filament at the equivalent
constant currents.[24]

McCracken noted that during extended operation of the filaments, a point
was reached where large crystals developed in the wire. Some of these
adjoining crystals had slips or dislocations along their grain boundaries, some
crossing the full diameter of the wire. However, they noted no failures as the
result of this crystal growth, as long as the filament current was kept below a
certain value (*i.e.*, 57 A).

Lawson and Woodward published the first definitive study of changes in the physical properties of these Ti/Mo filaments in time.[24] Their filaments measured ~1.6 mm ϕ with a length of 21.7 cm. They studied changes in wires initially comprising 85% Ti, 15% Mo by weight. First, they noted that none of the Mo was sublimed from the filament. They noted that 40% of the Ti in the wire "can be evaporated", presumably on reaching the end of filament life. Also, when operating the filaments at a constant voltage, 24% of the Ti can be evaporated at a constant, predictable rate. They indicated that the large crystal growth - which they termed a "macrocrystalline" structure - occurred on the wire reaching an alloy mixture of 74% Ti and 26% Mo. The onset of this structural transformation was independent of the evaporation temperature between the limits of 1550-1800° K. This result is somewhat confusing as if: 1) no Mo is sublimed from the filament, and, 2) the initial alloy mixture is 15% Mo and 85% Ti; then, 3) on reaching an alloy mixture of 26% Mo and 74% Ti through evaporation of the Ti, an alloy of 26% Mo and 74% Ti suggests that ~50% of the Ti has been sublimed at the onset of the macrocrystalline state. However, the total mass of the wire has decreased by ~42%.

Lawson and Woodward noted increases in the total emissivity of ~×2 in these wires as a consequence of the filament's crystalographic transformation. These changes in emissivity resulted in decreases in filament temperature for the same power. Wires were clamped in a holder, similar to that shown in Fig. 3.2.1, but comprising "fairly heavy copper clamps"; the implication being that these heavy clamps imposed a low-temperature thermal boundary condition at each end of the wire. Therefore, the central part of the wire initially operates at the highest temperature. This portion of the wire loses Ti at the greatest rate and therefore is the first portion of the wire to undergo the crystalographic transformation. It then, because of its higher emissivity, operates cooler than neighboring portions of the wire which have not yet gone through the transformation. The regions of transformation "grow" from the center of the wire towards the clamps. This fortuitous transformation mechanism prevents the central portion of the wire from overheating and premature filament failure. They also noted slight increases in the resistivity of the filament material (i.e., ~9%) as a consequence of this metallurgical transformation. However, for constant current operation, changes in the alloy resistivity are of much less importance to evaporation rates than are increases in wire resistance stemming from a decrease in the wire diameter due to sublimation of the Ti. They showed sublimation rate data for filaments operated under both constant current and constant voltage conditions. Under constant current conditions, under early stages of sublimation, the sublimation rate gradually increases, and eventually reaches a value of ~×4 that of the initial value.

Similar findings were noted by Chou and Lanni (~×6) under constant current conditions.[32] This increase in sublimation rate stems from an increase in the power dissipated in the wire (*i.e.*, constant current, but increasing wire resistance due to decreases in wire diameter). They too proposed that a *peaking* in sublimation rate was observed as additional Ti was sublimed, and the wire transformed over to the macrocrystalline state. They indicated that thereafter, the filament temperature decreased due to increases in emissivity, and the sublimation rate correspondingly declined. There is some disagreement in the published data regarding when, in the life of the filament, this maximum sublimation rate occurs. Chou and Lanni note, in their sublimation rate *vs.* time data, that the *peaking* in rate occurs after sublimation of ~10% of the usable Ti. Yet, they report that it occurs when the Mo content of the filament is ~26% "by weight". Their data does not seem to indicate this is the case. For example, if one integrates their sublimation rate vs. time data, the expended Ti at filament failure is ~1.2 gm. This closely agrees with my findings for such filaments. On the other hand, the *peaking* in sublimation rate in time, if occurring after sublimation of 10% of the Ti, one would expect that the alloy of the filament would be in the proportions of ~16% Mo and 84% Ti, rather than the suggested 26% Mo.

Integrating similar constant current data reported by Lawson and Woodward, we find that the *peaking* in sublimation rate occurs after expending ~50% of the total Ti sublimed, including sublimation results from increases in current late in the experiment. This *peak* corresponds to a filament alloy mixture corresponding to ~26% Mo. Their results are also shown in Fig. 3.4.2. Lawson and Woodward's constant current data were normalized to McCracken's data by taking into account the differences in the lengths, diameters and areas of the respective filaments.

Of course, the higher the sublimation rate with constant current operation, the sooner rate *peaking* occurs in time. However, I believe there is a second and equally important rate-effecting mechanism at play in these *constant current* experiments. That is, crystalographic changes in the wire occur as the consequence of temperature cycling. This effect has been reported in the literature with pure Ti sources.[42] For example, in order for both Chou and McCracken to make periodic measurements of weight changes in their *constant current* sublimation rate measurements, they had to cycle their filaments from RT up to the high temperatures associated with their respective *constant current* settings. From his data, we speculate that McCracken probably cycled his filaments 5 to 10 times. Chou implied that he cycled his filaments of the order of 70 times throughout his measurements. It is known from Ti/Mo alloy phase diagrams given by Brewer and Lamoreaux[37] that Ti/Mo alloys go through a crystalographic phase transformation when cycled from RT to temperatures ranging from ~600 to 900 C. A temperatures less than this, Ti/Mo alloys have a mix of both bcc and hcp crystal structure. However, above this temperature range, the alloy becomes exclusively bcc. If the temperature changes occur very slowly, the atoms of the alloy will reorder themselves into the crystal structure appropriate to the particular temperature. However, if these temperature changes are made very quickly, the filament will become a hodgepodge mixture of both the bcc and hcp crystals. This has the effect of increasing the surface roughness or area of the filament. As a consequence, filaments which are cycled numerous times through the crystalographic transformation temperature eventually develop a very rough surface texture. This in turn results in an increase in the effective surface area (or emissivity), and a reduction in operating tempera- ture, for the same input power. Lawson and Woodward's data were literally taken at a constant current (*i.e.*, without cycling the filaments). Therefore, I suggest that Chou and Lanni mistakenly interpreted the *peaking* in their sublimation rate data as being due to Ti depletion, as observed by Lawson and Woodward, rather than stemming from increases in surface roughness as a consequence of temperature cycling.

Constant Voltage Operation

Lawson and Woodward report that almost constant filament sublimation rates - in fact, slightly increasing rates - are achieved when operating filaments at constant voltage, until ~24% of the total Ti is sublimed.[24] They did not cycle the filaments during their tests.

I conducted a series of measurements on 2.0 mm ϕ, 85% Ti / 15% Mo filaments having lengths of ~17.5 cm. The usable filament length was of the order of 15 cm., as each end was clamped in a holder similar to that shown in Fig. 3.2.1. Sublimation rate measurements were made on a sample of six filaments. The filaments and holders were placed in a water cooled chamber, combination pumped with a 20 $\mathcal{L}$/sec sputter-ion pump. Two gauges were used to measure the indicated pressure. One was optically baffled from the Ti vapors, the second was deliberately put in line-of-sight with the Ti vapors. The intensity of the Ti vapors, monitored by the unshielded gauge, was used to determine the relative sublimation rate of each filament, in time. The second gauge was used to make sure that actual chamber pressure was sufficiently low so that errors did not occur in the inferred Ti flux readings taken with the first gauge.

Six filaments, with dimensions as described above, were used in the following tests. All were weighed prior to the experiment and had approxi- mate weights of ~2.7 gm. All experiments were conducted with a constant voltage setting, sufficient so that the initial current drawn by the filaments was 50 A. All of the filaments were outgassed prior to sublimation measurements. The first three filaments were each operated for one hour, and then removed and weighed. One of the three was inadvertently operated at a constant current of 50 A for one hour. The other two were operated at an initial current of 50 A, which decreased to 48.5 A at the end of one hour of oper- ation. The weight loss in the first filament corresponded to an initial equivalent sublimation rate, R_0, of 0.071 ± 0.017 gm/hour.

The remaining three filaments were operated in a manner similar to that of actual use. The current was initially set at 50 A, the voltage thereafter held constant and the current allowed to decrease in time as the resistance of the filaments increased in time. However, the voltage was cycled, 10 minutes on and 20 minutes off, throughout the life of the filaments. The sublimation rate, $R(t)$, throughout the life of the filaments followed the relationship:

$$R(t) = R_0 e^{-at} \qquad\qquad (3.4.1)$$

where R_0 = 0.073 ± 0.017 gm / hour,
        a   = 0.0455 / hour,
and t = time of sublimation in hours.

At end-of-life, the average weight loss of each filament was 1.16 ± 0.01 gm, or ~50% of the Ti in the wire had been sublimed. Of course, because of end effects, the high temperature portion of the filament had far greater depletion in Ti than this (i.e., ~73%). *Constant* voltage results are given in Fig. 3.4.3. Lawson and Woodward's data are also shown in this figure, but are normalized to the larger diameter filaments of my tests. Note that when *normalizing* all of the above data, ~5.0 cm was subtracted from the reported length of the filament. This was done because of end effects induced by securing the filament to the holders.

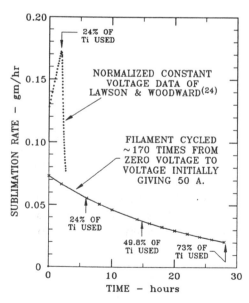

Figure 3.4.3. Sublimation rate data of Lawson and Woodward for a literally constant filament voltage, compared with Welch data where the filament was cycled on and off, between a voltage setting initially yielding 50 A current in the filament.

We must conclude from Fig. 3.4.3 that, as in the case of *constant current* operation, the cycling of the filament through a crystalographic phase transformation causes changes in the total emissivity of the filament. In my *constant voltage* experiments, the filaments underwent ~170 temperature cycles, where in the case of Lawson and Woodward the voltage was in fact literally held constant. Where Lawson and Woodward reported no decrease in sublimation rate up to a depletion of 24% of the total Ti, I observed sublimation rate decreases, in time, at the beginning of filament operation. Therefore, from the above *constant voltage* and *constant current* data, it is obvious that not only is the proportion of Ti and Mo in the filament important to predicting sublimation rates in time, but the number of temperature cycles through the crystalographic transformation temperature of the filament is of equal, if not greater importance.

Because of the perplexing problems posed in predicting sublimation rates of these filamentary sources in time, Kuznetsov[38] used the electron emission of the filament as an inference of its temperature. A refinement of this technique was later reported on by Strubin.[39] Electron emission from a hot filament can be predicted by using what is called the Richardson-Dushman equation.[40] Presumably, by knowing the temperature of the filament, one can thereby predict the sublimation rate. As you have by now observed, knowing the temperature is necessary, but not sufficient to the prediction of filament sublimation rates. That is, the rate of Ti sublimation at a given temperature depends on both the Ti and Mo proportions and the number of temperature cycles the filament has undergone.

Radiantly Heated Sources

These sources are heated by the thermal radiation of a secondary W filament. Herb and his colleagues were the first to report on this type of Ti source.[35] They reported on using a 0.5 mm ϕ W coil, inserted into a 1.6 cm ϕ tube with a length of 2.5 cm and a wall thickness of 3.3 mm. No sublimation rate data were given. Harra and Snouse reported on a refined radiantly heated source, which came to be known as the Varian TiBall® source.[42] A cross section of this source and another similar source, developed by me, is shown in Fig. 3.4.4. In the case of both the TiBall® and MiniTiBall® sources, current is passed through a W filament, completely surrounded by the Ti *ball*, and with a return path through the Ti sphere. Part of the inner surface of the TiBall® sphere is serrated for purposes of achieving a more uniform radiant heating of the Ti.

The *ball* assembly is mounted on a flanged holder, similar to that used with filament sources. For example, two such TiBall® sources could be coaxially positioned near the center of the cheveroned array depicted in Fig. 3.5.1, and each provide ~35 gm of usable Ti, prior to venting the system for source replacement. The TiBall® source was capable of providing sublimation rates of up to ~0.6 gm/hour with a maximum power of 750 watts. The smaller source had less uniform sublimation rates, and required the same power needed to operate the standard filament sources found in common use today (*i.e.*, ~380 W).

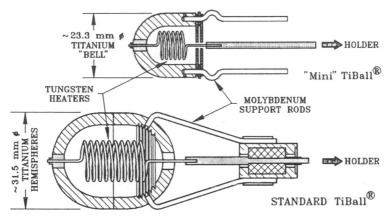

Figure 3.4.4. Two varieties of radiantly heated sublimation sources; the TiBall®, developed by Harra and Snouse[42] and a smaller, radiantly heated source developed by Welch.

Sublimation rates as a function of sublimed Ti for these two sources are given in Fig. 3.4.5. End-of-life of these *ball* or *bell* sources more often than not occurs when the Ti wall thins down to the point where a *window* to the outside develops. This window may be only ~1.0 cm^2 in area, but this is sufficient that radiation losses from the W filament, through the window, make further sublimation of the Ti negligible.

These types of sources have one major disadvantage. If they are not operated with some level of standby power, their life is significantly decreased. The standby power must be sufficient to keep the Ti source at a temperature in excess of ~880° C.[42] This is due to the fact that distortion occurs in the Ti source as a consequence of temperature cycling it through a crystalographic phase transformation which occurs in pure Ti at ~880° C. At temperatures $\leqslant 880°$ C, Ti exists in an α or hcp phase. At temperatures $\gtrsim 880°$ C, it changes into a β or bcc crystal structure. The temperature cycling of these pure Ti sources soon results in a knurled, distorted body of Ti. Martin encountered this distortion when using the TiBall® in a fusion-related program at Princeton.[26] Distortion usually leads to a shorting of the W filament to the Ti wall. Also, the very rough texture of the surface, as the result of the temperature cycling, results in a greatly increased surface area (or total effective emissivity). As a consequence of this increase in surface area, radiation losses increase, and sublimation rates diminish to unusable levels.

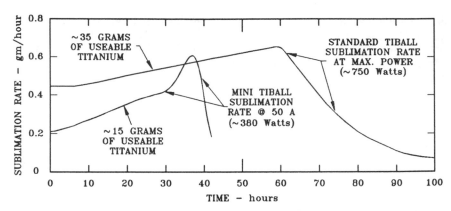

Figure 3.4.5. Sublimation rates, in time, of the standard and MiniTiBall® radiantly heated sublimation sources.[42]

I had the *bright idea* of making a source out of a β-stabilized material so as to minimize distortion of the *bell* (*i.e.*, the MiniTiBall®) on temperature cycling. I had a bell made of such an alloy. It worked magnificently. I was able to cycle the bell over 250 times from RT to operating temperature with only slight distortion of the configuration. Cycling of a pure Ti source a similar number of times led to failure due to shorting of the filament, and after only ~30% Ti utilization. The problem with the new material was that though the bell survived the temperature cycles, the sublimed film wouldn't stick to surfaces including $A\ell_2O_3$, stn. stl., Cu and glass. Film flakes, with a thickness of a only few μm, lay in piles on the bottom of the bell jar. This was due to residual stresses in the sublimed film.

In the final analysis, no solution to this phase transformation distortion problem has been discovered. Therefore, to date these sources must be operated with standby power. Standby power, for the smaller source shown above, is ~95 watts, and for the larger source ~200 watts. Though there is negligible sublimation at temperatures stemming from these power levels (theoretically in the millions of years of life) heat from the source may be troublesome to extreme high vacuum applications such as storage rings. Also, in applications such as storage rings, operating at pressures in the 10^{-10} Torr region and lower, TSP users are less concerned with the availability of large quantities of Ti than they are concerned about the reliability of the source. Filamentary sources usually number 3 to 4 on one holder. This provides redundancy in the system in the event of the failure of any one filament. Because of these two considerations, radiantly heated sources have not found wide acceptance in particle accelerator applications.

E-Gun Sources

Gould and Mandel, subsequent to a visit with Professor Herb at the University of Wisconsin, launched into an E-gun Ti source development program at Brookhaven National Laboratory.[43] The Alternating Gradient Synchrotron at Brookhaven was once pumped with wire-fed electrostatic ion pumps. It was evident, when *reading between the lines* of publications dealing with these electrostatic pumps, that they had serious reliability problems. Gould and Mandel developed an E-gun Ti source so as to eliminate the wire-fed Ti sources. It comprised a simple Ti plug, mounted on a Ta rod within the center of the pump. The Ti plug was bombarded by ~1.0 kV electrons emitted from a hot W filament, as shown in Fig. 3.4.6. Gould and Mandel noted at that time that "It is perfectly feasible to visualize a continuous feed rod (*i.e.*, Ti source) ..., depending only on the life of whatever filament or multiple filaments were installed." Gretz reported on an appendage pump having a similar configuration to Gould and Mandel's E-gun source.[53]

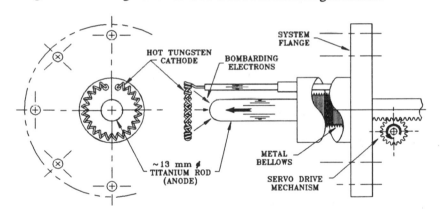

Figure 3.4.6. A form of "E-gun" titanium sublimation source developed and sold by the Ultek, Perkin-Elmer Corporation.

A variation of Gould and Mandel's proposed rod-fed source was reported on by Burthe and Munro, of the Ultek, Perkin-Elmer Corporation, only four years later.[44] Ti rods, with diameters of ~3.18 cm and of varying lengths, could be manually fed, as needed, into beams of electrons emitted from six or more short, heated W filaments. A later variation of this E-gun source, depicted in Fig. 3.4.6, made use of a toroidally wound, heated W electron source. This later version provided more uniform heating of the Ti rod.

The need of these sources arose with the need for higher sublimation rates than could be reliably achieved by filaments. E-gun sources were initially developed for semiconductor and decorative and optical coating applications.[45] The primary pumping application of these sources was in space simulation work. When pumping very large chambers with forms of combination pumping (see below), including TSP pumps, they were Ti limited. That is, even with multiple filament sources, they could not sublime enough Ti to take advantage of all of the available surface area. The above source could sublime Ti at a rate of up to 0.5 gm/hour, operated at less than one-third the power of the TiBall®, had ~55 gm of usable Ti, and did not require standby power. The Ti source rod was automatically advanced. This was controlled with a servo loop tied to the pervience of the electron gun (*i.e.*, the filament emission to the Ti anode, for a given voltage). Because of the need for less power than the TiBall®, Ultek advertised this source as the "cool sublimator".

A second form of E-gun source, reported on by Robertson, was developed and sold by Varian Associates.[54] Rather than having a continuously fed source of Ti, the source comprised layers of Ti sheet, as depicted in Fig. 3.4.7. A heated W filament served as the source of electrons.

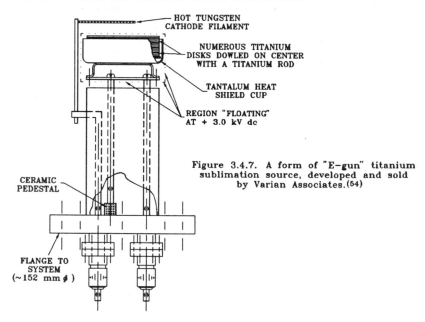

Figure 3.4.7. A form of "E-gun" titanium sublimation source, developed and sold by Varian Associates.[54]

This E-gun could sublime Ti at a rate of up to 7.0 gm/hour and provided ~200 gm of usable Ti. It operated at up to 3.0 kV, and required 3.0 kW of beam power at the maximum sublimation rate. It was bakeable to 450° C and did not require water cooling. For reasons noted below, these high sublimation rate, E-gun sources had rather limited use in pumping applications. Their present advantage is that, with the appropriate pumping arrays, they can provide pumping speeds of 50,000 - 200,000 $\mathscr{L}$/sec in very hostile environments (*e.g.*, high ionizing radiation fields).

3.5 Combination Pumping

As is now evident, some gases are not chemically pumped by a deposited layer of Ti. Also, some gases are synthesized on the hot filament, or liberated from the Ti film due to molecular dissociation. Because of this, various forms of *combination* pumping are used to pump these by-products of TSP pumping, and other gases such as He and Ar which are present in many vacuum systems. Three forms of combination pumping are often integrated into one system. These are cryopumping on LN_2 cooled surfaces (primarily to pump H_2O), TSP pumping, and sputter-ion pumping. An example of the configuration of such a system is shown in Fig. 3.5.1. In this system, the TSP pump affords speeds of the order of many thousands of liters per second for chemically active gases, the sputter-ion pump pumps gases including the CH_4 and the inert gases, and the LN_2 surfaces pump the H_2O.

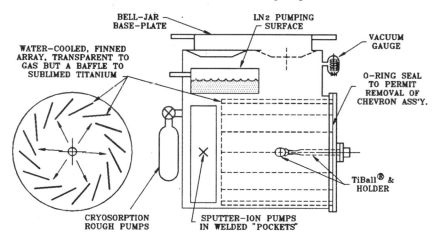

Figure 3.5.1. A combination pump configuration, reported on by Harra, and making use of TSP, sputter−ion and cryopumping.[20]

Pumping systems, having configurations similar to that shown above, were developed primarily for thin film deposition work in the semiconductor industry. As chip component densities became greater and line-widths smaller, the effects of contamination from the pumping systems became more significant. Diffusion pumped systems have problems of oil backstreaming. Also, if such systems are inadvertently let to air while the main isolation valve is open to the semiconductor deposition chamber - a process referred to as being *dumped* - this could result in system down time of weeks, and serious damage of the work in process. Because of this, use of diffusion pumped systems became less popular and systems such as shown in Fig. 3.5.1 came into brief prominence. Of course, even these systems became less popular with the advent of the gaseous helium cryopump.

Note that the system shown in Fig. 3.5.1 is rough pumped with sorption pumps. This too was a requirement of many of the semiconductor manufacturers. There was concern that possible oil backstreaming from mechanical roughing pumps might also pose a threat of system contamination. Essentially all of the UHV surface science equipment (e.g., Auger Electron Scanning systems) presently marketed use some form of TSP and sputter-ion pumping. A load lock or preparation chamber may be rough pumped with a cryopump of some form, but the UHV work is done in a chamber pumped with TSP (usually on RT surfaces) and sputter-ion pumps.

As an interim measure, TSP and sputter-ion pumping was also used in space simulation systems.[6,44,46,47] This was always augmented with LN2 pumping. This combination pumping was also prompted by oil backstreaming problems from diffusion pumps, which caused the contamination of work pieces. However, these large space simulation chambers are now almost exclusively pumped with closed loop, gaseous helium cryopumps, augmented with LN2 shrouds.

Combination TSP and sputter-ion pumping is also used extensively in particle accelerators and storage rings in operation throughout the world (e.g., see refs. 29,48,49). Many of the accelerators and storage rings noted in Section 2.10 use this form of combination pumping. Pressures in the 10^{-13} to 10^{-11} Torr range may be reliably achieved through this means.[48-51] TSP pumping surfaces are made comparatively large, so that the pump speed is large, and infrequent use of the sublimation pump is required to maintain the needed operating pressure. In such instances, the Ti films need not be cleaned off the pumping surfaces. This is because the operating *life* of the pump approaches the useable life of the accelerator and the accumulated thickness of the sublimed films is therefore negligible (see problems 5-11).

As a footnote to combination pumping, for reasons having to do with safety, I *strongly discourage* the use of TSP pumping in conjunction with cryopumps capable of sorbing (i.e., temporarily storing) large quantities of flammable gases. More on this is discussed in the following chapters.

Peeling of Titanium Films

The subliming of Ti onto LN2 cooled surfaces fell into disfavor in later years. It was determined that the advantages gained in values of sticking coefficients and film capacities with such use were outweighed by disadvantages of premature peeling of deposited films. This effect is not reported in the literature, though peeling effects of films on various surfaces have been reported. Harra reports the peeling of the sublimed films initiates after deposition of ~0.03 gm/cm^2.[20] This was a *working* system and the film was used intermittently (i.e., for thousands of cycles) to pump large quantities of air. Assuming, because of the sorption of gases, the film has a density of ~4.0 gm/cm^3, this corresponds to a film thickness of ~7.5 x 10^{-3} cm. Martin made measurements of the peeling of films deposited on surfaces of various materials, including sheets of Ti, Ta, Mo, Cu, Aℓ and stn. stl.[26] Half of each of the test sheets were sandblasted. He noted unsandblasted surfaces peeled with thinner films, compared to sandblasted surfaces. This work was of a qualitative nature, though he noted a "maximum" film thickness of 0.023 cm on peeling. Conservatively, a usable thickness of $\sim 5 \times 10^{-3}$ cm seems reasonable.

The peeling of films causes a deterioration in pump performance. As the film peels back from the surface of the pumping array, the film tends to operate at elevated temperatures, when the sublimation pump is turned on. This is because thermal conduction along the peeling film is poor, yet its surface area is comparatively large. As the result of increases in temperature of the peeling film, H_2 may be thermally desorbed, and result in unwanted increases in system pressure. Because of this peeling effect, it is necessary to clean the surfaces of the pump from time to time. Therefore, one must either have access to the deposition surface for cleaning, or provisions must exist for removing the pumping array from the system. The clever configuration shown in Fig. 3.5.1 served this purpose. Titanium was sublimed onto the inner surfaces of a cylinder comprising numerous cheveroned baffles. The baffles were transparent to the transmission of gas into the TSP pump, but completely cut off the sublimed Ti vapor. The cheveroned array could be easily removed from the system for periodic cleaning.

A word of caution: peeling films of Ti are *pyrophoric*. That is, they may spontaneous ignite, or spark when vented to air. Also, when scraping these films off of metal surfaces, if enough work is put into the film, it is possible to ignite the film and have a localized "flash fire". This is not a major safety hazard. But, we can often injure ourselves when responding to a surprise, and this is merely a word of warning. Secondly, film refuse should not be heaped together or discarded in containers having other flammable refuse.

3.6 Advantages and Disadvantages of TSP Pumping

There are disadvantages to the use of TSP pumps. First, in being capture pumps, they have a finite capacity. In high sublimation rate applications, the pump surfaces must be periodically cleaned (see section on film peeling). The power radiated from Ti sources, in the absence of water cooling, will raise the temperature of surfaces within the vacuum system. This causes system outgassing.

These pumps are low throughput devices (see problem 3). TSP pumps will not pump the inert gases, and gases including CH_4 are synthesized at the hot Ti source. Because CH_4 is a comparatively inert gas, it is virtually unpumped by the Ti film, and requires the use of combination pumping. Also, some form of rough pumping is required, as in the case of all high vacuum pumps.

The sublimed Ti vapor can cause problems in a vacuum system if adequate line-of-sight baffling does not exist between the Ti source and other vacuum equipment. For example, sublimed films can cause breakdown or leakage currents in high voltage insulators, the contamination of working films and optical surfaces, and damage to instrumentation within the system, without appropriate shielding from the Ti flux. The need for the shielding of surfaces from the Ti flux requires that gas impedances be interposed between the system working volume and the pump, thus reducing the available speed of the film surface to the system.

Combination pumping with TSP and sputter-ion pumps is the most reliable and cost effective method of achieving UHV pressures in bakeable, *lumped* systems. By *lumped*, I mean that the pumps are lumped elements rather than elements distributed along some great length, as in the application of NEG pumping in, for example, LEP.[52] TSP pumping is the most economical of all forms of pumping, in terms of operating power, or W/$\mathcal{L}$/sec at very low pressures. TSP pumps are bakeable, and with the exception of CH$_4$ and possible Ti vapors introduce no contaminants into the system. They are safe to operate, though when being serviced peeling Ti films pose potential hazards, as noted in Section 3.5. Systems in which they are used are simple to design. They may be operated in any position, are immune to high radiation fields, and present no source of vibration to the system.

Problem Set

1) Using the data of Fig. 3.2.5 and the pump configuration of Fig. 3.2.1, calculate the intrinsic speed of the pump for the RT gases H$_2$, N$_2$ and O$_2$, assuming a gas surface coverage 10^{13}, 10^{14} and 10^{15} molecules per cm^2.

2) Show that without a pumping synergistic effect, and the synthesis of CH$_4$, (3.3.1) should be true.

3) Assume a perfect stoichiometry of TiN is achieved in a continuously operated system. What would the throughput capability be, in Torr-$\mathcal{L}$/sec, of a TSP source having a sublimation rate of 0.5 gm/hour?

4) Use Table 3.2.1 to calculate the speed of a TSP pump for the gases H$_2$, CO, N$_2$ and O$_2$. Assume the temperature of the pumping surface is 77° K and the surface area is 10^3 cm^2, and the gas is at RT.

5) Assume that TSP combinations pumps are designed as shown in Fig. 3.6.1. Assume that the pump is operated at RT. For a freshly deposited film, what is the intrinsic speed of one pump for the gases H$_2$, CO, N$_2$, O$_2$ and CO$_2$?

6) Assume a system configuration similar to that of Fig. 3.6.1. The H$_2$ outgassing rate of the long beam pipe attached to this pump is ~10^{-12} Torr-$\mathcal{L}$/sec-cm^2. The distance between each pump is 2x cm. The outgassing of other gases from the beam pipe is negligible. Assume the values of α for

H_2 vary as shown in Fig. 3.2.5. Assuming an H_2 coverage of 10^{13} molecules/cm^2, what is the *average* pressure in the system if it is periodic in sets of the system shown in Fig. 3.6.1? Neglect outgassing from the pump and connecting manifold.

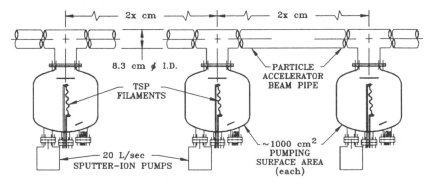

Figure 3.6.1. Particle accelerator storage ring pumping system similar to that described by Halama for the "ISABELLE" accelerator ring.[50]

7) What is the <u>average</u> pressure of the system in problem 6, if the beam pipe outgassing rate is 10^{-13} Torr-ℓ/sec-cm^2 H_2? What is the average pressure if the gas is CO and the outgassing rate is the same?

8) Assume that the H_2 sticking coefficient initially decreases as shown in Fig. 3.2.6, and extrapolate coverage data until a gas/film stoichiometry of TiH$_2$ is reached. Assume a film thickness of ~200 Å. Construct an analytical expression for the sticking coefficient as a function of coverage.

9) Using the results of problem 8, express the speed of the film as a function of coverage.

10) Assume that the diffusion of H_1 is not a limiting factor. Express the average pressure in the system as a function of time, assuming the H_2 outgassing rate of problem 7, the pumping speed found in problem 8, and until such time as a stoichiometry of TiH is reached.

11) Assume we operated the pump in problem 5 with a sublimation rate of 0.05 gm per hour, under batch sublimation pumping conditions. Conservatively, how long would we be able to operate the sublimation pump prior to the onset of film peeling?

References

1. Stout, V.L. and Gibbons, M.D., "Gettering of Gas by Titanium", J. Appl. Phys. 26(12), 1488 (1955).
2. Wagener, J.S., "Properties of Getters in Electron Tubes", Proc. 4th Nat. Conf. on Tube Techniques, 1958 (New York University Press, New York, 1959), p. 1.
3. Giorgi, T.A., Ferraio, B., Storey, B., "An Updated Review of Getters and Gettering", J. Vac. Sci. Technol. A3(2), 417 (1985).
4. Hayward, D.O. and Taylor, N., "The Adsorption and Absorption of Hydrogen by Evaporated Tantalum Films", Proc. 4th Int. Vac. Cong., 1968 (The Institute of Physics and The Physical Society, London, 1969), p. 115.
5. Okano, T., Iimura, K., Tominaga, G., "A Tantalum Evaporation Pump", Proc. 7th Int. Vac. Cong. and 3rd Int. Conf. on Solid Surfaces, 1977 (R. Dobrozemsky, F. Rüdenauer, F.P. Viehböck, A. Breth, Postfach 300, A-1082 Vienna, Austria, 1977), p. 81.
6. Prevot, F. and Sledziewski, Z., "The Titanium Evaporation Pump, its Application to Nuclear Fusion Experiments and Space Simulation", J. Vac. Sci. Technol. 9(1), 49 (1972).
7. Haque, C.A., "Simple Titanium and Nickel Sublimation Pump (TNSP)", J. Vac. Sci. Technol. 13(5), 1088 (1976).
8. Nazarov, A.S., Ivanovskii, G.F., Men'shikov, M.I., "Getter-Ion Pump With Filamentary Titanium and Chromium Evaporators", translated from Pribory i Tekhnika Éksperimenta, No. 5, pp. 157-161, 1963, Instrum. & Exper. Tech., 934 (1964).
9. Kuz'min, A.A., "Laboratory Ultrahigh-Vacuum Apparatus With Filamentary Solid-Phase Titanium Evaporators", translated from Pribory i Tekhnika Éksperimenta, No. 3, pp. 126-130, 1963, Instrum. & Exper. Tech., 497 (1963).
10. Clausing, R.E., "A Large-Scale Getter Pumping Experiment Using Vapor Deposited Titanium Films", Proc. 2nd Int. Vac. Cong., 1961 (Pergamon Press, Inc., New York, 1962), p. 345.
11. Herb, R.G., Davis, R.H., Divatia, A.S., Saxon, D., "Evapor-Ion Pump", Phys. Rev. 89, 897 (1953).
12. Hayward, D.O. and Taylor, N., "Some Fundamental Problems in the Measurement of Sticking Probabilities of Gases on Metal Films", J. Sci. Instrum. 44, 327 (1967).
13. Giorgi, T.A. and Pisani, C., "On the Concept of Pumping Speed in Gettered and Cryopumped Systems", Vacuum 16(12), 669 (1966).
14. Harra, D.J. and Hayward, W.H., "Sorption of Nitrogen by Titanium Films", Proc. Int. Symp. on Residual Gases in Electron Tubes, 1967, Supplemento Al Nuovo Cimento Serie I. 5(1), 56 (1967).

References

15. Kornelsen, E.V. and Domeij, B., "Simple Differential Pumping Stage for Connecting High to Ultrahigh-Vacuum Systems", J. Vac. Sci. Technol. 3(1), 20 (1966).

16. Harra, D.J., "Review of Sticking Coefficients and Sorption Capacities of Gases on Titanium Films", Proc. 22nd Nat. AVS Symp., 1975, J. Vac. Sci. Technol. 13(1), 471 (1976).

17. Grigorov, G.I., "Apparent and Real Values of Common Gas Sticking Coefficients on Titanium Films and Application to Getter Pump Devices With Periodic Active Film Renovation", Vacuum 34(5), 513 (1984).

18. Biryukova, N.E., Vinogradov, M.I., Efimov, M.N., "Sorption of Nitrogen by Continuously Deposited Titanium Films", Proc. 4th Int. Vac. Cong., 1968 (The Institute of Physics and The Physical Society, London, 1969), p. 184.

19. Elsworth, L., Holland, L., Laurenson, L., "The Sorption of N_2, H_2, and D_2 on Titanium Films at 20° C and −190° C", Vacuum 15, 337 (1965).

20. Harra, D.J., "Improved Titanium Sublimation Pumping Techniques for Long-Term, High-Throughput, Clean Pumping", J. Vac. Sci. Technol. 12, 539 (1975).

21. Gupta, A.K. and Leck, J.H., "An Evaluation of the Titanium Sublimation Pump", Vacuum 25(8), 362 (1975).

22. Reichardt, J.W., "The Kinetics of the Hydrogen-Titanium Reaction", J. Vac. Sci. Technol. 9(1), 548 (1972).

23. Steinberg, R. and Alger, D.L., "Access to Uncombined Titanium Through an Inhibiting Film in Sublimation Pumping of Deuterium", J. Vac. Sci. Technol. 10(1), 246 (1973).

24. Lawson, R.W. and Woodward, J.W., "Properties of Titanium-Molybdenum Alloy Wire as a Source of Titanium for Sublimation Pumps", Vacuum 17, 205 (1967).

25. Grigorov, G.I. and Tzatzov, K.K., "Theory of Getter Pump Evaluation. Sticking Coefficients of Common Gases on Continuously Deposited Getter Films", Vacuum 33(3), 139 (1983).

26. Martin, G.D., "Pulsed Gas Load Pumping of Hydrogen by Vapor Deposited Films", Princeton Plasma Physics Laboratory, Princeton, New Jersey, PPPL Report #MATT-1193, 1976.

27. Jepsen, R.L. and Francis, A.B., "Interactions Between Ionizing Discharges and Getter Films", Supplemento Al Nuovo Cimento 1(2), 694 (1963).

28. Klopfer, A. and Ermrich, W., "Properties of a Small Titanium-Ion Pump", Proc. 6th Nat. AVS Symp., 1959 (Pergamon Press, Inc., New York, 1960), p. 297.

29. Halama, H.J., "Behavior of Titanium Sublimation and Sputter-Ion Pumps in the 10^{-11} Torr Range", J. Vac. Sci. Technol. 14(1), 524 (1977).

References

30. Edwards, D., "Methane Outgassing from a Ti Sublimation Pump", J. Vac. Sci. Technol. 17(1), 279 (1980).
31. Edwards, D. and Lanni, C., "Ti Getter Study", J. Vac. Sci. Technol., 17(1), 1373 (1980).
32. Chou, T.S. and Lanni, C., "Lifetime of Titanium Filament at Constant Current", IEEE Trans. Nucl. Sci. NS-28(3), 3323 (1981).
33. Kornelsen, E.V., "A Small Ionic Pump Employing Metal Evaporation", Proc. 7th Nat. AVS Symp. 1960 (Pergamon Press, Inc., New York, 1961), p. 29.
34. McCracken, G.M. and Pashley, N.A.,"Titanium Filaments for Sublimation Pumps", J. Vac. Sci. Technol. 3(3), 96 (1966).
35. Herb, R.G., Pauly, T., Welton, R.D., Fisher, K.J., "Sublimation and Ion Pumping in Getter-Ion Pumps", Rev. Sci. Instrum. 35(5), 573 (1964).
36. Harra, D.J., "Predicting and Evaluating Titanium Sublimation Pumping", Presented at 6th Int. Vac. Cong., Kyoto, Japan, 1974, Varian Associates Report No. VR-88, 1974.
37. Brewer, L. and Lamoreaux, R.H., Molybdenum: Physico-Chemical Properties of its Compounds and Alloys, Special Issue No. 7 (International Atomic Energy Agency, Vienna, 1980, L. Brewer, Ed.), Chapter II, "Phase Diagrams", p. 333.
38. Kuznetsov, M.V., Nasarov, A.S., Ivanovsky, G.F., "New Developments in Getter-Ion Pumps in the U.S.S.R", J. Vac. Sci. Technol. 6(1), 34 (1969).
39. Strubin, P., "Study of a New Method to Control Precisely the Evaporation Rate of Titanium Sublimation Pumps", J. Vac. Sci. Technol. 17(5), 1216 (1980).
40. Spangenburg, K.R., Vacuum Tubes (McGraw-Hill Book Company, Inc., New York, 1948), p. 30.
41. Sweetman, D.R., "The Achievement of Very High Pumping Speeds in the Ultra-High Vacuum Region", Nucl. Instrum. Methods 13, 317 (1961).
42. Harra, D.J. and Snouse, T.W., "A Radiant Heated Titanium Sublimator", Proc. 5th Int. Vac. Cong., 1971, J. Vac. Sci. Technol. 9(1), 552 (1972).
43. Gould, C.L. and Mandel, P., "A Sublimation Pump", Proc. 9th Nat. AVS Symp. 1962 (The Macmillan Company, New York, 1963), p. 360.
44. Burthe, J.H. and Munro, D.F., "Bulk Sublimation of Titanium", Proc. 3rd Int. Vac. Cong., 1965 (Pergamon Press, London, 1966), p. 37.
45. Burden, M. St. J., Walley, P.A., "The Evaporation of Metals and Elemental Semiconductors Using a Work-Accelerated Electron Beam Source", Vacuum, 19(9), 397(1969).
46. Wolf, R.A., "The Application of Selective Pumping Techniques to Upgrade Vacuum Chambers", J. Environ. Sci., October 1966, p. 3.

References

47. Brothers, C.F., Tom, T., Munro, D.F., "Design and Performance of a 50,000 ℓ/sec Pump Combining Cold Cathode Ion Pumping and Active Film Gettering", Proc. 10th Nat. AVS Symp., 1963 (The Macmillian Company, New York, 1964), p. 202.
48. Ishimaru, H., "Ultimate Pressure of the Order of 10^{-13} Torr in an Aluminum Alloy Vacuum Chamber", J. Vac. Sci. Technol. $\underline{A7}$(3), 2439 (1989).
49. Johnson, J.W., Atkins, W.H., Dowling, D.T., McConnell, J.W., Milner, W.T., Olsen, D.K., "Heavy Ion Storage Ring for Atomic Physics Vacuum Test Stand for Pressure of 10^{-12} Torr", J. Vac. Sci. Technol. $\underline{A7}$(3), 2430 (1989).
50. Halama, H.J., "ISABELLE Vacuum Systems", Proc. 8th Int. Vac. Cong., 1980 (Supplé ment à la Revue, "Le Vide, les Couches Minces, No. 201", 1981), p. 115.
51. Halama, H.J., "Design and Construction of Vacuum Systems for Large Colliders Using Superconducting Magnets", Proc. 9th Int. Vac. Cong. and 5th Int. Conf. on Solid Surfaces, 1983 (Asociacion Español del Vacio y sus Aplicaciones, Madrid, 1984), p. 283.
52. Reinhard, H.P., "The Vacuum System of LEP", Proc. 9th Int. Vac. Cong. and 5th Int. Conf. on Solid Surfaces, 1983, (Asociacion Español del Vacio y sus Aplicaciones, Madrid, 1984), p. 273.
53. Gretz, R.D., "A High Vacuum Titanium Getter Pump", Vacuum $\underline{16}$(10), 537 (1966).
54. Robertson, D.D., "Evaluation of a High-Rate Titanium Getter Pump", Proc. 4th Int. Vac. Cong., 1968 (The Institute of Physics and The Physical Society, London, 1969), p. 373.
55. Holland, L., Laurenson, L., Allen, P.G.W., "The Formation of Hydrocarbon Gas by Titanium Getters Containing Carbon and Hydrogen Impurities", Proc. 8th Nat. AVS Symp. and 2nd Int. Vac. Cong., 1961 (The Macmillan Company, New York, 1962), p. 208.

CHAPTER 4

NONEVAPORABLE GETTERS

4.0 Introduction

There is a class of getters which effectively pump gases merely as a consequence of being activated by heating to a certain temperature, under vacuum. However, the getter activation temperatures, differing with different materials, are low compared to the operating temperatures of TSP sources. Therefore, appreciable sublimation or evaporation from these getter materials does not occur. Thus, they have been named nonevaporable getters, or *NEGs*. The NEG comprises a sintered-powder matrix of metal alloys which chemisorb gases. These alloys are typically operated at temperatures ranging from RT (room temperature) to ~400° C, though some require activation at slightly higher temperatures. A number of different alloys and pure metals possess this gettering capability. Giorgi, Ferrario and Storey review the properties and applications of some of these NEGs.[1] Kindl and Rabusin also provide a review of early NEG development work.[2] Most of the significant NEG development work to date has been done at the company SAES® Getter S.p.A. (*i.e.*, Società Apparecchi Elettrici e Scientifici, in Milan, Italy). Of course, studies of the pumping characteristics of these materials have involved the work of others throughout the world.

Two alloy mixtures have been widely reported on. These are the St 101® and St 707® materials. The St 101® material is an alloy of 84% Zr and 16% Aℓ by weight, where the St 707® material comprises an alloy of 70% Zr, 24.6% V and 5.4% Fe, by weight. In the development of these NEGs, some other NEG was used as a *benchmark* or measure of relative merit. For example, the St 101® material, first reported on by Della Porta, et al., in 1961,[3] was compared with NEGs comprising mixtures of Th and Aℓ. The high activation temperatures of the St 101® material (*101*, hereafter) prompted the development of the second popular NEG material, St 707® (*707*, hereafter), which was first reported on by Boffito, *et al.*, in 1981.[4] This chapter will primarily deal with the properties and applications of these two materials. I will first discuss some of the forms and mechanical configurations of these NEGs as presently used. After this, I will discuss general aspects of the pumping mechanisms of these NEGs, as presently understood. I will conclude by describing test results of a number of investigators using these NEGs in a wide variety of applications.

4.1 Mechanical Features

Boffito describes these materials as "sintered porous bodies".[4] Initially, the *101* material was available only in bulk form. That is, the ZrAℓ alloy was pressed into contained pellets having dimensions, for example, of 5-10 mm φ and a thickness of 2-3 mm. The pellets were then sintered. In spite of this sintering, the pellets retain their porosity. Kindl notes that this porosity is equivalent to ~0.015 ℓ/sec-cm² for CO (the alloy thickness and gas temperature were not given).[2] To increase the surface area of the pellets and facilitate ease of attachment, the pellets were also offered in the shape of washers. Pellets of this and other special shapes are now available in both materials.

New applications became possible with the discovery of a method of bonding thin layers of powdered *101* material to metal surfaces. This was first reported on by Kindl and Rabusin in 1967.[2] They reported on a "patented" method of bonding the material to metal substrates. The process did not involve the use of organic binders. The *101* material was reported by Boffito to effectively sinter to a Ni plated Fe substrate.[4] However, Kindl earlier reported problems with this method of bonding the ZrAℓ powder.[2] Zirconium forms a eutectic with Ni at ~960° C. Apparently, if the temperature (and perhaps the Ni plating thickness) was not controlled during sintering of the ZrAℓ powder to the substrate, the Ni would diffuse not only in the substrate, but throughout the bulk of the NEG material. This had the effect of significantly reducing the sorptive capacity of the NEG.

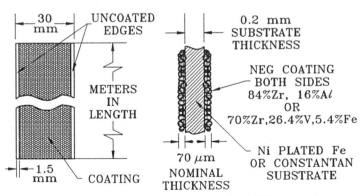

Figure 4.1.1. Example of strip NEGs used in distributed pumping applications.

In any event, both the *101* and *707* powder materials may be sintered to either Ni plated Fe substrates, or onto constantan (*i.e.*, a nonmagnetic alloy comprising 55% Cu and 45% Ni). The material is bonded onto strips of either of these two metals, as shown in Fig. 4.1.1. The thickness of the bonded material is ~70 μm, where the substrate may be any thickness; 200 and 400 μm thicknesses are typical values. The powdered alloy, prior to application, has particulate dimensions of the order of 50-100 μm. Therefore, the 70 μm sintered coating corresponds to about a one or two-particle NEG thickness. Apparently the powder, when pressed into the substrate - probably by high pressure rolling - sticks sufficiently well to itself and the substrate that it may be handled thereafter without falling off the substrate.

For reasons to be discussed, knowing the amount of pressed powder material on the substrate (*i.e.*, g/cm^2) is important in the prediction of the sorptive capacity of the NEG per unit area. This is also true as it relates to the relative density of the sintered powder. Benvenuti reported progressive improvements in the pumping capacity and speed of *101* materials supplied by SAES for use in the CERN LEP accelerator.[5] Part of these improvements may have stemmed from increases in the thickness of the sintered *101* material (*i.e.*, 82 - 106 μm thick *vs.* 70 μm) over the years. However, reported increases in both speed and capacity of the *101* materials may also have stemmed from increases in the density of the sintered powder. Apparently wide variations in powder density are possible. For example, using data from SAES published catalogs, we conclude that the density of sintered powder on substrates such as shown in Fig. 4.1.1 is ~2.65 g/cm^3 for the *707* material and ~2.3 g/cm^3 for the *101* material, assuming a 70 μm thickness in both cases. On the other hand, the density of sintered pellets of the *707* material is calculated to be 4.9-5.1 g/cm^3, where the density of *101* pellets is ~3.7 g/cm^3. Benvenuti reports that the porosity of newer batches of *101* powder, sintered onto substrates, is ~15%,[5] compared with a porosity of ~12% in earlier batches of the *101* material. I am not sure of the significance of these "porosity" numbers.

The bonded material retains sufficient ductility and adhesion, after sintering, that the strip of metal onto which the NEG is bonded may be bent or shaped into different configurations, depending on the application. One such application is in the construction of panel arrays such as shown in Fig. 4.1.2. Variations of such array configurations have been reported on for use in fusion applications.[6-10] These arrays are periodically *activated* by passing current through the metal substrate. The possible combinations of array configurations are limitless, and left to your imagination. Audi, *et al.*, report on the use of panels of NEG material installed in sputter-ion pumps.[11] For reasons noted in Section 2.9, this would have obvious advantages in the pumping of H$_2$ at extremely low pressures. Poncet and his colleagues have used variations of these NEG panel arrays, in both linear form and toroidally pleated form, in a particle beam pipe.[12,37]

Another such application is in the construction of modular pumps, such as shown in Fig. 4.1.3. In this case the pleated strips are formed into toroidal-shaped bands. Very high surface area densities may be created with this configuration. The modular pump is activated using independent heaters located within the center of the pump. The substrates of the modular pumps may be coated with either *101* or *707* material, each having advantages which will be discussed.

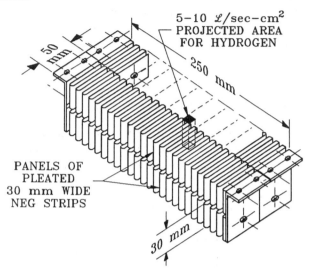

Figure 4.1.2. NEG panels arrayed in a plane as described by Ferrario and Rosai.[36]

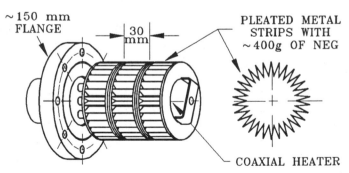

Figure 4.1.3. Example of modular, nude NEG cartridge pump manufactured by SAES.

The comparatively high pumping speeds per unit area of the strip material led to its use in applications of distributed pumping.[13] The use of NEGs in this capacity was first proposed by Benvenuti and Decroux at CERN.[14] They suggested the application of NEGs as distributed pumps for both electron and proton storage rings and accelerators. The most daring example of this application was its use in the LEP e^+e^- collider.[5,15,16] It was *daring* in that 27 km of accelerator was designed and constructed using distributed NEGs of the *101* material as the principal pumping mechanism. It proved most successful in this application.[17,18] Strips of NEG material were positioned in an antechamber in an extruded Aℓ beam pipe, in a manner similar to the distributed sputter-ion pump shown in Fig. 2.10.2 of Chap. 2.

Hseuh also successfully used a distributed NEG, comprising the *707* material, in a heavy ion beam transport line at Brookhaven.[19,20] Results of these and other applications studies will be discussed.

4.2 NEG Pumping Mechanisms

I will first discuss NEG pumping mechanisms in a general sense, as they are presently understood. With this background, I will then give specific findings reported in the literature of the characteristics of the *101* and *707* materials, in some of their present applications.

The Pumping of CO, CO$_2$, N$_2$, O$_2$

The sintered powder is very porous to the transport of gases between the voids in the bulk material. Because of this transparency to gas, the effective surface area of both NEG materials is of the order of ×70 to ×100 of its geometric surface area.[15,19] With exposure to air, or during the pumping of some gases, the surface of the granules become oxidized. This oxide layer passivates the surface; that is, it prevents further surface pumping of gases containing C, O and N. The primary ingredient responsible for surface chemisorption is the Zr. This was shown in the early work of Lampert, Rachocki and Lamartine in the study of St 171®, a NEG comprising a mixture of graphite and Zr.[35] Using ESCA (*i.e.*, soft x-ray photoelectron spectroscopy) they determined that the surface of the granules was covered with ZrO$_2$ and ZrC. However, when the material was activated by heating to ~900° C, the oxygen and carbon disappeared by diffusion into the bulk material. Miller, reporting on much earlier work of others, described this as also being characteristic of pure Zr.[21] He notes that the "protective" film of ZrO$_2$ dissolves into the bulk material at ~450° C, permitting further adsorption of O$_2$. He indicated that O$_1$ (*i.e.*, atomic oxygen) diffuses interstitially in a Zr lattice, with up to even 60 atomic percent oxygen. Also, atomic nitrogen, like oxygen, is absorbed in the interstices of the lattice, but has much less mobility. Because of the ability of all gases to enter voids in the porous body of the NEG, simplistic models involving diffusion processes through a semi-infinite plane are probably not appropriate to NEG pumping. On the contrary, gases are able to diffuse into the bulk nearly through a full 4π steradians of the particle surfaces. This should also be applicable to any

H2 pumping model (*e.g.*, see reference 2 2).

The above carbon and oxygen diffusion mechanism, applicable to the sorption of CO, CO_2 and O_2 by pure Zr and the St 171® material, is also applicable to the *101* and *707* materials. This was shown in the more recent work of Ichimura, Matsuyama and Wantanabe.[23] Using ESCA, they studied the measure of surface activation - this amounts to the freeing up of surface Zr - as a function of activation temperature for three materials (*i.e.*, the *101*, *707* materials and a ZrNi alloy). Through use of SIMS (*i.e.*, secondary ion microscopy scanning), they also determined the ratio of sputtered Zr^+ to ZrO^+ ions from the surface of the NEG to determine the relative amount of free Zr on the surface as a function of temperature cycling. ESCA and SIMS results were taken at RT after heating the NEG to successively higher temperatures under UHV conditions. They also conducted thermal desorption spectroscopy (TDS) measurements on the three materials. Their ESCA results are reproduced in Figures 4.2.1 and 4.2.2, below.

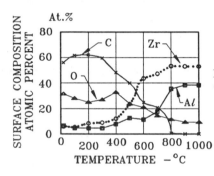

Figure 4.2.1. Changes in surface composition of the St 101® NEG vs. activation temperature.[23]

Figure 4.2.2. Changes in surface composition of the St 707® NEG vs. activation temperature.[23]

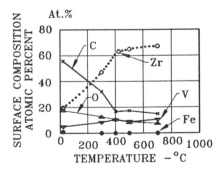

From their findings, including data in the two figures, they concluded the following. Surface oxygen in the *101* material, as-received, and at RT was attributable to the oxygen in Al_2O_3 and ZrO_2. After heating the material to ~800° C, the remaining "surface" oxygen was interstitial. The as-received surface Zr in the *101* material was actually tied up as ZrO_2. Their ESCA and SIMS measurements verified that the surface Zr began to appear in the metallic state only after heating the *101* material to temperatures >400° C. We see from Fig. 4.2.1 that complete activation of the *101* material occurs after the material is raised to a temperature of 800° C. The surface Zr in the as-received *707* material was also present as ZrO_2, though, as shown in Fig. 4.2.2, metallic Zr appeared at much lower temperatures, on heating the material. Using results of TDS, they observed only slight desorption of CO and CO_2 from the *101* material up to a temperature of 500° C. Above this temperature there was negligible desorption of these species. Yet, C and O disappeared from the surface of the material at higher temperatures. Therefore, they concluded that both the C and O diffused into the bulk of the material. A similar model applies to the *707* material. However, because of the apparent variation in the diffusivities of C and O in the two materials, activation of the *707* material is almost complete at 400° C, where activation of the *101* material asymptotes at ~800° C.

In summary, the pumping of gases such as CO, CO_2, O_2, N_2, and the oxygen component of H_2O, occurs on the surface of the powder granules. Carbon is tied up as ZrC and oxygen as ZrO_2. After a certain amount of gas has been chemically pumped on the surface, the free Zr sites become saturated. In order to activate the getter, or replenish these sites, the getter must be heated so that the C and O will diffuse into the lattice of the material. On the one hand, the diffusion of these elements into the material creates new surface pumping sites. On the other hand, the bulk material has a finite capacity for accommodating impurities diffusing therein. Because of this, NEGs, as all chemisorption pumps, have a finite capacity for gases. In the case of NEGs, we must define *two* capacities: 1) the capacity of the pump at saturation of the metal Zr surface sites (*i.e., surface capacity*); 2) the capacity of the bulk to accommodate diffusing impurities (*i.e., bulk capacity*).

Regarding bulk capacity, little independent experimental data are published on measurements clearly defining the total sorptive capacities of either the *101* or *707* materials for the gases CO, CO_2, O_2, H_2O, N_2, etc. In describing the impact of CO *loading* on the H_2 equilibrium pressure, Boffito, *et al.*, show H_2 equilibrium data for the *707* material which has presumably pumped ~15 Torr-$\mathcal{L}$ CO/g, and data for the *101* material which has pumped ~6 Torr-$\mathcal{L}$ CO/g.[24] Kindl and Rabusin report pumping CO to the point of saturating of *101* material.[2] They pumped "... 50 cm^3 Torr of CO at 800° C in one hour per mm^2 of geometric exposed surface ... under a CO pressure of 1 Torr." Assuming a film thickness of ~0.5 mm (*i.e.*, Fig. 3 of the reference) and a film density of ~4 g/cm^3, this would equate to a total RT capacity of the *101* material of ~6.8 Torr-$\mathcal{L}$ CO/g. Data published in SAES product catalogs indicate the total bulk capacity values shown in Table 4.2.1.

A note of caution. In analyzing the published data, it is often very difficult to reduce these data to a common base-line for engineering applications. For example, one such base-line might be the speed per unit area of NEG. In reformatting the data for your use, I have had to scale each datum of published figures which I thought to be salient. Excluding human error, the scaled data will be accurate to $\pm 5\%$. When all the conditions of a reported test were not stated, I have included the assumptions made in all data interpretation.

TOTAL "PRACTICAL" BULK CAPACITY $-$Torr$-\mathcal{L}/$g			
NEG	GAS		
	CO	N2	H2
St 101®	2.2	2.2	20.8
St 707®	9.0	9.0	20.0

Table 4.2.1. Total "practical" irreversible capacities of the "101" and "707" NEGs for N2 and CO, and maximum reversible H2 capacities prior to the onset of embrittlement problems.

Hydrogen Pumping

Unlike the gases noted in the prior section, the pumping of H2 is completely reversible in *101* and *707* NEGs. As noted in the previous chapters, H2 is pumped by first dissociating into an atomic state on the surface of the getter and diffusion at RT, thereafter, into the bulk of the material. We noted that sputter-ion pumps could be activated to have high pumping speeds for H2, if the surfaces were first *scrubbed* by sputtering with a noble gas. This scrubbing had the effect of sputtering away oxides and nitrides on the cathode surface which occupied hydrogen dissociation energy sites. These oxides also act as a barrier to the diffusion of hydrogen, though this latter effect is of less importance, as was noted in Chapter 3. Therefore, if the surface of the NEG has been activated so as to create fresh metallic Zr, H2 is readily dissociated and diffuses into the bulk of the material.

In fact, NEGs have such high capacities for H2, if certain H2 pumping limits are exceeded, the sintered powder material will become brittle and small pieces of the NEG will spontaneously *pop* off the surface of the substrate. This effect was noted by Malinowski, when conducting tests with the *707* material.[10] In a figure shown by Knize, Cecchi and Dylla, they indicate that this embrittlement threshold occurs after pumping ~18.5 Torr-$\mathcal{L}$ H2 per gram of *101* material.[22] Similar embrittlement data (*i.e.*, ~20 Torr-$\mathcal{L}$ H2/g) are reported for both the *101* and *707* materials in product literature distributed by SAES and in several publications by SAES staff.[36] In the pumping of gas mixtures, including CO, CO2, O2, H2O and H2, the former four gases eventually *take up* the active surface Zr and cause significant decreases in H2 speed.

Ferrario, et al., reported on the *plugging* effect or passivation of both *101* and *707* NEG materials to H2, as a consequence of the simultaneous pumping of the individual gases CO, CO2 and N2.[8] Results of these tests are shown in Figures 4.2.3 and 4.2.4. In these figures the decrease in H_2 speed is given as a function of the amount of each of the three gases sorbed. Ferrario used modular NEG panels in these tests, similar to that shown in Fig. 4.1.2.

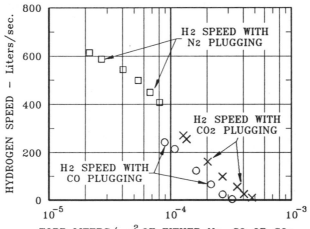

Figure 4.2.3. St 101® NEG plugging for the pumping of H_2 as a function of the amount of N_2, CO and CO_2 pumped.[8]

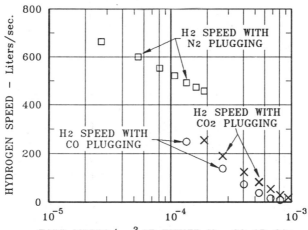

Figure 4.2.4. St 707® NEG plugging for the pumping of H_2 as a function of the amount of N_2, CO and CO_2 pumped.[8]

Speed measurements were conducted with the material at RT and with a system pressure ranging from 10^{-6} to 10^{-5} Torr. Actual gas conductances to all of the NEG material were unknown. Therefore, the total reported speed data are given. However, because of the comparatively low pumping speeds per unit area, reducing these data to units of $\mathcal{L}$/sec-cm^2 would give fairly reasonable results. The amount of *101* NEG material reported in each panel was ~58 g, where the amount of *707* NEG was ~70 g. From this, and knowing the reported approximate densities of *101* and *707* materials in strips, we conclude that the area of NEG material on these panels is ~0.37 m^2.

Though the NEG speed for H$_2$ can be drastically passivated by pumping the other gases, apparently some minute amount of H$_2$ pumping still occurs, and a perfect H$_2$ *dissociation barrier* is never created.[6,7,9,10] This fact may eventually prove fundamentally troublesome to fusion-related NEG applications. As noted in most of the above references, sorption of N$_2$ by these NEGs has a negligible effect on the reduction in H$_2$ speed. In fact, there is some hint of possible gas displacement effects of N$_2$ by, for example, O$_2$,[6] and perhaps CO$_2$ displacement by H$_2$O,[10] though I am unaware of specific studies to date in this area. Systematic studies of possible displacement effects might prove enlightening.

Miller reports that volume changes will occur in pure Zr as a consequence of H$_2$ sorption.[21] For example, sorption of 60 Torr-$\mathcal{L}$ H$_2$/g of Zr will result in an 8.2% volume change in the Zr. This volume change results from a change in the crystalographic structure as a consequence of sorbing the H$_2$. Similar to spalling effects in sputter-ion pump cathodes, stresses occur because of localized volume changes in the matrix. These stresses, due to volume changes, result in the NEG flaking or popping off the surface of the substrate. Because of this, limits are set on the permissible quantities of H$_2$ which may be pumped prior to coating embrittlement. These *limits* are more empirical than theoretical as the volume changes progress with increasing quantities of H$_2$ pumped.[21]

The equilibrium pressure of H$_2$ *vs.* the amount of H$_2$ sorbed by the bulk of both *101* and *707* materials is predictable, and said to follow Sievert's Law,[25] at least at high concentrations and temperatures. That is, the $\log_{10}$ of the equilibrium pressure is:

$$\log P = A + \log \rho^2 + B/T \tag{4.2.1}$$

where P = the equilibrium pressure of H$_2$ in Torr,
         ρ = the amount of H$_2$ sorbed in Torr-$\mathcal{L}$/g of getter,
and T = the temperature, ° K.

The constants A and B are 4.8 and −6116 respectively for the St 707® material and 4.82 and −7280 respectively for the St 101® material.[24] From (4.2.1) we are able to construct a family of curves, called adsorption isotherms, for the two materials. Fig. 4.2.5 shows some theoretical adsorption isotherms for the *707* material for two temperatures.

After pumping, at 450° K, a quantity of H$_2$ approaching some desired limit (*e.g.*, pressure), say point #1 of the curve in Fig. 4.2.4, an auxiliary pump may be attached to the system, and the getter heated to the temperature of point #2 of the figure (*i.e.*, 600° K). At this temperature, the equilibrium pressure is higher. Therefore, if one pumps on the NEG, the hydrogen may be removed from the system. The rate of outgassing will depend on the speed of the appended pump. That is, the pump must be able to achieve pressures lower than that of the equilibrium pressure of the NEG at point #2 or gas will not be removed from the NEG. Auxiliary pumping is maintained until equilibrium point #3 of the curve is established. After this, the auxiliary pump is valved out of the system, and the temperature of the NEG reduced to the original operating temperature, yielding the desired H_2 equilibrium pressure of point #4.

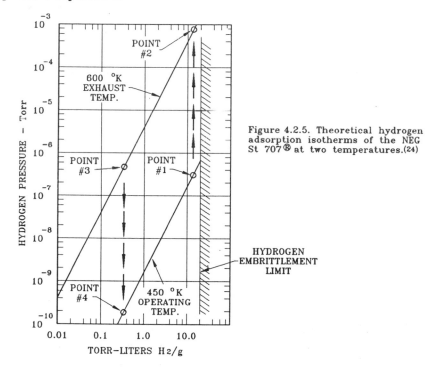

Figure 4.2.5. Theoretical hydrogen adsorption isotherms of the NEG St 707® at two temperatures.[24]

Because of the very high H_1 (and perhaps H_2) diffusivity of the NEG materials, with sufficient pumping speed, most of the H_2 can be desorbed from the NEG in a matter of ~30 minutes. This of course depends on the chosen activation temperature. Boffito, citing work of Ferrario, gives the time/temperature equation for H_2 desorption from both the *101* and *707* materials.[24] Ichimura, *et al.*, report on their own work and that of others relating to the activation energies and heats of absorption for all of the hydrogen isotopes, and NEGs including the *101* and *707* materials.[26]

It is noted by Dushman, at least as it applies to pure Zr, that with the

interstitial diffusion of oxygen, "... the volume of hydrogen sorbed at saturation is decreased by that of a volume equivalent of that of oxygen present."[27] For a given pumping array, this has the effect of the equivalent of
reducing the amount of getter material in (4.2.3) in direct proportion to the
amount of oxygen consumed per gram of material. However, Boffito and his
colleagues showed this increased H_2 equilibrium effect to be of no
consequence in NEGs sorbing, for example, CO far above their rated capacities.[24] However, only high temperature results (*i.e.*, $\gtrsim 475°$ C) were given.
Assuming Sievert's Law is applicable, we may use (4.2.1) to make a simple-
minded estimation of the relative H_2 capacities of the *101* and *707* materials.
Methods for calculating relative capacities are part of the problem set.
However, theoretical equilibrium pressures for three levels of H_2 loading of
the two materials are shown in Table 4.2.2.

Table 4.2.2. Theoretical equilibrium pressures for hydrogen pumping
assuming Sievert's Law is valid for the St 101® and St 707® materials.

TEMP. °K	NEG MATERIAL	HYDROGEN SORBED — Torr−ℓ/g		
		0.318	3.18	11.6
293	St 101®	9.5×10^{-22}	9.5×10^{-20}	1.3×10^{-18}
	St 707®	8.5×10^{-18}	8.5×10^{-16}	1.1×10^{-14}
400	St 101®	4.2×10^{-15}	4.2×10^{-13}	5.6×10^{-12}
	St 707®	3.3×10^{-12}	3.3×10^{-10}	4.4×10^{-9}
600	St 101®	4.9×10^{-9}	4.9×10^{-7}	6.5×10^{-6}
	St 707®	4.1×10^{-7}	4.1×10^{-5}	5.4×10^{-4}
753	St 101®	1.4×10^{-6}	1.4×10^{-4}	1.9×10^{-3}
	St 707®	4.8×10^{-5}	4.8×10^{-3}	6.4×10^{-2}

Hseuh, *et al.*, reported a departure from Sievert's law for H_2 sorption in the
707 material.[19] Hseuh's data were taken at temperatures $< 500°$ C, and
for H_2 concentrations up to ~50% of the embrittlement *limit*. The implication of his findings was that molecular hydrogen diffusion, rather than atomic
hydrogen diffusion, was occurring in the bulk material. I have total confi-
dence in Dr. Hseuh's data, and conclude that this may be a fundamental
discovery relating to NEGs. In Fig. 4.2.6, Hseuh's data are given in an
unorthodox format, showing isosteres of his data along with the theoretical
isosteres of the *707* material. Only data for coverages in excess of one
molecular layer are given. We see that at very high H_2 concentrations, the
data and Sievert's Law begin to converge. But, as pointed out by Hseuh, at
lower concentrations they diverge to the point of suggesting low temperature
equilibrium pressures many orders in magnitude higher than predicted by
Sievert's Law. Conceivably, the presence of C, N and O in the bulk material
might exacerbate this divergence from Sievert's Law at low H_2 concentrations. This is suggested by the data of Boffito, *et al.*[24]

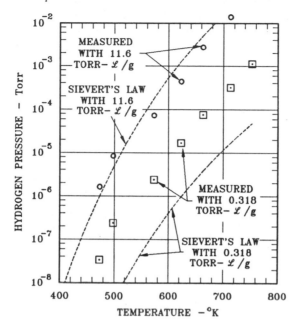

Figure 4.2.6. Comparison of measured and theoretical H₂ isosteres of the St 707® NEG for two concentrations.(19)

The Pumping of Hydrocarbons

It is not evident that CH_4 is synthesized on activation of either the *101* or *707* materials. However, it is evident that in the absence of an ionizing beam or hot filament,[28,29] there is little or no NEG pumping of this gas at RT by either material. This has been verified by numerous studies. Ichimura, et al., did note the presence of several hydrocarbon peaks as a consequence of heating both the *101* and *707* materials.[23] However, I suspect these hydrocarbons stemmed from sample contamination rather than hydrocarbon synthesis by the NEG.

Emerson, Knize, Cecchi and Auciello made pumping speed measurements of *101* material for the gases CH_4, C_2H_6, C_3H_8 and C_4H_{10}.[30] They used a panel coated with *101* material, which they were able to heat by passing current through the substrate material. Therefore, there was no dissociation of the above gases by some filament or heater used to indirectly heat the NEG (*i.e.*, such as is used in the modular pumps). They then developed a theoretical expression for the speed of these gases on the *101* material. Results of their test data are given in the below Table 4.2.3. I have reduced their data to units of Torr-$\mathcal{L}$/sec-cm^2, assuming the active surface area of the reported pump panel to be ~3600 cm^2.

Speed measurements were made at pressures of the order of 10^{-4} Torr using VdP/dt, as indicated by a total pressure gauge, as the measure of speed. This suggests that even though there may have been dissociative by-products created in the process of pumping the larger molecules, values of speed obtained, neglecting ionization cross section differences, were net values of speed for all gases. Similar data were reported by Emerson, et al., in the pumping of cycloalkanes.[31]

Because of the negligible pumping speed of all NEGs for CH$_4$ at reduced temperatures, some form of auxiliary (i.e., combination) pumping is usually used in conjunction with a NEG.[15,20]

Table 4.2.3. Pumping speed of the St 101® NEG for several hydrocarbons as a function of temperature.[30]

TEMP. °C	NEG SPEED – Liters/sec–cm^2			
	CH$_4$	C$_2$H$_6$	C$_3$H$_8$	C$_4$H$_{10}$
600	1.9×10^{-3}	$7.0^* \times 10^{-3}$	1.3×10^{-2}	1.1×10^{-2}
400	2.0×10^{-4}	2.5×10^{-3}	5.6×10^{-3}	6.1×10^{-3}
300	$3.8^* \times 10^{-5}$	1.2×10^{-3}	3.0×10^{-3}	3.9×10^{-3}
200	–	$4.2^* \times 10^{-4}$	1.3×10^{-3}	$2.2^* \times 10^{-3}$

* Data extrapolated.

Pumping Speeds for Gases and Gas Mixtures

It is of value to be able to predict intrinsic NEG speed per unit area vs. the amount of surface sorption at any time. If there is a decay in speed for a given gas on successive regenerations, we must also know the number of regenerations, and how these regenerations relate to the total bulk sorption of gas. With data such as these, one can then design a system using these NEGs. Much of the reported data, though having value in terms of determining surface and bulk capacities, is difficult to relate to intrinsic NEG speed.

Where gases such as CO, CO$_2$ and N$_2$ interfere with the pumping of H$_2$, the converse is not true. That is, H$_2$ does not interfere with the pumping of the other gases. Because of this, the data given by Ferrario for the speeds of the 101 and 707 materials for H$_2$, taken during the simultaneous pumping of each of the gases CO, CO$_2$, and N$_2$, (i.e., Figures 4.2.3 and 4.2.4), may be interpreted as being independent of the presence of H$_2$.[8] These data are shown in Figures 4.2.7 and 4.2.8.

Also, much data exists on the pumping speeds of these materials at elevated temperatures.[9,19] Sciuccati, et al., give speeds of the 707 material at several temperatures, including RT. However, there is some question regarding the amount of sintered NEG per unit area (i.e., noted to be 0.2 g/cm^2).[34] Similar 707 data, reported earlier by Boffito, et al., indicated a surface density of NEG material of ~0.3 g/cm^2.[4]

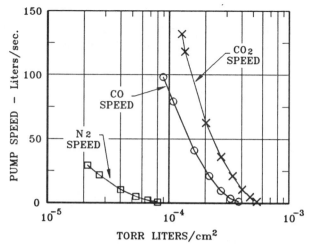

Figure 4.2.7. Speed of the St 101® NEG material for the gases N2, CO and CO2 as a function of the quantities pumped. (8)

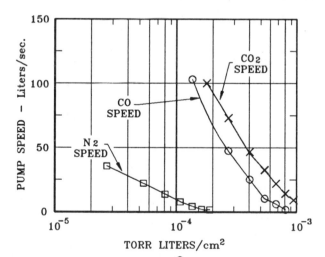

Figure 4.2.8. Speed of the St 707® NEG material for the gases N2, CO and CO2 as a function of the quantities pumped. (8)

The speed data of the *101* material used in the CERN LEP, published by Benvenuti[5,14,15] and Reinhard,[16] were most helpful in establishing intrinsic RT speed values for this material. These data, shown in Fig. 4.2.9, were taken for individual gases rather than a gas mixture. However, when used in conjunction with data in Fig. 4.2.3, they are useful in predicting mixed gas system behavior. In reducing the CERN data into the common units which I have chosen, I have assumed that the density of the sintered *101* material was ~2.3 g/cm^3, and that the thickness of the film was ~90 μm.[5]

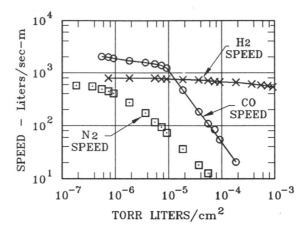

Figure 4.2.9. Speed of the St 101® NEG material for the gases H2, N2 and CO as reported by Benvenuti and Reinhard. (15,16)

The use of *101* has produced very successful results at CERN. During initial machine operation, a great deal of gas is desorbed from the walls of the beam pipe as a consequence of synchrotron radiation. However, the amount is predictable, and decreases in some relation to the product of machine beam current and operating hours. Because of these high initial gas loads, people at CERN predicted that it would be necessary to *recondition* the NEG 4 ±1 times during the first year of operation.(1 6) Benvenuti reported that they planned to *recondition* rather than *activate* the NEG.(1 5) This reconditioning amounted to heating the NEG to a temperature of ~400° C for a few minutes, where the *activation* process, as defined by Benvenuti, requires heating the NEG "... at 700° C typically for about half an hour ..." He reported that this reconditioning cycle was sufficient "... to restore pumping speed." We suspect from results of Fig. 4.2.1, that though the lower temperature *regeneration* might prove of value in H2 outgassing, it would have limited benefit in restoring CO and CO2 speeds.

Gröbner, *et al.*, indicate that during the first year of operation it was necessary to "regenerate" the LEP NEGs three times; two times within the machine operating time interval, and once at the conclusion of the first year of LEP operation.(1 7,1 8) The temperature of reconditioning, though not given, was implied to be in excess of 450° C. After the first two regeneration cycles, they indicated that "... the pumping speed for CO and CO2 had decreased from 500 ℓ/s/m to approximately 70 ℓ/s/m." The latter speed was observed after pumping ~0.12 Torr-$\mathcal{L}$ CO/m and 0.13 Torr-$\mathcal{L}$ CO2/m. Using my common units, these two speeds correspond to ~0.93 $\mathcal{L}$/sec-cm² and ~0.13 $\mathcal{L}$/sec-cm², respectively. The former seem ~×4 lower than the original CO speed data reported by Reinhard and Benvenuti. Normalized CO and

CO_2 loading corresponds to ~0.011 Torr-$\mathcal{L}$ CO/g and ~0.012 Torr-$\mathcal{L}$ CO_2/g of NEG material. Ferrario indicates that the *101* material reaches a point of surface saturation after pumping ~0.03 Torr-$\mathcal{L}$ CO/g and ~0.04 Torr-$\mathcal{L}$ CO_2/g.[8] However, in this case, speeds for the two gases approached zero, rather than the finite minimum speed needed to afford reasonable beam life-times in the LEP collider.

Regarding exposure of the *101* material to atmospheric pressures of various gases, Reinhard reports that "After 5 air exposures, the pumping speed of the reactivated NEG starts decreasing, and after 30 exposures to air, it is only about 30% of the initial value."[16] Regarding N_2, Reinhard reports that "... 30 ventings would result in a pumping speed reduction of 10% only." These results make evident the advantages of venting the "101" NEG material to N_2.

Results of these air exposures provide some valuable insight into the total capacity of the NEG for gases such as O_2, CO, CO_2, etc. That is, assume that each time the NEG is exposed to air, a layer of ZrO_2 is formed on the surface. On subsequent pumpdown, the only way the ZrO_2 can be removed is through activation. That is, the getter must be heated and the ZrO_2 reduced to metallic Zr through O_1 diffusion into the bulk of the getter. Assume the *101* material has an effective surface area of ~$\times 100$ of that of the geometric area.[15] On the initial exposure to air, the total amount of O_2 sorbed will be ~7×10^{16} molecules/cm^2.[32] As a rough approximation, assume that the O_2 speed of the NEG, immediately after activation, is directly proportional to the number of available fresh Zr sites, and that the number of sites decays linearly with each exposure. This being the case, if after 30 air exposures the NEG speed is ~$\times 0.3$ that of its original speed, then the average amount of O_2 sorbed as a consequence of each air exposure is ~4.5×10^{16} O_2 molecules/cm^2 of effective surface area. Therefore, the total amount of O_2 sorbed after 30 air exposures is ~1.35×10^{18} molecules per cm^2 of effective area. With a density of ~2.3 g/cm^3 and a 90 μm thickness of NEG, we conclude that the speed of the NEG was reduced to 30% of its original value as a consequence of pumping ~2.0 Torr-$\mathcal{L}$/g. These calculations are probably conservative. For example, Halama and Guo have shown, using *707* material, that even after cycling the material to air as much as 68 times, and activating between each cycle, they were able to achieve repeated surface coverages of >60% for CO and CO_2, and still observe measurable speeds for these two gases (*i.e.*, see Fig. 4.2.10, below).[33] Therefore, it is possible the *101* NEG may have sorbed as much as ~3.0 Torr-$\mathcal{L}$/g as a consequence of 30 air exposures. These results are in reasonable agreement with *101* data of Table 4.2.1.

The most definitive evaluation of the *707* material to date has been done by Halama and Guo.[33] Their interest in this study was to determine if the *707* material also possibly had applications in electron storage rings and accelerators. Because of this, they conducted speed measurements for this NEG material, using a gas mixture comprising 50% H_2, 35% CO and 15% CO_2. Speed measurements for these individual gases were also conducted. The pump comprised a strip of 3.0 cm wide, 2.2 m long NEG (see Fig. 4.1.1) coated on both sides with 707 material. The strip was mounted in an 8.8 cm ϕ, 2.5 m long stainless steel tube. Before each measurement, the entire system was baked at 200° C for 48 hours, and the NEG activated by heating it to a temperature of 400° C - 700° C. However, they reported the optimum activation temperature for this material as being 400° C - 450° C, for a duration of ~30 minutes. Speed and capacity measurements were taken at RT. In normalizing Halama's data, I have assumed a total NEG surface area of 1188 cm^2, a coating thickness of 70 μm, and a density of 707 material of 2.65 g/cm^3. The speeds for the individual gases H_2, CO and CO_2 were comparable to their respective speeds when pumped in a gas mixture. They noted that when pumping H_2, the speed "... remains high (after pumping) up to several Torr-$\mathcal{L}$/m." Results of pumping the above individual gases are shown in Fig. 4.2.10.

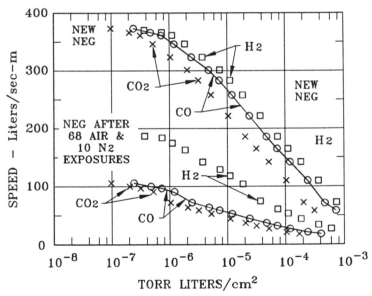

Figure 4.2.10. Speed of the St 707® NEG material for the individual gases H2, CO and CO2 as a new NEG and after exposed to both air and N2, and each time activated. [33]

Halama and Guo showed a spread in speed data for a freshly activated, unused NEG (*i.e.*, equivalent to error bars). This variation in speed ranged approximately ±25% about some average. I have reported the average speeds in Fig. 4.2.10. You will note that there was little difference in the initial speeds for the three gases.

In a second series of tests, Halama and Guo measured the speed of an NEG for the mixed gas, as a function of number of exposures to air and N_2, for a total of 68 air exposures and ten N_2 exposures. This was an outstanding series of experiments, and the results provided significant insight into the properties of the 707 NEG. After each air or N_2 exposure, the NEG was reactivated, the system again baked, and NEG speed measured for the gases H_2, CO and CO_2. Data, showing the initial speeds at the end of the 68th regeneration cycle, are also given in Fig. 4.2.10. Also, after each regeneration, speed measurements were conducted until a quota equivalent of $\sim 9.26 \times 10^{-4}$ Torr-$\mathcal{L}$/cm^2 of H_2, $\sim 6.48 \times 10^{-4}$ Torr-$\mathcal{L}$/cm^2 CO and $\sim 2.79 \times 10^{-4}$ Torr-$\mathcal{L}$/cm^2 CO_2 was pumped. I call this the *used* NEG. Therefore, they were able to establish the effect of both air exposure and the pumping of substantive quantities of the above three gases, on NEG performance.

In a third series of tests, they measured only the initial speeds of a *new* NEG for H_2, CO and CO_2 after exposure to air, activation and a rebake of the system. By conducting these measurements, they were able to distinguish between differences which existed in NEG speed when exposed only to O_2, H_2O and N_2 (*i.e.*, the *new* NEG) *vs.* both air exposures and the pumping of the gases CO and CO_2 (*i.e.*, the *used* NEG). If a major difference existed between results of these two tests, then changes in NEG speed as a consequence of exposure to air would not be sufficient to characterize the irreversible bulk sorption capacity of the NEG.

This proved to be the case, as is shown in Fig. 2.4.11. That is, the speed of the NEG which had pumped the first quota of CO and CO_2 (*used*) was irreversibly reduced by $\sim \times 5.6$ over that of a NEG (*new*) which had simply been exposed to air, regenerated and only briefly pumped H_2, CO and CO_2. Also, the difference in speeds diverged to $\sim \times 20$, on further cycling. The implication in this is that oxygen diffuses much more readily into the NEG, than carbon. One wonders how much CO and CO_2 had to be pumped by the NEG with *no gas loading* prior to establishing somewhat steady-state speeds. That is, would O_2, alone, have had any significant effect in decreasing the NEG speed?

The capacity of the 707 NEG material to sorb gases is remarkable. For example, because the NEG evidences speeds at what correspond to $>60\%$ surface coverage at the end of each test sequence shown in Fig. 4.2.11, we must conclude that on air exposure, the NEG must sorb something equivalent to a molecular layer of gas. At the end of the 68th test cycle of Fig. 4.2.11, the NEG had consumed the equivalent of ~ 2.4 Torr-$\mathcal{L}$ CO/g and ~ 1.0 Torr-$\mathcal{L}$ CO_2/g. Assuming 68 air exposures and a consequential 100% coverage, we calculate that *used* NEG, at the conclusion of the 68th exposure, had also sorbed the equivalent of ~ 2.1 Torr-$\mathcal{L}$/g of O_2. Of course, the *new* NEG pumped the equivalent of ~ 5.5 Torr-$\mathcal{L}$ O_2/g at which time the CO speed was

~24% of the initial speed.

Halama and Guo noted that exposure of the 707 NEG to N_2 had "... only a small negative effect on NEG behavior (speed)." Because of this, it is suggested that systems containing NEGs be vented to dry N_2, rather than air. Benvenuti and Reinhard made similar observations regarding the 101 material.[15,16]

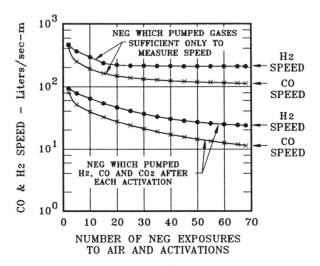

Figure 4.2.11. Speed of an St 707® NEG for the gases H_2 and CO;[33] one NEG merely as a consequence of exposure to air; the second NEG for the same reason and as the result of each time having pumped 0.5, 0.35 and 0.15 Torr−ℓ/m H_2, CO and CO_2 respectively.

One must be careful in interpreting initial, low-coverage, speed data in any baked system having a significant surface area. Also, in the case of the "101" material, if the geometric surface area of the system proves to be ×5 to ×10 larger than the geometric surface area of the NEG, it is conceivable that Aℓ, which is probably sublimed from the NEG during a high temperature activation (e.g., 800° C), might cover the vacuum system surfaces and lead to erroneously high initial NEG speed data for CO and N_2 after each activation (e.g., see problem 12).

4.3 Advantages and Disadvantages of NEGs

NEGs are effective room temperature pumps for all the chemically active gases, excluding hydrocarbons such as CH_4. They do not pump the inert gases. Also, they retain and even have improved sorptive capacities for many gases when operated at elevated temperatures. They do not require the use of high voltage, will function in very hostile radiation fields, and have no preferred operating orientation with respect to gravity. NEGs present no vibration problems to a system. They are neither a source of nor perturbation to magnetic fields when operated at RT. The *707* material is not a source of sublimed material to the vacuum system. However, some investigation is needed regarding possible vaporization of $A\ell$ from the *101* NEG during activation.

On the one hand, they are low throughput devices. On the other hand, when activated, they provide very high speeds per unit area while at extreme high vacuums. Materials are supplied in both sintered tablet form and bonded to metallic surfaces. Metal strips coated with NEG material may be bent and formed into a wide variety of shapes to suit a particular application. Strips of these materials have been successfully used in both modular and distributed pumped forms.

Based on published data, it appears that gases containing carbon cause far greater reductions in speed, on regeneration, than do O_2 and N_2. The reversible pumping of H_2 proves to be an asset, where the embrittlement of these materials, when pumping in excess of ~ 20 Torr-ℓ H_2/g of NEG, presents a risk of problems if not heeded. Based on SAES data sheets, there is some indication that thermal cycling of the material at excessive temperatures for numerous cycles may cause NEG peel-off problems. This has not been reported as a problem in the literature. However, the spontaneous *popping* off of small pieces of NEG from the metallic substrate, due to H_2 embrittlement, has been cited.

The need for high temperature activation requires attention to heat transfer design details and differential expansion considerations, when integrating these NEGs into systems. Activation of the *101* material requires temperatures of the order of 700° C - 800° C, where the *707* material activates at temperatures of the order of 400° C - 450° C. Assuming similar and constant (*i.e.*, *vs.* temperature) spectral emissivities for the two materials, the relative power requirements to activate the two materials differ by $\sim \times 4$. A 3 cm wide strip of the *101* material, in an extruded $A\ell$ chamber, requires ~ 900 W/m power to achieve 700° C.[15] A method for calculating power radiated from a NEG *vs.* temperature is given in problem 10.

Theoretically, it appears that the *101* material has greater H_2 capacity than the *707* material (*e.g.*, see Table 4.2.2). This may be true at very high H_2 concentrations. However, it has been shown that Sievert's Law is not applicable low H_2 concentrations in the *707* material (*i.e.*, UHV applications). This may also be true for the *101* material. No such data were found. The cost of the *707* material is slightly higher than that of the *101* NEG. However, this difference is minimal, and it appears that the *707* material has decided advantages in terms of irreversible bulk capacity and post-activation speed.

Problem Set

1) Assume you wish to construct a pump with St 101® material. The pump is constructed in a fashion similar to that shown in Fig. 4.1.3. Assume that the density of the *101* material is ~2.3 g/cm^3, and that the NEG coating is 70 μm thick. What is the minimum geometric surface area of NEG required to pump 1.2×10^4 Torr-$\mathcal{L}$ H$_2$ while at the same time not exceeding the embrittlement limit of the NEG material?

2) Assume that you have a substrate panel coated on both sides with *101* material, of the density and thickness noted in problem 1, and measuring 2 m $\times$ 2 m. Using the data of Fig. 4.2.9, what would the approximate *zero loading,* panel speeds be for the individual gases CO, N$_2$, and H$_2$?

3) Using data of Figure 4.2.3, what would the approximate speed be for H$_2$ after pumping ~1.6 Torr-$\mathcal{L}$ N$_2$; ~8.0 Torr-$\mathcal{L}$ N$_2$; 8 and 25 Torr-$\mathcal{L}$ CO; 8 and 40 Torr-$\mathcal{L}$ CO$_2$?

4) Using data of Figure 4.2.7, what would the approximate speed be for the individual gases after pumping ~1.6 and 6.4 Torr-$\mathcal{L}$ N$_2$; ~13.4 Torr-$\mathcal{L}$ N$_2$; 16 Torr-$\mathcal{L}$ CO; 16 Torr-$\mathcal{L}$ CO$_2$?

5) Assume the panel in problem 2 is exposed to air 30 times and activated each time thereafter. If the panel initially has a speed of ~1.0 $\mathcal{L}$/sec-cm^2 for CO, what would the total CO speed of the panel be after the 30th activation?

6) Assume a $\times$100 effective surface area and 100% O$_2$ coverage each time the NEG in problem 5 is exposed to air. How much O$_2$ would have been sorbed by the NEG after the 30th activation. Express your answer both in total Torr-$\mathcal{L}$ O$_2$ and in Torr-$\mathcal{L}$/g of NEG.

7) Using data of Halama and Guo, and assuming the use of *707* material with a density of ~2.65 g/cm^3, calculate the answers to problems 2 - 6 for this material, excluding N$_2$. Assume a $\times$70 effective surface area for the *707* material.

8) The power radiated between two parallel plates is given by:

$$W = \sigma A [(1/\epsilon_n) + (1/\epsilon_w) - 1]^{-1} [(T_n)^4 - (T_w)^4],$$

where W = power in Watts radiated per cm^2,
   σ = 5.67×10^{-12} Watts/°K^4-cm^2,
   A = the area of the panel in cm^2,
   ϵ_n = ~0.6, the total spectral emissivity of the NEG,
   ϵ_w = ~0.1, the total spectral emissivity of the wall,
   T_n = the temperature of the NEG in °K,
and T_w = the temperature of the wall in °K.

What would be the power, radiated to the RT walls of a vacuum envelope, from a NEG 30 mm wide and 1.0 m long for the temperatures 400° C, 700° C and 800° C?

Problem Set

9) Assuming the same amount of *101* and *707* material, that Sievert's Law is applicable, and that P_{101} and P_{707} are H_2 equilibrium pressures of the *101* and *707* materials respectively, what is the numerical value of the ratio P_{101}/P_{707} at RT, 200° C and 400° C?

10) Using data of Hseuh and Lanni, speculate on the possible RT equilibrium pressure of the *707* material with a loading of ~10 Torr-$\mathscr{L}$/gm.

11) Assume the panel in problem 2 is located at the center of a very large space simulation chamber. Because of the size of the chamber, we may assume ϵ_w ~1. How much power would be required to activate the NEG at temperatures of 400° C and 700° C?

12) Find a copy of Dushman's book (*i.e.*, Reference 25). Look up the definition of Raoult's Law. Use this to calculate the vapor pressure of Aℓ in the St 101® material as a consequence of heating it to 800° C.

References

1. Giorgi, T.A., Ferrario, B., Storey, B., "An Updated Review of Getters and Gettering", J. Vac. Sci. Technol. $\underline{A3}$(2), 417(1985).
2. Kindl, B., Rabusin, E., "A Large-Surface Nonevaporating Getter", Supplem. Al Nuovo Cimento $\underline{V}$, 36(1967).
3. Della Porta, P., Giorgi, T., Origlio, S., Ricca, F., "Investigations Concerning Bulk Getters from Metals of the IVth A Group and Thorium", Proc. 2nd Int. Vac. Cong. and 8th Nat. AVS Symp., 1961 (Pergamon Press, Inc., New York, 1962), p.229.
4. Boffito, C., Ferrario, B., della Porta, P., Rosai, L., "A Nonevaporable Low Temperature Activatable Getter Material", J. Vac. Sci. Technol. $\underline{18}$(3), 1117(1981).
5. Benvenuti, C., Francia, F., "Room-Temperature Pumping Characteristics of a Zr-Aℓ Nonevaporable Getter for Individual Gases", J. Vac. Sci. Technol. $\underline{A6}$(4), 2528(1988).
6. Cecchi, J.L., LaMarche, P.H., Dylla, H.F., Knoze, R.J., "Technique for In Vacuo Passivation of ZrAℓ Alloy Bulk Getters", J. Vac. Sci. Technol. $\underline{A3}$(3), 487(1985).
7. Emerson, L.C., Mioduszewski, P.K., Simpkins, J.E., "Performance of Zr-Aℓ Getter Pumps Under Transient Load Conditions", J. Vac. Sci. Technol. $\underline{A2}$(4), 1583(1984).
8. Ferrario, B., Boffito, C., Doni, F., Rosai, L., "Zr Based Gettering Alloys for Hydrogen Isotope Handling", Presented at the 13th SOFT, Varese, Italy, 1984. (This paper may be obtained through the authors at SAES.)
9. Knize, R.J., Cecchi, J.L., Dylla, H.F., "Compatibility of the Zr-Al Alloy with a Tokamak Plasma Environment", J. Nuc. Mat. $\underline{103}$ & $\underline{104}$, 539(1981).
10. Malinowski, M.E., "Decreases in Deuterium Pumping by St707® Getter Alloy Caused by Carbon Dioxide Preexposure", J. Vac. Sci. Technol. $\underline{A3}$(3), 483(1985).
11. Audi, M., Dolcino, L., Doni, F., Ferrario, B., "A New Ultrahigh Vacuum Combination Pump", J. Vac. Sci. Technol. $\underline{A5}$(4), 2587(1987).
12. Brouet, M., Girasini, M., Poncet, A., Wolf, A., Hütten, L., Poth, H., Habfast, C., "An Ultra-High Vacuum System for Coolers", CERN Technical Report No. CERN/PS/85-4 (ML), CERN, Geneva, Switzerland 1984.
13. Moenich, J.S., "Nonevaporable Getter for Ion Beam Fusion", J. Vac. Sci. Technol. $\underline{18}$(3), 1114(1981).
14. Benvenuti, C., Decroux, J-C., "A Linear Pump for Conductance Limited Vacuum Systems", Proc. 7th Int. Vac. Cong. and 3rd Int. Conf. on Solid Surfaces, 1977 (R. Dobrozemsky, F. Rüdenauer, F.P. Viebö ck, A. Breth, Postfach 300, A-1082 Vienna, Austria, 1977), p. 85.
15. Benvenuti, C., "A New Pumping Approach for the Large Electron Positron Collider (LEP)", Nucl. Instrum. Methods $\underline{205}$, 391(1983).

References

16. Reinhard, H.P., "The Vacuum System of LEP", Proc. 9th Int. Vac. Cong. and 5th Int. Conf. on Solid Surfaces, 1983 (Asociacion Española del Vacio y sus Aplicaciones, J.L. de Segovia, Editor, 1984), p. 273.

17. Gröbner, O., Laurent, J-M., Mathewson, A.G., "The Behavior of the Synchrotron Radiation Induced Gas Desorption During the First Running Period of LEP", 2nd European Particle Accelerator Conf., 1990 (CERN Technical Report No. CERN-AT-VA-90-10, CERN, Geneva, Switzerland, 1990).

18. Gröbner, O. "The Performance of the Ultra-High Vacuum System of LEP and the Experience Gained During the First Year of Operation", 2nd European Particle Accelerator Conf., 1990 (CERN Technical Report No. CERN-AT-VA-90-09, CERN, Geneva, Switzerland, 1990).

19. Hseuh, H.C., Lanni, C., "Evaluation of Zr-V-Fe Getter Pump for UHV System", J. Vac. Sci. Technol. $\underline{A1}$(2), 1283(1983).

20. Hseuh, H.C., Feigenbaum, I., Manni, M., Stattel, P., Skelton, R., "Ultra-high Vacuum System of the Heavy Ion Transport Line at Brookhaven", IEEE $\underline{NS-32}$(5), 391(1985).

21. Miller, G.L., Metallurgy of the Rarer Metals, 2 Zirconium (Academic Press, Inc., New York, 1954).

22. Knize, R.J., Cecchi, J.L., Dylla, H.F., "Measurement of H_2, D_2 Solubilities in Zr-Aℓ ", J. Vac. Sci. Technol. $\underline{20}$(4), 1135(1982).

23. Ichimura, K., Matsuyama, M., Wantanabe, K., "Alloying Effect on the Activation Processes of Zr-alloy Getters", J. Vac. Sci. Technol. $\underline{A5}$(2), 220(1987).

24. Boffito, C., Ferrario, B., Martelli, D., "Equilibrium Pressures of H_2 and D_2 for Different Getter Materials and the Effect of CO Impurities", J. Vac. Sci. Technol. $\underline{A1}$(2), 1279(1983).

25. Dushman, S., Scientific Foundations of Vacuum Technique (John Wiley and Sons, Inc., New York, 1958), p. 554.

26. Ichimura, K., Matsuyama, M., Watanabe, K., Takeuchi, T., "Absorption/Desorption of Hydrogen Isotopes and Isotopic Waters by Zr-Alloy Getters", J. Vac. Sci. Technol. $\underline{A6}$(4), 2541(1988).

27. Dushman, S., op. cit., p. 584.

28. Della Porta, P., Ferrario, B., "A Magnetless Gauge Appendage Pump Utilizing Non-Evaporable Getter Material", Proc. 4th Int. Vac. Cong., 1968 (Institute of Physics and Physical Society, London, 1968), Part 1, p. 369.

29. Shen, G.L., "The Pumping of Methane by Ionization Assisted Zr/Al Getter Pump", J. Vac. Sci. Technol. $\underline{A5}$(4), 2580(1987).

30. Emerson, L.C., Knize, R.J., Cecchi, J.L., Auciello, O., "Dissociative Pumping of the Alkanes Using Nonevaporable Getters", J. Vac. Sci. Technol. $\underline{A4}$(3), 297(1986).

References

31. Emerson, L.C., Knize, R.J., Cecchi, J.L., "Pumping of Hydrocarbons Using Nonevaporable Getters", J. Vac. Sci. Technol. A5(4), 2584(1987).
32. Redhead, P.A., Hobson, J.P., Kornelsen, E.V., The Physical Basis of Ultrahigh Vacuum (Chapman and Hall, Ltd., London, 1968), p. 41.
33. Halama, H.J., Guo, Y., "Non-Evaporable Getter NEG Investigation at the National Synchrotron Light Source", Presented at the 37th Nat. AVS Symp., Oct. 1990, Toronto (Proceedings to be published in J. Vac. Sci. Technol.).
34. Sciuccati, F., Ferrario, B., Gasparini, G., Rosai, L., "In Situ Pumping with NEG (Non-Evaporable Getters) During Vacuum Processing", Vacuum 38, 765(1988).
35. Lampert, W.V., Rachocki, K.D., Lamartine, B.C., Haas, T.W., "Study of the Activation of Nonevaporable Getters", J. Vac. Sci. Technol. 18(3), 1121(1981).
36. Ferrario, B., Rosai, L., "New Types of Volume Gettering Panels for Vacuum Problems in Plasma Machines", Proc. 7th Int. Vac. Cong. and 3rd Int. Conf. on Solid Surfaces, 1977 (R. Dobrozemsky, F. Rüdenauer, F.P. Viebö ck, A. Breth, Postfach 300, A-1082 Vienna, Austria, 1977), p. 359.
37. Pace, A., Poncet, A., "Monte Carlo Simulations of Molecular Gas Flow; Some Applications in Accelerator Vacuum Technology Using a Versatile Personal Computer", Vacuum 41(7-9), 1910(1990).

CHAPTER 5

CRYOPUMPING

5.0 Introduction

Cryopumping is the most elementary of all of the forms of capture pumping. By now you have learned most of the fundamentals needed to understand how cryopumps work. In Chap. 1, you learned the definitions of pressure and temperature; additionally, the concepts of vapor pressure, conductance, speed, throughput, and the *counting of molecules* were all introduced. In Chap. 3, I introduced the concept of the sticking coefficient, α, of a gas on a chemisorbing surface. It was therein noted that there was a counterpart in cryopumping. This counterpart is called a *capture coefficient*, and is sometimes denoted by the symbol c. It has the exact same meaning as α. Therefore, I will stick with α. Assuming that you have mastered these definitions and concepts, excluding hardware considerations, there is little new to be learned in achieving an understanding of cryopumping.

From a physics standpoint, there are two simple effects which describe the mechanisms of cryopumping. One is defined as *cryocondensation* pumping and the second as *cryosorption* pumping. These two effects are respectively related to two physical phenomena: one is the *vapor pressure* of gases as a function of temperature; the second is called an *adsorption isotherm* of a gas. With a few minor exceptions, these are the only new physical concepts introduced in this chapter. One exception deals with the concept of *thermal transpiration*; a second exception deals with the phenomenon of *cryotrapping*; a third exception deals with the use of *molecular sieve* materials in cryosorption pumping.

I will first discuss the above concepts, and then how they relate to the more practical considerations of cryopumps such as their speed, capacity, and blank-off pressures. After this, I will discuss the various forms of cryopumps, including cryosorption roughing pumps (*sorption pumps*) which operate at LN_2 (*i.e.*, liquid nitrogen) temperatures, the elements of liquid He cryopumps (*LHe pumps*) and then conclude with a discussion of closed-loop, gaseous helium cryopumps (*CLGHe pumps*). References given in the literature on various aspects of cryopumping number in the thousands. Those references given in the tables and text herein are given as representative work. They were selected on this basis and are few in number.

251

5.1 Cryocondensation vs. Cryosorption

Cryopumping is merely the temporary storage of gases on cooled surfaces. There are two defined extents, or measures, of this pumping. One measure is called *cryocondensation*; the second is called *cryosorption*. These are merely words used to distinguish between two extents of pumping.

> *Cryocondensation is the pumping of gas to the extent that comparatively thick layers of gas condensate are accumulated on the cooled surface.*

> *Cryosorption is the pumping of gas under circumstances where the extent pumped is limited by the population of gas on the surface.*

Both cryocondensation and cryosorption pumping exists in part because of the comparatively weak van der Waals' (*i.e.*, *dispersion*) forces at and near the pumping surface. At reduced temperatures, the thermal energy retained by a molecule (*i.e.*, translational, rotational, vibrational, *etc.*) is less. In fact, the energy retained by a gas molecule as a function of temperature is one of the methods used to arbitrarily define temperature (*e.g.*, Section 1.4 of Chap. 1).

A molecule becomes *thermalized* when impinging on a cold surface. That is, it gives up heat energy to the surface and in doing so is itself cooled to reduced temperatures. The amount of heat energy which a molecule or atom gives up with each collision with a cooled surface is called the *accommodation coefficient*. It will vary from 0.01 to 1.0, depending on the gas and surface and their respective temperatures.[1] It is generally believed that for the same van der Waals' force, the likelihood of a molecule being captured on a surface increases both with the decreasing temperature of the gas being sorbed and the surface temperature.[2,3] Exceptions have been reported regarding the temperature dependency of the gas being pumped.[4,5]

Intuitively, we would speculate that the heavier gas molecules, when giving up sufficient thermal energy to be cryopumped, because of their very mass would be less readily *bumped* off the surface by another gas molecule once initially sticking to that surface. Therefore, heavier molecules would tend to be more readily cryosorbed on surfaces, including *ice* of the same and different gas species. Conversely, the lighter the gas molecule (*e.g.*, He, H_2 and Ne), the more difficult it is to cryopump at a given temperature. With some notable exceptions, this proves to be the case. The exceptions to this mass-dependency occur when a gas molecule has a measure of *polarity*, as in the case of H_2O. In this case, other forces come into play, and the *polar* molecules are more readily pumped at comparatively higher temperatures.

Of course, there are degrees in molecular polarity. Being *polar* means that the gas molecule, because of its average internal electrical charge distribution, distorts the electrical charge distribution of surface molecules (*e.g.*, the *ice* or metal substrate). These charge distortions cause an attractive force between the molecule and the surface, which is additive with the van der Waals' forces.

Smooth, round atoms having filled electron shells are nonpolar. This is the case with all of the inert gases. Therefore, primarily van der Waals' forces cause these gases to stick to surfaces. This qualitatively explains why at the same temperatures H_2 is more readily cryopumped than He, CH_4 more readily than Ar, and H_2O more readily than Ne; the former gases are the lighter, yet more *polar*, gases.

A more quantitative measure of the relative strengths of these forces between molecules of the same and other species is found from the boiling points (and critical temperatures) of the respective species. For example, Dushman, reporting on work of others, gives the near-RT (*i.e.*, room temperature) capacity of a carbon (charcoal) surface for different gases as a function of their respective boiling points.[6] These data, in modified form, are given in Table 5.1.1. We see from the table that to first order, the lighter the molecule, the lower the possible surface coverage of the gas at a given temperature. More correctly, the lower the boiling temperature (or critical temperature) of the gas, the lower the possible surface coverage of the gas at a given temperature. Note that the equilibrium pressure of the gases in Table 5.1.1 was 750 Torr, and the *cryosorbing* surface was maintained at ~288° K. Nevertheless, this correlation between possible surface coverage and boiling points of gases persists even at very low pressures and temperatures.

Table 5.1.1. The extent of coverage of gas molecules on a 288°K charcoal surface, vs. the respective boiling points of the gas.[6]

GAS	Torr-$\mathcal{l}$ /gram	MOLECULES[1] /cm^2	COVERAGE σ_m	BOILING POINT °K
COCl$_2$	330.0	1.3 x 10^{15}	–	281.3
SO$_2$	285.0	1.1 x 10^{15}	–	263.0
CH$_3$Cl	207.8	7.9 x 10^{14}	–	249.3
NH$_3$	135.8	5.2 x 10^{14}	1.140	239.7
H$_2$S	74.3	2.8 x 10^{14}	–	211.2
HCl	54.0	2.1 x 10^{14}	–	189.3
N$_2$O	40.5	1.5 x 10^{14}	0.309	183.5
C$_2$H$_2$	36.8	1.4 x 10^{14}	0.176	184.5
CO$_2$	36.0	1.4 x 10^{14}	0.258	194.5
CH$_4$	12.0	4.6 x 10^{13}	0.088	111.5
CO	6.8	2.6 x 10^{13}	0.032	81.0
O$_2$	6.0	2.3 x 10^{13}	0.026	90.0
N$_2$	6.0	2.3 x 10^{13}	0.028	77.3
H$_2$	3.8	1.4 x 10^{13}	0.009	20.1

1) Assumed ~900 m^2/g of charcoal.

5.1.1 Cryocondensation Pumping

We observe the effects of cryocondensation frequently in our everyday lives. For example, from time to time during the winter I find that overnight a thick layer of ice has accumulated on the windshield of my automobile. This is simply the cryocondensation of H_2O on the window. Water accumulates on the surface of a cocktail glass. This too is the cryocondensation of H_2O. The limiting consideration in the measure of the thickness of the ice on the car window is the surface temperature of the ice on the window and the water vapor content in the atmosphere. As long as there is a source of H_2O in the atmosphere and the ice on the windshield is sufficiently cold, ice will continue to build up on the windshield through cryocondensation.

Similarly, assume the vessel represented in Fig. 5.1.1 is vacuum baked so as to remove the internal surface gases. After this, assume that it is enclosed in a special refrigerator which is used to vary the temperature of the vessel walls to a uniform value, and which is set at 30° K. Assume that we valve out the pump and inject sufficient RT N_2 gas into the volume to build a thick layer of N_2 *ice* on the vessel walls. On turning off the gas source, what will the eventual equilibrium N_2 pressure be once it has thermalized to the wall temperature? It is the vapor pressure of N_2 at 30° K.

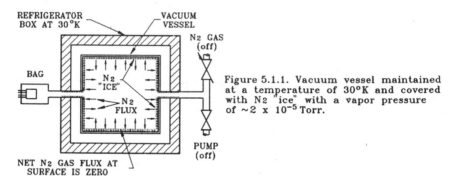

Figure 5.1.1. Vacuum vessel maintained at a temperature of 30°K and covered with N_2 "ice" with a vapor pressure of ~2 x 10^{-5} Torr.

Referring to Honig and Hook's data (*i.e.*, Fig. 1.7.3 of Chap. 1),[7] we note that the N_2 pressure in the volume would be ~2 x 10^{-5} Torr. As long as the temperature of the N_2 *ice* on the inner surfaces of the volume remains at 30° K, the pressure in the volume will be ~2 x 10^{-5} Torr. If we now change the temperature of the vessel to 25° K, the N_2 pressure will start to decrease and eventually equilibrate at a value of ~10^{-7} Torr, the vapor pressure of N_2 at 25° K. Reducing the temperature to 20° K would result in an eventual equilibrium pressure of ~10^{-11} Torr. In this case, to be at *equilibrium* means that the rate at which gas is departing from and impinging on (and sticking to) the N_2 ice is constant and the same. Therefore, there is a net gas flux of *zero* at the surfaces. In fact, if we know the pressure, we can calculate this flux using (1.9.6) of Chap. 1. Analytical expressions for the vapor pressure of all gases may be derived from the famous Clausius-Clapeyron equation.

5.1.2 The Clausius-Clapeyron Equation

The saturation vapor pressure of a gas may be analytically derived as follows.[8] The origin of this derivation is the Clausius-Clapeyron equation which describes the slope of the saturation vapor pressure curve as follows:

$$\frac{dT}{dP} = \frac{v_g - v_f}{s_g - s_f} = \frac{T(v_g - v_f)}{h_{fg}}, \qquad (5.1.1)$$

where T = the temperature of the surface,
         P = the pressure over the liquid,
         v_g = the saturated-vapor specific volume,
         v_f = the saturated-liquid specific volume,
         s_g = the saturated-vapor entropy,
         s_f = the saturated-liquid entropy,
and h_{fg} = the heat of vaporization.

Note that $s_g - s_f \triangleq h_{fg} / T$, and that $v_g \triangleq \Re T / P$. If we assume that $v_g \gg v_f$, and that h_{fg} may be expressed as some linear function of the temperature (i.e., $h_{fg} = k_1 + k_2 T$, where k_1 and k_2 are some constants), then (5.1.1) takes the form:

$$\frac{dT}{dP} = \frac{\Re T / P}{(k_1 + k_2 T)} = \frac{\Re T^2 / P}{(k_1 + k_2 T)}. \qquad (5.1.2)$$

On separating the variables and integrating both sides, we derive the simple algebraic equation for the pressure of a gas as a function of temperature:

$$\log_{10} P = A - B T^{-1} + C \log_{10} T \qquad (5.1.3)$$

where P = the pressure in Torr,
         T = the temperature in $^\circ$ K,
         A = $C_3 / 2.303$,
         B = $k_1 / 2.303 \Re$,
         C = $k_2 / 2.303 \Re$,
and C_3 = the combination of the constants of integration.

The values of A, B and C differ for the specific gas and units of measurement. Also, when reading the literature be aware that one person's A may be another person's B, etc.[7,9-12] Also, h_{fg} might have initially been represented as an infinite series in T, and lead to an expression similar to (5.1.3), but containing many more higher order terms in T. Table 5.1.2 gives values for A, B and C for some of the common gases.

Table 5.1.2. Clausius–Clapeyron constants for H2 and D2 from Lee,[12] and others from Honig & Hook,[7] and Redhead, *et al.*[14]

GAS	A	B	C	ΔH	ΔHv
H2	3.881	42.730	0.50	–	196
D2	4.234	62.970	0.50	–	289
CH4	7.123	482.871	–	2208.7	1955
H2O	10.285	2637.006	–	12061.7	–
Ne	6.889	109.270	–	499.8	431
N2	8.362	387.428	–	1772.1	1333
CO	8.732	445.831	–	2039.2	1444
O2	9.082	480.871	–	2199.5	1630
Ar	7.703	419.566		1919.1	1558
CO2	9.734	1349.809	–	6174.0	4041
Kr	7.797	581.967	–	2661.9	2158
Xe	7.775	801.641	–	3666.7	3021

The data for H_2 and D_2 were taken from the work of Lee.[12] Data for the remainder of the gases were calculated by "force-fitting" data of Honig and Hook into (5.1.3),[7] while at the same time assuming $C \sim 0$. The value of ΔH in this table has different interpretations depending on the temperature at which the data are taken.[13] For example, if the data are taken on the *liquid side* of the melting point, B would be proportional to ΔH_s, the *heat of vaporization*, and $\Delta H = \Delta H_s$. If the data were taken on the *solid side* of the melting point, B would be proportional to ΔH_v, the *heat of sublimation*, and $\Delta H = \Delta H_v$. In either instance at constant temperature,

$$\Delta H_s = \Delta H_v + \Delta H_f, \tag{5.1.4}$$

where ΔH_v = the *heat of vaporization* (*i.e.*, liquid to gas),
  ΔH_s = the *heat of sublimation* (*i.e.*, solid to gas),
and ΔH_f = the *heat of fusion* or melting (*i.e.*, solid to liquid).

In applications involving cryocondensation pumping, the heat of sublimation is usually applicable. In cryosorption pumping, the heat of vaporization is usually thought to apply. The term *heat of sorption* is often used synonymously with heat of vaporization. Data for the heat of vaporization of several gases, taken from Redhead, *et al.*, are given to the far right of Table 5.1.2.[14] There are obvious differences in the values of ΔH and ΔH_v appearing in this table. These in part stem from the approximations used in arriving at (5.1.3), as well as the ΔH_f component in (5.1.4).

Condensation at Higher Pressures

Practical cryocondensation pumping involves the pumping of gases at pressures greater than their respective equilibrium vapor pressures. In the previous physical example, we might have continuously introduced gas into the cooled volume, but at a rate which did not appreciably alter the temperature of the N_2 *ice*. The introduced gas might have a temperature higher than that of the *ice*. The continuously introduced gas causes there to be an imbalance in the surface flux. When this flux imbalance exists, there is a net pumping by the N_2 ice. That is, N_2 gas is cryocondensing on the N_2 ice. This is illustrated in Fig. 5.1.2. As long as the induced pressure is higher than this vapor pressure, there will be a net flux of gas into the ice. (However, we will shortly see that there is at least one exception to this.) This is simply surface pumping by cryocondensation. Also, as long as the surface temperature of the *ice* remains constant, the rate at which gas is pumped when cryocondensation exists does not vary with the amount of gas pumped. That is, there is no surface gas population density dependency with cryocondensation pumping, only a temperature-species dependency.

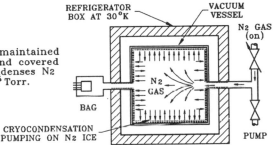

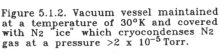

Figure 5.1.2. Vacuum vessel maintained at a temperature of 30°K and covered with N_2 "ice" which cryocondenses N_2 gas at a pressure $>2 \times 10^{-5}$ Torr.

The sticking coefficient, α, of a gas for cryosorption or cryocondensation is merely the ratio of the average number of molecules which stick, when impinging on a cooled surface, divided by the total number of molecules impinging per unit area. For reasons to be noted, it proves necessary to separate the surface gas fluxes due to the vapor pressure or some other equilibrium pressure, from those fluxes due to induced gases at higher temperatures. Therefore, the sticking coefficient disregards effects of vapor or equilibrium pressure and is expressed by:

$$\alpha \quad = \quad (\nu_i - \nu_v)/\nu_i \tag{5.1.5}$$

where ν_i = the impinging flux rate per cm^2,
and ν_v = the departing flux rate per cm^2.

The speed of a cryocondensation pump is obviously area limited. For example, assume that we continuously introduce gas into the vessel so that the flux impinging on the N_2 *ice* is x1000 greater than the introduced flux of gas departing from the ice. This corresponds to a sticking coefficient for N_2 of $\alpha \sim 0.999$; *i.e.*, ~ 1.0. That is, on average 99.9% of the introduced gas impinging on the surface sticks. If we were to reduce the temperature of the N_2 ice so that the impinging flux of gas sticking to the surface was now 99.99%, the new sticking coefficient would be $\alpha \sim 0.9999$; *i.e.*, ~ 1.0. For the reduced operating temperature, there is essentially no perceivable change in the speed of the pumping surface (*i.e.*, see (5.1.8)). Therefore, if we wanted to increase the speed of the pump, further reduction in the *ice* temperature would be of no avail. We could only increase the speed of the pump by increasing the geometric area of the pumping surface. This is called *surface conductance limited* pumping.

If the *ice* is held at a constant 20° K, the rate at which N_2 leaves the ice is equivalent to that of its equilibrium vapor pressure *plus* a second component of gas flux. This second component is that portion of the continuously introduced gas which when impinging on the surface, rather than sticking, is reflected back from the surface.

5.1.3 Thermal Transpiration

An interesting phenomenon occurs when two chambers are joined together through a given conductance and the chambers are maintained at different temperatures. The pressure of gas in each chamber will be different even when the net flux of gas passing between the two chambers is *zero*. This phenomenon is called *thermal transpiration* and is variously coined in the literature as the *thermomolecular effect*.[15-17] As you will see, this has applications in the design and evaluation of cryopumps.

We must often indirectly deduce the behavior of gases being pumped on cold surfaces in one chamber, through the direct measurement of throughput, Q, and pressure, P, observed in a second chamber maintained at a different temperature. For example, assume that the two vacuum chambers shown in Fig. 5.1.3 are maintained at different temperatures, and that they are separated by an aperture of area A_a. Assume that one chamber is extremely cold and maintained at temperature T_2, and the second is at $T_1 = 293°$ K (*i.e.*, RT). Assume that the gas in the RT chamber comprises air at a pressure of 10^{-6} Torr, and that there are no sources or sinks (*i.e.*, pumps) of gas in either chamber. If the pressure in both chambers is constant, what is the pressure in the cold chamber? If the pressure is neither rising nor falling in either chamber and there are no sources or sinks of gas, the net flux of gas passing through the aperture of area A_a must be *zero*. Using (1.9.6) of Chap. 1, we can express the flux of gas leaving the warm chamber as:

$$\nu_1 = A_a \, 3.51 \times 10^{22} \, P_1 \, / \, (MT_1)^{1/2}, \qquad (5.1.6)$$

where A_a = the area of the aperture in cm^2,
          P_1 = the pressure in the warm chamber in Torr,
          M = the *molecular weight* of air (*i.e.*, ~ 28.7),[22]
          T_1 = the temperature of the warm chamber in ° K.

An identical equation may be constructed for ν_2, the flux leaving the cold chamber and entering the warm chamber. In that $\nu_1 = \nu_2$, we arrive at the relationship:

$$P_1 (T_1)^{-\frac{1}{2}} = P_2 (T_2)^{-\frac{1}{2}}, \tag{5.1.7}$$

or $\qquad P_2 \qquad = P_1 (T_2 / T_1)^{\frac{1}{2}}.$

We could have just as easily invoked (1.12.3) and (1.12.7) of Chap. 1, and arrived at the same result. In this case though, we must also take care to note that the conductance separating the two chambers differs depending on the origin of the gas (*e.g.*, see problem 2).

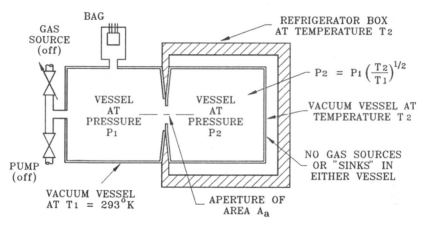

Figure 5.1.3. Thermal transpiration effect.

Now assume that there is some sort of mysterious pumping process going on in the cold chamber. We need not understand the process for now. We will only assume that the surfaces in the cold chamber are *sticky* to the gas being introduced in the warm chamber and passing through the aperture. Using an equation identical to (3.2.1), we can express the pumping speed of the surface in the cold chamber as:

$$S_s = \alpha C_s, \tag{5.1.8}$$

where $\quad \alpha \quad$ = the sticking coefficient of the cold surface,
and $\quad C_s \quad$ = the conductance at the surface (*i.e.*, dependency on molecular weight and temperature of the gas).

Assume that gas is introduced through a variable leak into the warm chamber at a rate Q. Clearly, $Q = S_s P_2 = \alpha C_s P_2$. From this result, and by invoking conservation of mass, the following is found:

$$Q \quad = \quad C_1 P_1 - C_2 P_2 = \alpha C_s P_2, \tag{5.1.9}$$

where $\quad C_1 \quad = k A_a (T_1 / M)^{1/2},$
$\qquad \quad C_2 \quad = k A_a (T_2 / M)^{1/2},$
$\qquad \quad C_s \quad = k A_s (T_2 / M)^{1/2},$
and $\quad\; k \qquad$ = a constant of proportionality (see section 1.12.1).

From these simple equations we may derive S_{eff}, the effective speed produced in the warm chamber due to the pumping of the cold chamber, assuming we know the value of T_1, α, A_s and A_a (see problem 3); α, assuming we know Q and P_1 (see problem 4); or α, assuming we know P_1 and P_2 (see problem 5). As a word of caution, one should be careful in indiscriminately invoking (5.1.7) as an identity. It is only an identity when the net flux of gas through the separating conductance is *zero*. That is, (5.1.7) is true only when there is no net flow of gas. One frequently encounters this error in the literature. From this simple model, one finds the speed delivered to the warm chamber due to pumping by the cold chamber as:

$$S_{eff} \quad = \quad \frac{\alpha C_1 A_s}{(A_a + \alpha A_s)} . \tag{5.1.10}$$

Note that the temperature of the cold chamber does not appear in (5.1.10), though it is implicit in the value of the sticking coefficient, α, of the gas on the cold surface. Therefore, we see that α is all important in predicting the behavior of cryopumps. There are numerous excellent publications on the subject of the sticking coefficients of gases on cold surfaces, and some I feel are *classics*. One of these includes the work of Dawson and Haygood.[18] This work contains a comprehensive listing of sticking coefficients of the gases N_2, CO, O_2, Ar, CO_2 and N_2O, at temperatures ranging from 77° K to 400° K and on cryofrosts of the same gas at temperatures ranging from 10° K to 77° K, and similar data for 13 other RT gases on 77° K surfaces. It also contains a most lucid treatment of elementary kinetic theory in the definition and interpretation of the sticking coefficient of a cryosurface in one chamber adjoined by a second at RT.

In another early *classic*, Levenson used a quartz crystal microbalance to measure the sticking coefficients of molecular beams of Ar, CO_2, Kr and Xe as a function of gas and substrate temperature.[3] The work of Brown, Trayer and Busby is also frequently cited in publications dealing with the sticking coefficients of gases on *ices* of the same species (*i.e.*, cryocondensation pumping).[4] Through use of collimated molecular beams, they measured the sticking coefficients of the gases N_2, Ar and CO_2 as a function of gas temperature, cryofrost temperature and angle of incidence of the gas beam with the surface. The results indicated only slight variations in α for the three gases over gas temperatures ranging from 300° K to 1400° K. The implication of this finding is that cryopumps may be used to cryopump very hot gases. This is not a surprising result (*e.g.*, see problem 6). The marked dependency of α on ν_j, which they noted for 300° K CO_2 gas on CO_2 *ice* of varying temperatures, is shown in Fig. 5.1.4.

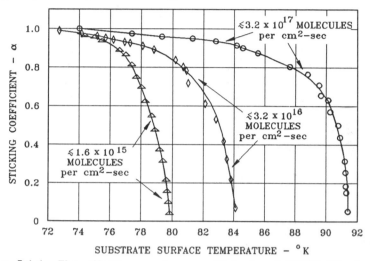

Figure 5.1.4. The effective capture coefficients of 300°K CO_2 beams of varying intensities and impinging on CO_2 "cryofrosts" of varying temperatures, as reported by Brown, Trayer and Busby.[4]

Results of this work (*i.e.*, Fig. 5.1.4) proved troubling to many readers, as there is a strong temptation to lend significance to the *zero* intercept of the various flux intensities. That is, we would intuitively assume that the sticking coefficient would vanish at that flux intensity equivalent to the vapor pressure of the cryofrost. Brown, *et al.*, reported that for CO_2, Ar and N_2 this was not the case. That is, the sticking coefficient of these gases vanished at surface flux intensities far lower than that equivalent to the vapor pressure.

I have modified their data, in Fig. 5.1.4, to reflect beam intensities, as they reported on total flux. The inequality signs for the flux values result from my assumption of no beam divergence at the cryosurface. We can qualitatively state that a cryopumping surface with a sticking coefficient of $\gtrsim 0.5$, whatever the conditions, would pump a significant quantity of gas. With this criterion, if we assign a pressure equivalency to the beam flux of $\leq 1.6 \times 10^{15}$ CO_2 molecules/sec-cm^2, and vapor pressure equivalency of the *ice* at the point of $\alpha \sim 0.5$ shown in Fig. 5.1.4, we determine that, at least for CO_2, effective cryocondensation pumping occurs when the *pump operating pressure* is $\sim \times 100$ that of the vapor pressure of the *ice*. Note that as the CO_2 beam intensity was increased to 3.2×10^{17} molecules/sec-cm^2, this would require a *pump operating pressure* margin of $\sim \times 2000$ that of the vapor pressure.

In molecular beam experiments similar to those of Brown, *et al.*, Bentley and Hands reported similar findings for the sticking coefficient of CO_2 on cryofrosts of same, but disagreement in results for the gases N_2 and Ar.[5,19] For the latter gases, they concluded that α vanishes at or very near that equivalent to the vapor pressure of the cryofrost. They proposed a simple model for the anomalous cryocondensation behavior of CO_2 based on the tendency of CO_2 to nucleate and to stick to other surface CO_2 molecules more readily than to a *clean* cryopanel. Bentley and Hands noted that the effect became much less pronounced once several monolayers of gas had accumulated on a cryopanel.[19] This phenomenon would result in an hysteresis effect in sticking coefficient data of the nature shown in Fig. 5.1.4, as α would depend on the thickness of the cryofrost. This strange phenomenon seems unique to CO_2, though it may have been observed by Hobson with the gas Xe,[20] and Chubb, *et al.*, with H_2 condensation on 3.7° K surfaces.[87] It is referred to as a *super saturation* effect.

Büttner measured the cryocondensation speeds of H_2, D_2 and a 1:5 mixture of $H_2:D_2$ on a surface cooled to 4.2° K (*i.e.*, cryofrosts of the gases). He determined that the speed for H_2 decreased to *zero* very near the equivalent of the vapor pressure of the H_2.[21] Chubb and Pollard reported near-unity capture coefficients for RT H_2 on *ice* of H_2 at a temperature of 3.6 - 3.9° K, and for sorbed H_2 fluxes of the order of ×15 to ×160 that of the surface efflux due to the vapor pressure of H_2.[2] They reported that "... there was no dependence of (the) sticking coefficient upon gas flow rate ..." for the indicated fluxes.

5.1.4. Adsorption Isotherms

The second form of cryopumping involves cryosorption which in turn relates to an *adsorption isotherm*. You frequently encounter this term and the terms *isostere* and *isobar* in the literature. Mathematical definitions of these terms are as follows:

an *isotherm* is defined as $\sigma = f(P,T_c)$ at constant T,
an *isostere* is defined as $P = f(T,\sigma_c)$ at constant σ,
an *isobar* is defined as $\sigma = f(T,P_c)$ at constant P,

where σ = the density of molecules of gas on a surface per cm^2,
        P = the equilibrium pressure in the system,
and T = the system temperature.

Note that we could have just as well used the volume of gas, V_{stp}, at standard temperature and pressure, in place of σ. This appears in some places in the literature. Regarding *adsorption isotherms*, assume that we have a vacuum vessel identical to that shown in Fig. 5.1.1, which we bake under vacuum to remove all of the surface gases. Assume we fix a gauge to this system to enable measurement of the pressure within the vessel, while knowing both the vessel and gauge temperatures and thereby compensating for their differences using (5.1.7). The vessel surface area is A_v and the volume V_v.

In this instance, after cooling the system, we will introduce a very limited, known amount of gas. Some of the gas will land on the surface, stick there for a while, and then come off again as *free* gas. If all of the gas is continually in a gaseous state, the pressure in the vessel, P_v, will behave according to the *ideal gas law (i.e., $P_v = n\Re T / V_v$)*. However, if for example, at any instant in time half of the gas is resident on the walls, then the pressure in the vessel will be half of that predicted by the ideal gas law, *etc.* Therefore, there is a net pumping on the walls. If we know the amount of gas introduced, and the subsequent equilibrium pressure in the vessel, we can calculate the average population density of gas on the surface of the vessel at any time. A measure of the surface population density is usually given the symbol σ in the literature, where σ represents the extent of surface coverage in molecular or atomic layers. For example, σ = 0.1 indicates a 10% coverage; σ_m is usually reserved to note a surface coverage of one molecular layer or a *monolayer*, σ = 2 indicates coverage of two monolayers, *etc.* Sometimes the symbol θ is also used in the literature to represent surface coverages relative to the total number of possible adsorption sites, and sometimes as the ratio of σ / σ_m. As a *rule of thumb*, a molecular layer comprises $\sim 10^{15}$ molecules (atoms) per cm^2. See Redhead, *et al.*, for more precise values of σ_m.[23]

When plotting experimental data, it is usually the convention to display the independent variable (*i.e.*, the *cause*) along the abscissa or x-axis and the dependent variable (*i.e.*, the *effect*) along the ordinate or y-axis. There are two ways one might perceive an adsorption isotherm: 1) one might assume that the surface population density *causes* the equilibrium pressure; or, 2) one might assume that sustaining the equilibrium pressure over the surface *causes* the surface population density. Each are reasonable scenarios, and I will show data both ways.

Assume that we successively introduce additional known quantities of type A gas into the system, and after each introduction we wait for the pressure to equilibrate and then make a pressure measurement. From this sequence of measurements we are able to plot a curve of the equilibrium pressure P_e *vs.* the average surface coverage, σ. Data of P_e *vs.* σ at constant temperature, T_1, are called an *adsorption isotherm*. An *isotherm* merely means that pressure *vs.* coverage data are taken at the same temperature. Such data might take the form represented in Fig. 5.1.5a, where the equilibrium pressure is given as a function of surface coverage at a given temperature. If we were to repeat the measurement all over again, but this time at a different temperature, T_2, where $T_2 > T_1$, we would observe a distinctly different pressure for a given coverage. We note that as the temperature is increased, the equilibrium pressure is higher for the same gas coverage.

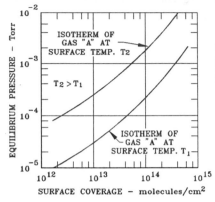

Figure 5.1.5a. Variations in equilibrium pressures, for the same coverage, but at different temperatures.

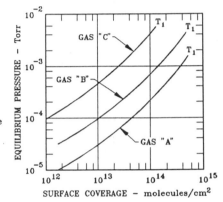

Figure 5.1.5b. There are unique adsorption isotherms for each gas and for the same surface temperature.

Were we to repeat this series of measurements for three distinct gases, identified as types A, B and C in Fig. 5.1.5b, each gas would have its own distinct T-dependent adsorption isotherm. That is, at the same surface temperature, each type of gas would have a different equilibrium pressure for an identical surface coverage, as illustrated in this figure. For all gases, the measured pressures of the adsorption isotherms are in each case less than the vapor pressures of the given gas at the particular temperature. However, as surface coverage increases to the point of several atomic layers, the equilibrium pressure of an adsorption isotherm converges to the much higher value of the vapor pressure of the species at that temperature.

In the case of adsorption isotherms of gas A, the cold surface has a finite speed for gas A as long as the pressure of the adsorption isotherm is less than the operating pressure of gas A. This effect is graphically illustrated in Fig. 5.1.6. Assume an adsorption isotherm for gas A as shown in Fig. 5.1.6a. With greater and greater surface coverage, the equilibrium pressure over the surface continues to increase.

Clearly, after pumping σ_1 amount of gas, the blank-off pressure of the pump is *zero* at P_1; *i.e.*, the speed of the pump at P_1 has decreased to *zero*. However, at higher pressures, and for the same σ_1 coverage, the speed increases as we depart from the base pressure of the pump, P_1. Eventually, the pump speed asymptotes at some S_{max} value, which is dependent on the area of the cooled surface and the sticking coefficients of the gas being pumped. This amounts to *surface conductance limitation* for cryosorption pumping. This effect is shown in Fig. 5.1.6b.

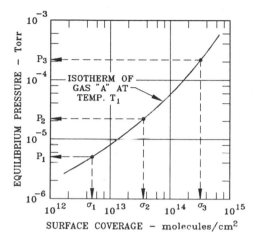

Figure 5.1.6a. Variations in equilibrium pressure for surface coverages σ_1, σ_2 and σ_3 and temperature T_1.

Figure 5.1.6b. Variations in pump speed due to changes in base pressure because of gas surface coverage, where $\sigma_1 < \sigma_2 < \sigma_3$ for gas "A", and a surface temperature T_1.

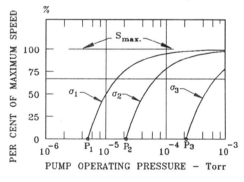

After pumping enough gas to increase the surface coverage to σ_2, the speed of the pump becomes *zero* at P_2. With increasing surface coverage, the *zero* speed point of the pump progressively increases. This results in a family of speed curves such as shown in Fig. 5.1.6b. Each speed curve is dependent on the respective surface coverage. We conclude that when discussing the speed of a cryopump used for cryosorption pumping, we must have knowledge of the quantity of gas pumped up to that point in time and the pressure. This is similar to requirements in the use of NEG pumps, noted in Chap. 4. Also, when making claims about the capacity of a pump to cryosorb gases (*i.e.*, the amount of gas which it can pump), these claims only have meaning if the pressure at which the speed measurements are taken is also given. For example, if the speed measurement for a gas A was taken at pressures somewhere on the *knee* or flat portion of the speed curve *belonging* to the σ_2 curve, the speed for that gas would be *zero* or negative at pressures $\leq P_2$.

The rate at which the *zero* speed of the pump increases depends on the rate at which gas is pumped and is the true measure of a cryopump's capacity. In practical applications, when the system pressure increases to some predetermined limit as a consequence of the *zero* speed point of the pump increasing in pressure, we conclude that pump capacity has been exceeded and the cryopump needs a *regeneration*. I will later return to the subject of cryopump regeneration.

Classifications of Adsorption Isotherms

In Fig. 5.1.5b I showed the adsorption isotherms of fictitious gases as being smooth functions and devoid of inflection points. An *inflection point* is a point along some curve (*e.g.*, the *P vs. σ* curve) where the curve seems to depart from a trend and change directions (*i.e.*, $d^2\sigma / dP^2 \neq 0$). All adsorption isotherms have a minimum of one inflection point and some as many as three or four. To further complicate their theoretical modeling, all evidence hysteresis effects which may or may not be time dependent functions.[24-26] In addition, Stern and DiPaolo report on a *conditioning effect* occurring with some sieve materials.[29] That is, an isotherm of a gas on a given material may differ for the same temperature after, in an interim period, having been returned to RT. An in-depth treatment of the physical process resulting in these complex *P vs. σ* functions is beyond my complete understanding. However, ...

Adsorption isotherms often have a unique and repeatable form, which is attributable to the gas, temperature and sorbing surface. Robens provides an excellent qualitative review, with a comprehensive bibliography, on the classification of about 13 different types of adsorption isotherms.[27] Dushman discusses several types of adsorption isotherms reported in the early literature. I liberally use this reference below.[28]

No unified theoretical model presently exists which fits the peculiarities of all of the variously shaped isotherms, though some theoretical models are very accurate for specific gases on specific types of materials. For example, one model might be found to agree with experimental results in a pressure range varying ~13 orders in magnitude for certain gases on a specific material,[20] where the experimental results with a different gas on another material were found to diverge from the model.[29] I will briefly note some of the historical developments in this area to provide you with a familiarity with some of the terms and their origins. To avoid repetition, I will use the terms σ, P and T as defined in the previous section.

There are several popular semi-empirical adsorption isotherm models which we vacuum technicians find frequently cited in the literature. One such model is called *Henry's Law*. It is sometimes applicable only at very low

pressures,[29,30] while at other times it is applicable at high pressures.[31]
It is a variant of Henry's Law dealing with the solubility of gases in liquids.
This law states that the amount of gas which will go into solution with a liquid
is directly proportional to the pressure of the gas over the liquid. The vacuum
variant of Henry's Law predicts that σ will vary linearly with pressure. Or,

$$\sigma = k_1 P, \tag{5.1.11}$$

where k_1 is some constant of proportionality. In this case, if Henry's Law is
applicable, when the P-σ data are plotted on log-log paper, the slope of the
data will fall on a 45° line.

The *parabolic* or *Freundlich* isotherm is also frequently cited.[32] In this
case,

$$\sigma = k_1 P^m, \tag{5.1.12}$$

where k_1 is a constant and $m \leq 1$. Freundlich's model includes Henry's Law.

Langmuir was the first to attempt to model adsorption processes in terms
of gas-surface physics relating to potential occupancy sites, variations in the
number of sites possibly occupied by different gas species, and how all of this
affected the amount of surface gas which could be accommodated. Because
of Langmuir's work, his insightful approach started vacuum technicians
thinking in these sort of terms. In honor of him the term *Langmuir* is often
used in publications to denote a gas coverage of one monolayer. His model
dealt with surface coverages where $\sigma \leq$ one *Langmuir*. The equation which
he developed for an isotherm is sometimes called the *hyperbolic* isotherm, and
takes the form:

$$\sigma = \frac{\sigma_s (k_1 P)^{1/m}}{1 + (k_1 P)^{1/m}} \tag{5.1.13}$$

where k_1 is a constant, σ_s denotes the total number of available sites per
cm^2, and m represents the number of surface sites which is occupied by a
particular gas species.

Almost a quarter of a century later another important adsorption model
was derived by Brunauer, Emmett and Teller.[33] They were troubled by
published data indicating that cryosorption could include the build-up of
surface gases to an extent exceeding a monolayer of gas - perhaps to the
extent of several monolayers as in the case of cryocondensation. Their
interest was in the use of molecular sieve materials (see Sieve Materials,
below) as catalysts. It was noted that the shape of the P vs. σ isotherms of
gases on these sieve materials was concave downward at low pressures and at
higher pressures became convex. They coined this effect an *s-curve* isotherm.
They expanded on Langmuir's theory to include its applicability to gas

coverages exceeding a monolayer, with the added assumption that the same forces which came into play in cryocondensation were applicable to cryosorption at higher temperatures. They derived the following equation:

$$\sigma \;\; = \;\; \frac{\sigma_m \, k_1 \, P}{[P_v - P][1 + (k_1 - 1)(P/P_v)]}, \qquad (5.1.14)$$

where k_1 is some constant much greater than unity, σ_m is a *Langmuir* and P_v is the vapor pressure of the test gas. Equation (5.1.14) may be rearranged into the form:

$$\frac{P}{\sigma(P_v - P)} \;\; = \;\; \frac{1}{\sigma_m \, k_1} + \frac{(k_1 - 1)}{\sigma_m \, k_1} P/P_v. \qquad (5.1.15)$$

Therefore, assuming the theory was correct, a plot of $P / \sigma (P_v - P)$ as a function of P / P_v should produce a straight line, the ordinate intercept being $1/\sigma_m \, k_1$ and the slope having the value $(k_1 - 1)/\sigma_m \, k_1$. This enables one to determine the values of both k_1 and σ_m. They expanded on (5.1.14) to include coverages exceeding one monolayer. However, (5.1.15) proved so successful in characterizing the *s-curve* adsorption isotherms of materials that it developed into an industrial standard for specifying surface areas of sieve materials used for catalysis and in vacuum applications. In recognition of Brunauer, Emmett and Teller's work, it became known as the *BET method* for determining the area of sieve materials.[34,35] Nitrogen is usually used as the standard gas in this application and the sorbing material is cooled to LN_2 temperatures.

A decade later, Dubinin and Radushkevich published what is now abbreviated in the literature as the *DR* adsorption isotherm.[36,98] This equation takes the form:[29]

$$\log \sigma \;\; = \;\; \log \sigma_m - k_1 (\log P_v / P)^2 , \qquad (5.1.16)$$

where k_1 is some constant and P_v is the vapor pressure of a *liquid* of the gas. Stern and DiPaolo have found fairly good agreement in this equation when pumping N_2 on sieve materials (Linde 5A) cooled to LN_2 temperatures.[29] However, the *DR* theory and their results diverged at pressures $\lesssim 5 \times 10^{-4}$ Torr. Troy and Wightman could find no general experimental agreement with either the *BET* or *DR* theory. Their experiment involved the pumping of CH_4, N_2, Ar and Kr on stn. stl. surfaces ranging in temperature from 77 to $90°$ K. Halama observed similar departures in the *DR* theory with measured results for He on 5A sieve, but at pressures $\lesssim 10^{-10}$ Torr.[25]

Hobson found close agreement with the *DR* theory and measured isotherms of Ar, Kr and Xe on porous silver, and over a *very* broad pressure range. This latter experiment must have been a test of both scientific technique as well as endurance. The results are shown in Fig. 5.1.7.

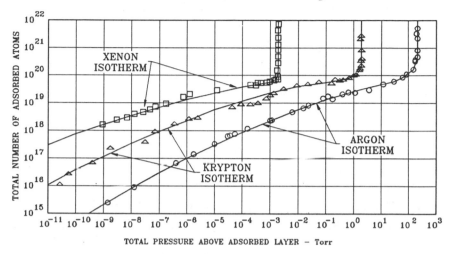

Figure 5.1.7. Adsorption isotherms of Xe, Kr and Ar on a porous silver adsorbent at a temperature of 77.4 °K, as reported by Hobson.[20]

We see from this figure that in order to conduct this experiment, Hobson had to measure equilibrium pressures ranging 13 orders in magnitude. Detailed attention to UHV technology was a prerequisite to the experiment. Also, achieving equilibrium pressures at times requires elapsed times of an hour or more for each datum. With all this in mind, this has to be one of the *classic* experimental papers on the measurement of adsorption isotherms.

Hobson predicted in an earlier theoretical paper that the isotherms of gases would depart from the *DR* theory at very low pressures, and begin to converge with Henry's Law.[30] The theoretical model which he presented in ref. [20] accommodated predictions of the onset of Henry's Law at low pressures. However, he did not observe this effect. Also, it was not observed in similar experiments which he conducted with He on porous silver at LHe temperatures.[37] In this case, the lower pressure limit of the experiment was $\sim 5 \times 10^{-11}$ Torr He. Hseuh, *et al.*, noted that the Henry's Law limit was in no manner applicable for He on activated coconut charcoal at temperatures ranging from 4.2 to 18.2° K.[38]

Defining a universal model to predict an adsorption isotherm becomes very complex, as they are critically dependent on the type of gas (*e.g.*, polar *vs.* nonpolar) and the sorbing material (*e.g.*, the nonpolar carbon materials *vs.* the highly polar zeolites).[100] Existing models are still somewhat flawed. For example, using graphical (or analytical) methods similar to those described relating to the *BET* method, Hobson found a disparity between that of the *DR* method and *BET* methods in defining σ_m (*i.e.*, equivalent to the effective area of the porous silver) for the gases Ar, Kr and Xe on porous silver at LN_2 temperatures. There was reasonable agreement for Xe (*i.e.*, within 30%), but a divergence in data for lighter molecules to the point of a disparity of ×4.3 for the *BET vs. DR* methods. The effective surface areas of his porous bed, as predicted by the *BET* and *DR* methods, were 1.1 × 10^5 cm^2 and 2.4 × 10^4 cm^2, respectively. Secondly, there was disagreement in implied surface areas for the different gases when comparing results of one method for the three gases. These were minor in the case of the *BET* method (*i.e.*, 25%), but significant in the case of the *DR* method (*i.e.*, ~×2.5).[20]

In a paper published three months later, Hobson reported measurement of He adsorption isotherms, presumably on the same porous silver bed, but this time at LHe temperatures.[37] He made a direct comparison of these results with that of Danilova and Shal'nikov, who measured the isotherms of He on a smooth Cu plate also maintained at LHe temperature.[39] From this, Hobson was able to conclude that the *effective* area of his porous silver bed was ~×1000 that of a flat Cu surface, or ~3.5 × 10^5 cm^2 for He. Therefore, for the same bed, and different gases, there is a variation of over an order in magnitude in the predicted effective surface area using these three different techniques. Recognizing Hobson's exacting technical skills, one therefore must conclude that much about this theory is still found wanting.

Hobson suggests that we classify cryosorption pumping as a process where $\sigma \lesssim 5 \sigma_m$, and cryocondensation pumping when surface coverages are such that $\sigma \gtrsim 5\sigma_m$.[40] This is a reasonable scenario if we envision that van der Waals' forces extend into free space something greater than the equivalent of three molecular layers (*e.g.*, see Fig. 2.4.7). However, $\sigma \lesssim 5$ is a statement about the probable upper limit of the adsorption process as, for example, He on LHe cooled surfaces has a coverage σ ~3 × 10^{-2} σ_m at 10^{-6} Torr.[39] The effectiveness with which gases are cryosorbed relates to equivalent energy sites created by surface-gas and surface gas-gas forces. The forces between gas molecules and the parent surface tend to be greater than the forces between gas-gas molecules on the surface. Therefore, the heat of sorption will initially be greater than the heat of vaporization of a gas on a liquid of the same species. However, as the surface coverage increases, the heat of sorption decreases and finally converges to the heat of vaporization (or sublimation).[41]

Table 5.1.3 lists representative examples of both cryocondensation and cryosorption pumping experiments reported in the literature. It covers a wide variety of gases and surfaces, including cryofrosts of similar and different gases. It is somewhat chronological.

Table 5.1.3. Examples of some of the referenced literature on cryocondensation, cryosorption and cryotrapping pumping.

SURFACE TEMP. °K	APPROX. PRESS. Torr	GASES	REF.	SURFACE [1]
4.2	10^{-11}– 10^{-6}	H2, He	45	CT by Ar.
10–77	varied	N2,CO,O2,Ar,CO2,N2O	18	CF of S.
77	varied	13(e.g., H2O–CH3COCH3)	18	CF of S.
3.5–99	not indicated	H2, Ar	2	CF of S.
77 & 87	10^{-9} – 10^{-4}	Kr, Xe	107	Mo films & glass.
3.8–77	molecular beam	Ar, CO2, Kr, Xe	3	CF of S; quartz crystal.
4.2	2×10^{-11}– 10^{-5}	He	39	Cu plate.
2.18–3.68	7×10^{-10}– 7×10^{-8}	H2, D2	87	CF of S.
20–92	molecular beam	N2, Ar, CO2	4	CF of S.
77–90	10^{-9} – 10^{-4}	CH4, N2, Ar, Kr	108	304 stn. stl.
2.5–4.2	10^{-12}– 7×10^{-7}	H2	12	Cu and CF of S.
2.5–4.2	10^{-12}– 4×10^{-7}	H2	12	CF of S & CF of Ar.
3.7–5.2	10^{-12}– 5×10^{-8}	D2	12	Cu & CF of S.
2.3	8×10^{-12}– 4×10^{-11}	H2	91	Ag plated stn. stl.
4.2	10^{-6} – 10^{-4}	H2, D2	96	CF of S & stn. stl.
2.5–4.3	2×10^{-14}– 10^{-6}	H2, HD, D2	92	CF of S & Ne,Ar,N2,Kr.
4.2	<1.3×10^{-4}	D2	103	CF of S & Al.
4.2	<1.3×10^{-4}	D2	103	CF of D2.
3–300	molecular beam	N2, Ar, CO2	19	CF of S & Cu.
2.23–4.2	not indicated	He, H2, N2	42	CF of S & N2 SCT.
3–300	molecular beam	N2, Ar, CO2	5	CF of S & Cu.
7.3–15.3	8×10^{-12}– 8×10^{-6}	H2	48	CF of CO2.
4.1–4.9	3×10^{-5} – 10^{-4}	H2, D2, N2	94	CF of S and stn. stl.
4.2	10^{-7} – 6×10^{-5}	H2, D2, H2+D2	21	CF of S & CCT of H2.
4.2	2×10^{-6} – 2×10^{-4}	D2, D2+He	46	CF of S & CCT of He.
4.2	not indicated	H2	61	CF of S and stn. stl.
3.7–4.3	4×10^{-6} – 4×10^{-5}	H2	97	CF of S and Cu.
6.0–11	5×10^{-8} – 7×10^{-4}	H2	50	CF of CO , Kr & Xe.

5.1.5 Speed and Capacity of Cryopumps

In a previous section, we noted that when the pressure of N2 in the vessel was greater than the equilibrium vapor pressure of the N2 *ice*, the net flux of gas was into the N2 *ice*, and the surface of the *ice* was a cryocondensation pump. Also, similar mechanisms applied to cryosorption pumping. That is, if the equilibrium pressure related to a given adsorption isotherm was less than the pressure due to introduced gas, the surface would pump gas as the system sought a new equilibrium. In both cases, the system attempts to equilibrate at a state where the net flux of gas impinging on and leaving the surface is *zero*. In most cases of cryocondensation pumping, the surface *ice* has *zero* speed for a gas A at a pressure equivalent to the vapor pressure of the *ice* A, and it has finite speeds for gas A at pressures greater than the vapor pressure of *ice* A. An exception to this was noted in the case of CO2. Therefore, we must recognize and compensate for the fact that the values α and $(1 - P_e / P_o)$ may stem from two different effects. The equation for the pumping speed of a cryocondensing or cryosorbing surface is as follows:

$$S \quad = \alpha\, C\,(1 - P_e/P_o)' \tag{5.1.17}$$

where C = the *conductance* of the pumping surface, in $\mathcal{L}$/sec.,

P_e = the equilibrium vapor pressure of the condensed ice,
or the adsorption isotherm equilibrium pressure,

and P_o = the operating pressure above the *ice* or surface.

This is simply a variation of (1.15.1) of Chap. 1. In this case the *base pressure* of the pump is simply the equilibrium pressure of the *ice* when cryocondensation pumping, and the adsorption isotherm equilibrium pressure in the case of cryosorption pumping. In (5.1.17) we have assumed that the gas impinging on and leaving the cold surface is at the same temperature as the surface. This is implicit in the term C. In order to compensate for possible differences in the *free* gas temperature, T_g, and the temperature of gas departing the surface, T_s (assuming a unity accommodation coefficient), we must modify (5.1.17). Remember that P_e/P_o corresponds to the equilibrium fluxes of gas leaving and entering the surface. In other words, $P_e/P_o = \nu_v/\nu_i$. Making this substitution in (5.1.17) gives us a new equation for the speed of a cryosurface which includes the effects of gas temperature. That is:

$$S \quad = \quad \alpha\, C_g\,[1 - (P_e/P_o)(T_g/T_s)^{1/2}], \tag{5.1.18}$$

where α = the sticking coefficient (5.1.5),

P_e = the equilibrium pressure of the cryosorbed gas,

P_o = the pressure (induced gas) above the cryosurface,

C_g = the surface conductance for gas at temperature T_g,

T_g = the induced gas temperature,

and T_s = the cryosurface temperature.

When cryocondensing or cryosorbing gases, (5.1.18) is always applicable. We may now combine (5.1.8) and (5.1.18) to predict the behavior of a cryopump of the more complex configuration such as shown in Fig. 5.1.8. In this case there is a cryopanel of area A_3 located in the colder chamber. The temperature of this additional panel is T_3, where $T_3 < T_2 < T_1$. If a cooled chevron was placed over the aperture connecting the two chambers, this would fairly well describe the simplest form of multi-stage cryopumps. More will be discussed later on the purpose of such cryochevrons and cryopump *stages*.

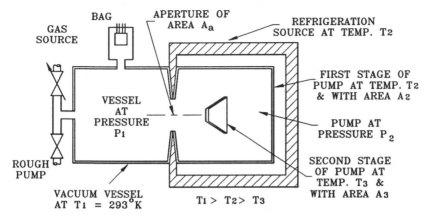

Figure 5.1.8. The elements of a two-stage cryopump.

5.1.6 Cryotrapping

Cryotrapping is the concurrent or sequential cryopumping of two or more gases for the purpose of trapping a less readily pumped gas in the cryodeposits of a more readily pumped gas.

For example, define D as a less readily pumped gas and E as a more readily pumped gas at some temperature. We might pump the D gas in a mixture rich in proportions of the E gas. The vapor pressure of the E gas might be very low at the given temperature, but that of the D gas very high. If the D gas becomes trapped in the cryofrost matrix of the E gas during its cryocondensation, we say that the D gas has been *concurrently cryotrapped.*

In another example, assume that we have cryocondensed a thick cryofrost of D gas. However, at the cryopanel temperature the vapor pressure of the D gas is higher than desired. In this case we might introduce sufficient amounts of E gas so as to cover over the D gas with a cryofrost of E gas and thereby depress the observed system partial pressure of D gas. This would be *sequential cryotrapping.* Overlays of E-type gases have also been used to shield D-type gases from desorption effects due to energetic particles, as Hilleret reported doing with N_2 on cryodeposits of H_2.[42]

Several requirements become obvious if cryotrapping is to work. First, there must be no surface gas displacement effects. For example, suppose Hobson tried to depress the 2×10^{-3} Torr Ar partial pressure, in his experiment of Fig. 5.1.7, by the subsequent introduction of Xe. At this pressure, σ was $< \sigma_m$. Indeed the Xe would have been more readily pumped than the Ar, but it probably would also have displaced Ar molecules off the surface as it successfully competed for physisorption energy sites. Hseuh, *et al.,* observed this gas displacement effect when pumping H_2 and He on activated coconut charcoal.[38] We observed a similar effect with TSP pumping. However, in this case gases were displaced in competition for chemisorption sites (*i.e.*, see Table 3.3.1). Therefore, cryotrapping probably has no significant application in cryosorption pumping. Also, this competition for energy sites may explain why cryocondensation pumping of certain gas mixtures leads to unexpected results. For example, Büttner reported that the cryocondensation speed of a mixture of $H_2:D_2$ (1:5) on supercooled He surfaces was less than their respective speed components when individually pumped at the same temperature.[21] Some of these anomalous effects, when sieve materials are used, may also be attributable to sieve *plugging* or unplugging effects (see below).

Secondly, if cryotrapping is to work, the D gas must be somewhat insoluble in the E gas. For example, if the D gas is soluble in the E gas, it will readily diffuse in the cryofrost matrix of the E gas. If on reaching the surface of the E gas, the heat of sorption of the D gas on the E gas is less than the D gas heat of vaporization, the D gas will outgas into the vacuum system and eventually equilibrate at a pressure corresponding to its equilibrium isotherm on the E gas. Halama, *et al.*, saw this very effect when attempting to cryotrap He with H_2 on LHe cooled surfaces.[43,44]

There have been many successful examples of the applications of cryotrapping. For example, in an early paper Hengevoss and Trendelenburg reported on the cryotrapping of both He and H_2 with the concurrent pumping of Ar on 4.2° K surfaces.[45] Chou and Halama concurrently cryotrapped both H_2 with D_2 in the proportions of $D_2:H_2$ of ~10:1.[44] Batzer, *et al.*, used a directed stream of Ar to pump He and isotopes of hydrogen on 4.2° K surfaces in a fusion-related application.[46] The flux of Ar:He had to be maintained in the proportions of ~30:1. They resorted to this method of pumping He because the H_2 plugging of sieve materials decreased its capacity for He. They also pointed out the need for UHV conditions when studying the cryopumping of one gas. They indicated that cryotrapping of the study gas by system background gases could lead to errors. Longsworth and Webber conducted similar concurrent cryotrapping experiments for Ar and H_2 in a CLGHe cryopump.[24]

Also, a number of sequential cryotrapping experiments have been conducted to determine the effectiveness with which D gases are cryosorbed on cryofrosts of E gases (*e.g.*, see (12,47-50,90)). These studies were undoubtedly in part aimed at determining the heats of sorption of D gases on the E gases as they relate to the mentioned solubility-diffusion considerations. Also, it has been concluded that D gases are less readily desorbed from metal surfaces if a substrate of E gas is first deposited on the surface.[12,90] It is believed that the E gas serves as a phonon barrier between the D gas and the metal substrate. Some of these works and others related to cryotrapping are referenced in Tables 5.1.3 and 5.1.5.

5.1.7 Sieve Materials

In cryosorption pumping it is obvious that the equilibrium pressure within the vessel is critically dependent on σ, the measure of surface coverage. Further, for a fixed pressure, σ depends on the temperature of the pumping surface. As will be discussed, there are certain practical limits to the temperatures which can be readily achieved in various forms of cryopumps. Because of this, cryocondensation pumping has limited application in the UHV pumping of the gases He, H_2 and Ne. For example, the practical lower temperature limit of CLGHe cryopumps is $\sim 10°$ K. We see from the data of Haefer[52] (i.e., Fig. 1.7.4 of Chap. 1) that the gases He, H_2 and Ne would not be effectively cryocondensed at these temperatures (i.e., at UHV pressures). Therefore, these gases can only be pumped at lower pressures by cryosorption.

Sorption roughing pumps are operated at LN_2 temperatures (i.e., $\sim 77°$ K). Therefore, if these pumps are to be used to rough pump substantive quantities of gases, cryosorption pumping must come into play. In fact, the vapor pressure of LHe (liquid He) is ~ 750 Torr at $\sim 4.3°$ K. At this temperature, a LHe cooled surface would be of marginal value in the pumping of H_2 in UHV applications and useless in the cryocondensation pumping of He. However, I have noted examples of the effective cryosorption pumping of both of these gases at temperatures $\leq 20°$ K.

Therefore, many gases, including the *difficult* gases He, H_2 and Ne, must be pumped in most applications by cryosorption pumping. But we have seen that cryosorption pumping has an inherent capacity limit established by the effective area and temperature of the cold surface. As the surface coverage of gas increases, the equilibrium pressure rises. I model this in my *mind's eye* as being due to mobile surface gases. That is, because some molecules temporarily residing on the surface still have sufficient thermal energy to move about on the surface, from time to time they *bump* into their neighbors and give these neighbors sufficient energy to escape the surface, as gas. It is reasonable that the consequences of this unneighborly behavior (i.e., higher pressures) would depend on the surface population on the gas. Solid state physicists explain the higher pressures, with higher surface coverage, as stemming from *phonons*. A phonon is a *quantum bundle* of heat energy propagating through the crystal lattice. These phonons, residing in the cooled substrate, bump into the surface gas and cause its ejection. If, for a given temperature, the phonon concentration is constant in the substrate, it is reasonable that with increasing surface coverage more gas would be ejected. No matter which model you pick, the consequences are the same, the greater the surface population, the higher the equilibrium pressure. Therefore, the only hope of increasing the cryosorption capacity of a pump for He, H_2 and Ne is to increase the effective surface area of the pump.

You will recall that charcoal was used in making the measurements leading to the data in Table 5.1.1. This material has very strange and unique properties. That is, for a comparatively small mass of the material, charcoal evidences very large *effective* surface areas. Materials with this unique property have come to be known as *molecular sieve* materials, or just *sieve* materials. Actually, the first reported vacuum use of charcoal was in a paper read by Professor Sir James Dewar before the Royal Society of Edinburgh in 1874. I discovered its existence when it was referenced in a vacuum application note appearing in the journal *Nature* a year later.[53] In this second paper he reported heating the charcoal while pumping on it with what he called "a mercury pump". He then sealed off the charcoal in the glass vile, while it was still hot. The *cryogen* was the RT air which cooled the vile after bakeout. He reported on various electrical and mechanical experiments conducted in the sealed vile which served as verification of the improved vacuums achieved by the cooled charcoal.

Over the last three-quarters of a century, much work has been done in studying the properties of these sieve materials at both RT and cryogenic temperatures. The variety of materials studied has included almost everything imaginable (*e.g.*, cement, coal, rubber, glass, *etc.*).[54] Study of their RT properties was given impetus because of the development of gas masks before and during WWI.[55]

Unfortunately, these sieve materials must still be used as *adsorbents* in this application. An *adsorbent* is a material on which gas is adsorbed. An *adsorbate* is the gas which is adsorbed on an adsorbent. These are often abbreviated as *sorbents* and *sorbates* respectively. A review of much of the early work on the pumping properties of various sorbents for various sorbates, and over wide temperature ranges, is given by Dushman.[56]

A second extremely important RT application of these sieve materials is in the area of catalysis.[33,57] Use of catalysis on an industrial scale requires large surface areas. Because of this, a variety of man-made sieve materials were specifically developed for catalysis applications. Some are called *artificial zeolites*. Zeolite is an aluminosilicate, crystalline matrix. It occurs naturally in some forms, the man-made material being distinguished as *artificial*, and usually comprising some form of calcium or sodium aluminosilicate.[58]

Because of their large surface areas, these sieve materials have been investigated for vacuum applications. Some of these materials are listed in Table 5.1.4, an embellishment of a summary published by Stern, *et al.*,[34] and Redhead, Hobson and Kornelson.[59] I find the results of this table to be nothing short of remarkable. For example, a gram of activated coconut charcoal has the equivalent effective surface area of ~890 m^2 (*i.e.*, using the *BET* method). The Linde 5A artificial zeolite has an effective surface area of ~600 m^2/g, and the 13X artificial zeolite ~515 m^2/g, *etc.*

Table 5.1.4. Effective areas of some sieve materials.[34,59]

SIEVE MATERIAL	FORM	AREA [1] m^2/gm	POUR SIZE–Å	REF.
Porous Silver	sintered matrix	0.22 [2]	–	20
Porous Silver	sintered matrix	0.71 [3]	–	37
Alumina	spheres, 3 mm ϕ	287	50	34
Silica Gel, Type ID	granular	311	150	34
Linde 13X (artificial zeolite)	powder	514	9–10	34
Linde 5A (artificial zeolite)	pellet	600	5	34
Silica Gel, Type R	granular	784	22	34
Coconut Charcoal	granular	889	10–30	34
Sanan Charcoal, Type S–85	powder	1170	–	110
Coal	granular	1050–1150	–	35
Coconut Charcoal, Type PBC	granular	1150–1250	–	35

NOTES: 1. Measured using the BET method.[33]
2. Measured using Ar rather than N2.
3. Measured using He rather than N2.

These results imply that if there was some convenient way to bond these materials onto cryogenically cooled surfaces, it would be possible to cryosorption pump many orders of magnitude more gas than would be possible on similarly cooled metal substrates of the same geometric surface area. For example, Danilova reported the cryosorption pumping of only ~6 × 10^{-7} Torr-$\mathcal{L}$ He/cm^2 on a Cu surface cooled to 4.2° K and before an equilibrium pressure of ~10^{-7} Torr was observed.[39] Halama, by cooling the surface onto which a sieve material was bonded to 4.2° K, was able to pump ~23 Torr-$\mathcal{L}$ He/cm^2 of geometric surface area and achieve the same equilibrium pressure.[26] That is, for the same projected surface area, ~4 × 10^7 more He could be accommodated.

Similarly, Sedgley, et al., epoxy-bonded activated coconut charcoal to a cryopanel.[35] When the surface was cooled to 4.2° K, and the equilibrium He pressure reached ~10^{-5} Torr, they reported the pumping of ~2.0 Torr-$\mathcal{L}$ He/cm^2 of bonded material. This corresponds to ~×1.2 × 10^6 more He than could be accommodated on a Cu panel at equivalent temperatures.

The term *activated* coconut charcoal (*i.e.*, *vs.* less active) has to do with the relative sorptive capacity of the charcoal. This relative sorptive capacity varies with the refinement of impurities out of the charcoal and vacuum and air roasting procedures. Use of proper procedures and recipes leads to *activated* charcoal.[55]

The mysterious properties of these materials are qualitatively explained by Fig 5.1.9. In Fig 5.1.9a, the cryogenic pumping surface is a flat, metal plate. On an atomic scale, the surface is comparatively smooth, so that gas can randomly move about the surface, and suffer collisions with neighboring molecules on the surface. Gas on the surface of the plate is pumped gas, where gas which is *bumped* off contributes to the pressure in the system.

Bulk sieve material comprises a complex matrix of *caverns* or cavities, with interconnecting apertures. For example, Bolton notes that the unit cells of the Linde 13X zeolite matrix comprise "... eight large cavities, each with a free diameter of about 13 Å".[5 7] These cavities are interconnected by apertures with diameters of ~7.4 Å. The Linde 5A zeolite has internal cavities with diameters of the order of 11 Å, which are interconnected by apertures having diameters of ~4.2 Å. Because of these caverns within the bulk of the sieve material, copious quantities of gas can be pumped.

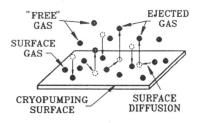

Figure 5.1.9a. Cryosorbing plate.   Figure 5.1.9b. Cross section of small chunk of sieve material.

The mechanism for gas pumping, depicted in Fig. 5.1.9b, is as follows. Gas impinges on the geometric surface area of the sieve. Because of a build-up of surface concentration gradients, the gas diffuses on the surface of the sieve and as *free* gas down into the caverns within the bulk material. If a molecule, when residing on the wall of one of these inner caverns, is bumped off by a neighboring gas molecule, it lands on a nearby cavern wall. This struggle for real estate between molecules occurs within the innards of the sieve and therefore has no effect on the pressure over the geometric surface area of the sieve. I have coined this process as a *labyrinth effect*. Note that if the interconnecting apertures are large, gas is able to more readily diffuse down into the bulk of the material. However, this accessibility argument works both ways. That is, it will be easier for free gas within the bulk to find its way out of the labyrinth. For this reason, both the sieve surface area and the size of the interconnecting apertures are important.

Table 5.1.5 is a chronological summary of some of the publications relating to the use of different sieve materials for the pumping of a variety of gases and in applications involving sorption, LHe and CLGHe cryopumps.

Plugging of Sieve Materials

The preponderance of the surface area of a sieve material is located within the body of the sieve. Therefore, it is important that the entrance apertures to the body of the sieve remain open so that gas has access to the inner surfaces. Because of the relatively small sizes of these entrance apertures, they can become plugged by certain gases. The phenomenon associated with the plugging of sieve materials is called *persorption* .[3 3] It merely means that the bigger or more polar the sorbate molecule, the greater the likelihood that it will plug the entrance apertures.

Table 5.1.5. Examples of some of the referenced literature on the use of sieve materials and forms of cryotrapping.

SURFACE TEMP. °K	APPROX. PRESS. Torr	GASES	REF.	SIEVE MATERIAL[1]
–	not indicated	He,H2,Ne,N2,O2,Ar,CH4	60	Charcoal & porous glass.
77.3	10^{-4} –750	Air, H2O	80	MS–5A, ACC.
77	4×10^{-6} – 4×10^{-4}	N2, Ar	104	MS–13X, ACC.
4.2	10^{-11} – 10^{-6}	He, H2	45	CCT with Ar.
20.2	10^{-8} – 10^{-4}	para–H2, ortho–H2	34	Seven sieve materials.
20.2	10^{-5} –1.0	He	34	MS–13X, silica gel.
20.2–40.4	10^{-7} –10.0	He	34	ACC.
4.2	5×10^{-8} – 7×10^{-4}	He	68	MS–5A.
20.2	7×10^{-9} – 10^{-4}	H2	67	MS–5A.
4.3	8×10^{-10} – 4×10^{-7}	He, H2	101	MS–5A.
77.3	10^{-6} – 10^{2}	He, Ne, N2	29	MS–5A, MS–10X, MS–13X.
10.7	3×10^{-7} – 2×10^{-5}	He, H2	79	ACC.
4.2	5×10^{-11} – 2×10^{-3}	He	37	Porous silver.
9	not indicated	H2	49	On Ar cryofrost.
77.4	10^{-11} – 10^{2}	Ar, Kr, Xe	20	Porous silver.
77	10^{-6} –750	Air	99	MS–5A.
4.2*,20	8×10^{-8} – 8×10^{-5}	He*, H2	86	MS–5A.
12.4–21.5	5×10^{-8} – 3×10^{-4}	H2	47	On CO2 cryofrost.
≥4.2	10^{-11} – 10^{-7}	He	25	MS–5A.
4.2	10^{-11} – 10^{-7}	He	28	MS–5A.
4.3, 20	3×10^{-8} – 4×10^{-3}	He, H2, D2	63	MS–5A.
4.2	≤1.3 10^{-4}	H2	103	On D2 cryofrost.
~20	10^{-6} – 7×10^{-4}	H2	106	ACC.
4.2	not indicated	H2+He, D2+He, H2+D2	44	CCT.
≥4.2	10^{-8} – 10^{-3} (H2)	He, H2, D2, D2+He	65	MS–5A.
2.23–4.2	not indicated	He, H2, N2	42	SCT with N2.
7–20	≥10^{-6}	D2	85	MS–5A.
7.5–15.3	10^{-11} – 10^{-5}	He	48	On CO2 cryofrost.
4.2	2×10^{-7} – 4×10^{-4}	He, H2, He+H2	62	MS–5A.
4.2–18.2	10^{-9} – 2×10^{-6}	He, He+H2	38	ACC.
4.2	10^{-7} – 6×10^{-5}	H2, D2, H2+D2	21	CCT of H2 with D2.
2.18, 4.2	8×10^{-10} – 10^{-5}	He	51	On Ar cryofrost.
4.2	varied	He, H2, D2, T2	43	ACC.
10–20	varied	He	43	ACC.
4.2	6×10^{-5} – 4×10^{-4}	He, He+D2	46	CCT with Ar jet stream.
4.2	not indicated	He, H2	61	ACC.
18–22	10^{-7} – 10^{-4}	H2	105	ZX–15 charcoal (PRC).
6.0–11.0	8×10^{-10} – 8×10^{-8}	H2	50	On CO2 cryofrost.
9.5	not indicated	He, H2	70	ACC.
10–18	10^{-10} – 10^{-9}	He	84	ACC.
4.2–10	not indicated	He	69	ACC.
7.5–8.0	4×10^{-6} – 8×10^{-5}	He	35	Variety of charcoals.
4.3	not indicated	He, H2, He+H2	64	ACC.
4.3	not indicated	He+Ar	64	CCT of He w/ jet of Ar.
5(?)–23	1.2×10^{-5} – 2.3×10^{-4}	H2	109	ACC; CLGHe cryopump.
4.3	4×10^{-6} – 8×10^{-5}	He	64	Six varieties of charcoal
10.4–17.5	10^{-7} – 10^{-4}	He	102	ACC., CLGHe cryopump.
11–25	2×10^{-8} – 10^{-4}	H2	102	ACC., CLGHe cryopump.

1 Key: SCT: Sequential cryotrapping of one gas with another.
CT : Cryotrapping. ACC: Activated coconut charcoal.
CCT: Concurrent Cryotrapping with more readily pumped gas.
CF of S: Pumped on cryofrosts of same gas or substrate.
MS–5A, 10X and 13X are zeolite products of the Union Carbide Corporation.

The phenomenon of persorption is used to determine the size of the entrance apertures in sieve materials. This plugging of entrance apertures can cut off all gas access to the inner sieve labyrinth.

Gases which prove particularly troublesome in the plugging of sieve materials include oil vapors and water vapor. Also, any gas which might cryocondense on the outer surface of the sieve, so as to bridge over the entrance apertures, will cause plugging of the sieve. For example, Stern, et al., reported on the plugging of Linde 5A sieve by N_2, when pumping H_2 at ~20° K.[67] Even H_2 and D_2 have been noted to plug sieve materials for the pumping of He, when the sieve was operated at ~4.2° K.[46,61-65] This has been observed with both activated coconut charcoal[61] and the artificial zeolites.[62] There may also be some gas displacement effects as well as plugging effects in these cases.

In the early days of CLGHe cryopumping, these pumps were often used to replace diffusion pumps on vacuum systems. Diffusion pumps are notorious for oil backstreaming into traps, valves and systems to which they are appended. If, when replacing the diffusion pump with a CLGHe cryopump, the remainder of the system is not first completely cleaned of traces of the pump oil, eventually this oil will find its way to the sieve material and cause irreversible plugging. Problems due to the sieve plugging (e.g., negligible subsequent H_2 capacity) might take several months to become evident. Within time the system might have been shut down and the pump cycled back to RT several times. However, eventually the oil will get to the sieve material by creeping over RT surfaces.

Sieve materials are frequently used in devices to trap out possible oil backstreaming from mechanical pumps. These *sieve traps* may be periodically baked to high temperatures (i.e., ~300° C) to get rid of the oil accumulation, and refresh the sieve trap. However, baking to this extent is not possible with many forms of cryopumps. For example, if the sieve material of a CLGHe cryopump becomes plugged with hydrocarbons, the element to which the sieve material is bonded must be discarded, and replaced with a new element. This is for two reasons. First, the sieve material is usually bonded to the surface of a cryopanel using some form of nonbakeable epoxy-resin. Secondly, the CLGHe mechanisms would be damaged by high temperature bakeouts.

Sieve materials are also plugged by H_2O. This gas is present in most systems. Also, if it is initially cryocondensed on some surface other than the sieve, on warming up the pump at some later time, the water vapor will eventually find its way to the sieve material. Because of the tenacious manner in which H_2O sticks to many of these sieve materials, the only way in which it can be removed is through bakeout. This is particularly true of the artificial zeolites,[29,99] and there is some suggestion that it also applies to porous silver sieve materials.[20] For this and another reason, these zeolites are primarily confined to use in sorption pumps and bakeable LHe cryopumps.

The above H_2O sieve plugging problem would seem to present an insoluble obstacle to the use of these materials in CLGHe cryopumps. Fortunately, activated coconut charcoal does not require baking to remove most of the pumped water vapor. In fact, very pure charcoal is said to be almost *hydrophobic* in nature.[66] That is, at the conclusion of some number of pumping cycles, the pump may be brought back to RT. Water vapor consequently *finding its way* to the charcoal may be subsequently pumped away at RT using some form of trapped roughing pump. This proved most fortuitous, as it made possible the practical use of mechanical refrigerators, capable of achieving temperatures of 10-20° K, in the cryosorption and cryocondensation pumping of all gases. I will return to this subject.

Surface Bonding of Sieve Materials

Developing methods, free of organic binders, for the bonding of sieve materials to surfaces has been a challenge. The artificial zeolites are ceramic-like and have the form of grains or nuggets. Coconut charcoal may be purchased in small, regularly shaped cylinders of ~4 mm ϕ × 7 mm, or in irregularly shaped chunks varying in size from < 1 mm to > 4 mm ϕ.

I surmise that technologists used some form of glue or epoxy to bond sieve materials to metals prior to 1964, as Hemstreet, in a patent application at that time, noted the disadvantages of this practice.[58] He noted that the epoxy or glue tends to diffuse into part of the bulk of the sieve and cause plugging. He described in the final patent a number of methods used to successfully bond both charcoal and artificial zeolites to metal surfaces. Some involved the formulation of a slurry comprising inorganic binders, sieve material (< 1 mm ϕ) and metal particles (< 5 mm ϕ). This slurry was applied to a roughened, pretreated surface and then fired in either N_2 or Ar, so as to sinter the matrix together and to the metal surface. Other methods, also noted in the patent and elaborated on by Stern, *et al.*, involved putting the bulk slurry into cross-milled grooves machined into the metal surface.[67] Thereafter, the plate was baked to remove the inorganic binders. Presumably, it was still required that the matrix be sintered to the grooves, cut in the plate with a circular end-mill, though Granier and Stern also used this same method with an Aℓ plate.[68] This technique was used by the Excalibur Corporation to bond sieve materials to the third stage of a bakeable LHe cryopump (*e.g.*, see below).

Hseuh, *et al.*, report using a 6.3 mm thick bed of 3.5% AgSn solder to bond coconut charcoal.[43,61] The charcoal was pressed half way into the alloy while it was heated to a molten state under vacuum. He indicated that when subsequently inverting the bed and tapping on the metal plate, ~10% of the charcoal fell off the surface. However, the surface of the solder could not be seen in the vacant sites as some of the broken charcoal remained stuck to the surface. Tobin, *et al.*, report the successful bonding of coconut charcoal to Cu plates with the use of a "silver-based braze alloy".[69] According to Tobin, this was only applicable to the coating of small, flat surfaces. It was applied to a copper plate using a sheet of Sil-Fos (*i.e.*, 5% Ag, 15% P, 80% Cu) and an oxyacetylene torch.[88]

Coupland reported on the use of sintered Ni as a sieve.[70] Hobson reported on a technique of flame-spraying an 8.5% CaAg alloy onto a stn. stl. surface.[20] The Ca was thereafter oxidized with steam, and leached out of the matrix with 20% acetic acid, leaving a 1.25 mm thick bed of porous silver bonded to the surface. The bed could be baked to ~500° C.

All commercial manufacturers of CLGHe cryopumps use some form of epoxy to bond the activated coconut charcoal to the second stages of the cryopumps. It appears that this practice was *rediscovered* some time in the early to mid 1970s. I recall, when visiting CTI Incorporated in August of 1975, that they were producing and selling three varieties of cryopumps at that time. All featured the use of activated coconut charcoal on the second stages. Turner and Hogan, when reporting on a modified Taconis-cycle cryopump in 1966, were unaware of the technique.[71] Cryopumping papers published as late as 1977 reported the need of the use of turbo-molecular pumps to augment CLGHe cryopumps in the pumping of He and H_2.[72] This same year Visser, *et al.*, reported on the use of epoxy to bond charcoal to the second stage of a CLGHe cryopump. In May of 1976, Longsworth read a paper (unpublished) on CLGHe cryopumps before a joint meeting of the New York and Philadelphia Chapters of the AVS. The implication of this paper was that he had used epoxy to bond charcoal to the second stages of cryopumps on or before 1975. This technique has also been successfully used in LHe cryopump applications.[38]

5.2 Sorption Roughing Pumps

Sorption roughing pumps are not, as thought by some, scientific novelties which are confined to use in some laboratory. On the contrary, they are widely used in a number of very practical applications. For example, with few exceptions, all UHV surface science apparatus comes equipped with sorption roughing pumps. All large molecular beam epitaxy systems are sorption roughed. On a much larger scale, Neal instituted the use of sorption pumps to rough the sectors of the two-mile SLAC accelerator.[73] Each sector comprised a volume of $\sim 7 \times 10^3$ $\mathcal{L}$ with a surface area of $\sim 3 \times 10^6$ cm^2 of stn. stl. and Cu. Prior to the advent of the use of CLGHe cryopumps in the semiconductor industry, there was a brief period when many coating systems were combination-pumped systems. All these systems were cryosorption roughed. The primary motive in using sorption pumps in all the above cases was to avoid possible catastrophic problems stemming from back-diffusion of oil from mechanical roughing pumps. A definitive work on the practical application of sorption rough pumps was published in 1972 by Frederick Turner of Varian Associates.[74] I liberally reference his work herein.

If one were to pour a quantity of sieve material into a small volume with flanges on each end, such as shown in Fig. 5.2.1a, the bulk sieve would afford an effective trap for gases including water vapor and oils. These are called *sieve traps* and are often used in conjunction with mechanical pumps to protect, to some extent, the system being rough pumped from backstreaming pump oils.[75] As will be later discussed, these RT sieve traps are also used to filter out impurities in gas streams. For similar reasons, bulk sieve materials have also been used, both at RT and reduced temperatures, as traps over mercury and oil diffusion pumps, as shown in Fig. 5.2.1b.

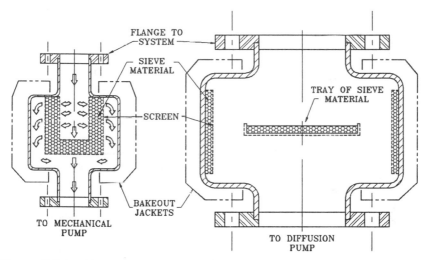

Figure 5.2.1a. Mechanical pump sieve trap.[75]

Figure 5.2.1b. Sieve trap used to limit oil or Hg backstreaming.

The first commercial use of bulk sieve materials for sorption roughing came about as a consequence of the use of sputter-ion pumps. These pumps and the devices to which they were appended (*e.g.*, klystrons) could be damaged by backstreaming oils from mechanical roughing pumps. Jepsen submitted patent application for a commercial sorption roughing pump in 1959.[76] This pump made use of bulk coconut charcoal. A cross section of the pump is shown in Fig. 5.2.2. To effect pumping, the vessel containing the charcoal was cooled on the exterior by filling a dewar in which it resided with LN_2.

This type of sorption roughing pump was widely adopted in many applications requiring the preservation of clean surfaces and UHV conditions. However, the use of coconut charcoal had a couple of disadvantages. First, the charcoal tended to fragment into carbon dust within the pump. This dust could accidentally be carried over into a UHV system if the isolation valve between the pump and system was inadvertently opened while the system was under vacuum and the pump at atmospheric pressure.

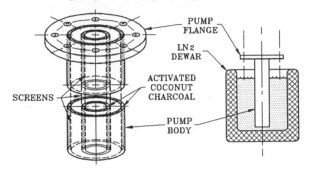

Figure 5.2.2. Representation of the first commercial sorption roughing pump as described by Jepsen.[76]

Secondly, the presence of LN₂ suggests that under certain circumstances there may be LO₂ (*i.e.*, liquid oxygen) present. Carbon dust and LO₂ comprise a potentially explosive mixture. This was recognized, and the artificial zeolites were subsequently used in sorption pumps of similar configuration.

These pumps are extremely simple devices. Bulk sieve material is simply poured into the volume of the pump. A coaxial screen is located down the center of the pump so that gas will have access to the sieve material the full length of the pump. Single sorption pumps are available which accommodate from ~1.4 to 4.2 kg of sieve material. The pump, and accompanying LN₂ dewar, are simply made longer if greater capacity is needed. It is important that the sieve material be uniformly cooled. Various commercial designs were developed to facilitate this uniform cooling of sieve material. Two such schemes are shown in Figs. 5.2.3 and 5.2.4. The design in Fig. 5.2.3, a sorption roughing pump manufactured and sold by Varian Associates, makes use of extruded Aℓ, finned quadrants. The finned quadrants are longitudinally welded to form a tubing having an inner finned structure. A blank Aℓ cap is then welded on one end of this cylinder and a flanged cap, with an Aℓ to stn. stl. transition joint, is welded to the other end. This simple configuration, with a screen and some sieve material, comprises the Varian sorption pump. The fins, radially protruding into the innards of the cylindrical pump, afford a means of cooling even that sieve material located near the center of the externally cooled pump.

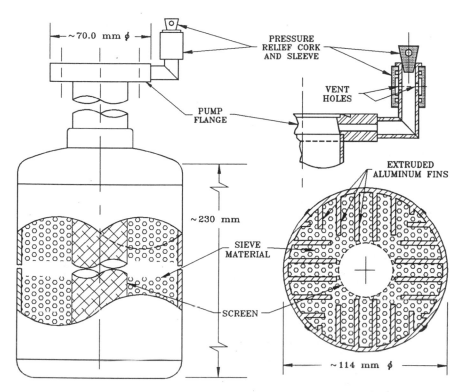

Figure 5.2.3. Representation of a single−length sorption pump, with ~1400 g sieve charge, offered by Varian Associates, Incorporated.

Perkin-Elmer commercially manufactures and sells a pump of slightly different design. This sorption rough pump is represented in Fig. 5.2.4. In this configuration, cooling of the more centrally located sieve material is accomplished by the use of six stn. stl. tubes, extending the full length of the sorption pump and serving as convective passages for the flow of LN2. The visible top of the sorption pump is in fact not an integral member of the vacuum vessel. Rather, it serves as a splash shield for the LN2 which is vigorously circulating about the outer shell and within the six convective tube passages. Copper fins are attached to the six tubes, radially extended out into the volume of the pump, and thus provide the needed sieve temperature uniformity during cooldown. These two pump designs are merely given as representative examples of commercial sorption pumps. Other companies are known to have manufactured and sold these pumps in the last few decades (*e.g.*, see (7 7)).

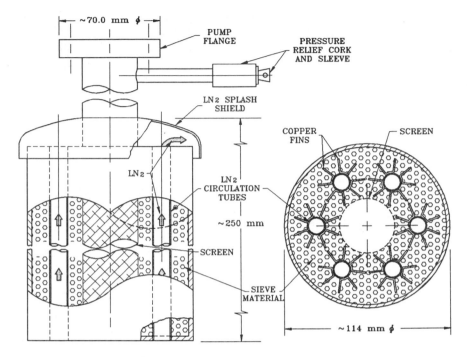

Figure 5.2.4. Representation of a single-length sorption pump, with ~1400 g sieve charge, offered by Perkin-Elmer, Inc.

5.2.1 Staging of Sorption Pumps

The *staging* of sorption pumps is defined as the sequential use of more than one pump to rough a volume, but with no more than one pump evacuating the volume at the same time. For example, assume you have three sorption pumps appending your UHV system for the purpose of roughing the system. Such a configuration is schematically represented in Fig. 5.2.5.

The reasons for this *staging* of pumps will be clarified, after first clarifying the meaning of staging. The sorption pumps each have their associated valves leading to a common manifold. These may be elastomer sealed valves. The roughing manifold is in turn attached to the UHV system through one valve, which may perhaps be of all-metal construction. In some applications, the first roughing stage of this three-stage roughing is accomplished by some form of bulk roughing pump such as a carbon-vane, gas-aspirator, or even a metal-bellows pump.[78] By means of the first-stage roughing, anywhere from 50 to 80% of the bulk gas is removed. After this, the sorption pumps come into play. Though these pumps might be simultaneously cooled with LN2, roughing is accomplished by the sequential valving in and out of each pump.

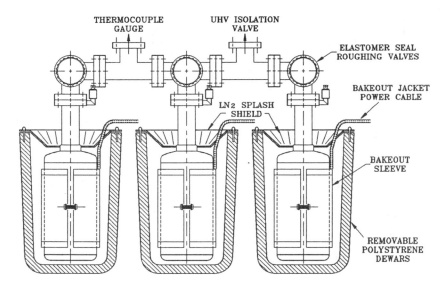

Figure 5.2.5. Typical configuration of a multi-stage sorption system.

This will lead to a pump-down curve such as given by Turner,[74] and shown in Fig. 5.2.6. In this instance, Turner used three sorption pumps to rough pump a 200 $\mathcal{L}$ vessel, initially at one atmosphere of air. No first stage bulk roughing pump was used.

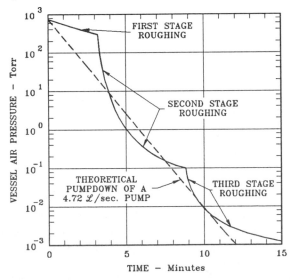

Figure 5.2.6. The staged roughing of a 200 liter volume using three single–high sorption pumps filled with Linde 5A sieve material.[74]

In this measurement, each sorption pump was used individually to pump on the vessel, and it was then valved out prior to use of the next sorption pump. This is what is meant by the staging of sorption pumps. Note that though the first *knee* of this pressure *vs.* time curve occurs at $\leq$ 3 minutes into the pumpdown, over half of the bulk gas is pumped in this interval of time. For comparison purposes, the theoretical pumpdown curve of a 4.7 $\mathcal{L}$/sec. (*i.e.*, 10 ft^3/min) mechanical pump is also shown in this figure.

These sorption pumps were charged with ~1400 g of artificial zeolite. Several of the artificial zeolite materials were exhaustively characterized by Stern and his colleagues.[34] Typically, the Linde 5A sieve is used. An adsorption isotherm for the gas N_2 and one of these pumps, as taken by Turner, is given in Fig. 5.2.7.[74]

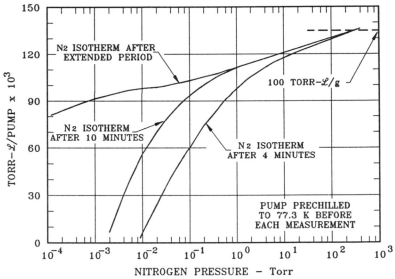

Figure 5.2.7. Variations in N_2 adsorption isotherms, with time, in a sorption pump charged with 1350 g of Linde 5A sieve material.[74]

The pump was charged with 1350 g of Linde 5A sieve material and refrigerated with LN_2. Because of the thermal conductivity limitations of the sieve material, and the aforementioned diffusion-related processes associated with molecular sieve pumping, achieving the equilibrium pressures of an adsorption isotherm is a long-term process. This is evident in the data given in Fig. 5.2.7. For example, Hobson reported that when pumping both Kr and Xe on porous Ag at LN_2 temperatures, it often took up to an hour for pressures to equilibrate.[20] Similar equilibrium pressure, time-dependent effects were reported by Stern, *et al.*, for the gases H_2 and He on a variety of sieve materials at 20.2° K and 77.3° K;[29,34] also see, for example, Halama and Aggus[25,26] for He on Linde 5A sieve at $\geq$4.2° K; Sedgley, *et al.*, for the pumping of He on six varieties of charcoal;[35] and Longsworth for H_2 on charcoal and *supercarbon*.[24,79] Because of this time dependency, one must use judgement in applying any adsorption isotherm data to the calculation of pumpdown performance of sorption pumps.

Contrary to common belief, the gases He, H_2 and Ne are sorption rough pumped. Adsorption isotherms for the separate gases He, H_2 and Ne, taken by Turner, are given in Fig. 5.2.8.[74] These isotherms are given for one sorption pump, with a charge of 1350 g of Linde 5A sieve material at LN_2 temperature (*i.e.*, ~77.3° K). Of course, the data in this figure were taken after the extended times needed for true equilibrium to be achieved.

To qualitatively appreciate the importance of staging, let us conduct a simple *gedanken* experiment (*i.e.*, a mental exercise). Assume that we have a vessel with a volume of 1.0 $\mathcal{L}$ and that it is pressurized with H_2 to 750 Torr. Assume that a valved manifold is constructed which enables us to simultaneously pump on the vessel with 750 sorption pumps, each having 1350 g of Linde 5A, and each chilled to LN_2 temperatures. In that we have simultaneously valved in all 750 sorption pumps, it is necessary that each sorption pump, on reaching equilibrium, pump the equivalent of only ~1.0 Torr-$\mathcal{L}$ of H_2. Referring to Fig. 5.2.8, we can therefore predict that on achieving equilibrium, the total system H_2 pressure will be ~4×10^{-4} Torr.

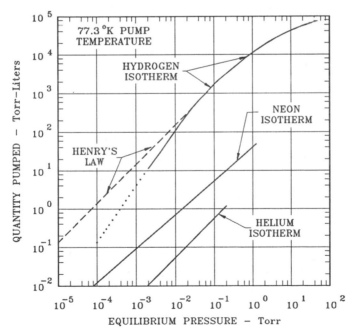

Figure 5.2.8. Adsorption isotherms for He, Ne and H_2 pumped with a sorption pump containing ~1359 g of Linde 5A sieve material. [74]

Now let us attach only two of the same sorption pumps to the system, and repressurize the 1.0 ℓ volume with 750 Torr of H_2. This time, let us valve in only one of the two sorption pumps. The pressure of the first sorption pump, on achieving equilibrium, is again found from Fig. 5.2.8, and determined to be $\sim 5 \times 10^{-2}$ Torr as a consequence of pumping ~ 750 Torr-ℓ H_2. Now, let us valve out the first sorption pump. The H_2 pressure in the 1.0 ℓ volume is still $\sim 5 \times 10^{-2}$ Torr. Now let us valve in the second, unused sorption pump. This sorption pump must pump the equivalent of $\sim 5 \times 10^{-2}$ Torr-ℓ of H_2. Again referring to Fig. 5.2.8, we see that the final equilibrium pressure probably falls somewhere in the mid 10^{-5} Torr range.

We have learned two things from the exercise: 1) that sorption rough pumps, even if cooled to only LN_2 temperatures, very effectively pump H_2; 2) that we can obtain better ultimate H_2 pressure by the staging of two sorption pumps than could be achieved with the simultaneous pumping of 750 such pumps.

Effects of Neon When Rough Pumping

We have seen the obvious benefits of staging pumps when pumping a single, difficult gas such as H_2. Similar benefits are gained when using these pumps to rough-pump vessels filled with air. For example, Fig. 5.2.9 shows pump-down data taken by Turner for two conditions: the first gives total vessel pressure, as a function of time, when using two sorption pumps to simulta-neously pump down a 100 ℓ vessel at an atmosphere of air; the second gives the total vessel pressure, in time, when using the sorption pumps to sequen-tially pump down the vessel under identical conditions.

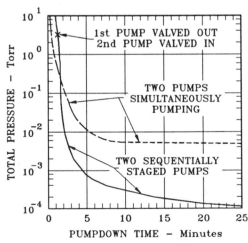

Figure 5.2.9. Pumpdown of a 100 liter volume, initially at atmosphere, with two sorption pumps. In one case the pumps are sequentially staged, in the other, used simultaneously.[74]

The results of Fig. 5.2.9 are explained by the presence of Ne in the air. That is, the difference in the final pressures of the two pumpdown curves is due to the partial pressure of Ne in the air. We recall from Table 1.8.1 that the partial pressure of Ne in the atmosphere is ~1.4 × 10⁻² Torr. The total pressure of Ne in the vessel which was simultaneously sorption pumped is less than its partial pressure in the atmosphere. This merely indicates that there was some *cryotrapping* of the Ne in the sorption pumps.

The reason one is able to achieve better total pressures with staged pumping is modeled in Fig. 5.2.10. It is suggested that substantive quantities of Ne are pumped in the first of the two staged pumps as a consequence *viscous drag* effects. That is, the Neon is swept along with the O_2 and N_2 evacuated from the system. If the first pump is valved out of the system in sufficient time, (*i.e.*, at time t_1 of Fig. 5.2.10) and the second pump valved in, the Ne is retained in the first pump. However, if one dawdles in valving out the first pump (*e.g.*, to time t_2 of Fig 5.2.10), Ne will back diffuse from the first pump, and eventually contaminate the system with a Ne partial pressure equivalent to that found at atmospheric pressure. As a *rule of thumb*, one should attempt to valve out the first pump on achieving a roughing pressure of ~0.1 Torr.

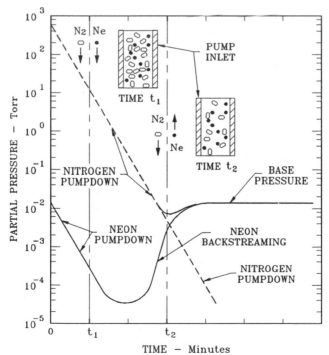

Figure 5.2.10. Limitation in observed base pressure of sorption roughed system as a consequence of Neon backstreaming.[74,99]

Stern and DiPaolo noted the benefits of staging sorption pumps and referred to the above effect as *entrainment pumping*.[99] Three important lessons are learned by these results: 1) gas displacement effects probably exist when sorption roughing; 2) though the component of Ne is only ~1.8×10^{-5} of the total atmospheric pressure, only ~65% of this small Ne component could be removed from the system by cryotrapping; 3) the staged roughing of systems will also aid in the pumping of the more *difficult* gases.

5.2.2 Dewars and Bakeout Regeneration Heaters

The artificial zeolites must be baked to remove sorbed water vapor. If this is not done, though other air gases are liberated on warming up the pump, H_2O will remain behind and plug the sieve material on subsequent use. This is also required on initially charging a pump with new sieve material. Because of this, some form of sorption pump bakeout provision must exist. This does not mean that each time a sorption pump is used it must be baked. On the contrary, in UHV system applications, they are often used a number of times between regenerations (bakeouts). This is totally dependent on the partial pressure of H_2O in the gases being pumped. The regeneration process merely involves heating up the valved-off pump for an hour or so. Released gases escape through a pressure relief valve which appends all of these sorption pumps. It is not necessary to pump on the sorption pump during regeneration, though associated *bulk* roughing pumps are often valved in to the sorption pumps during the last ten minutes or so of bakeout.

Figure 5.2.11 shows H_2O isotherms on the Linde 5A sieve material.[74,80] Four H_2O isotherms are shown in this figure. In this case, a reduction in the capacity of the sieve material for H_2O at higher temperatures is interpreted as a measure of the thoroughness of the bakeout regeneration. Stern and DiPaolo have shown that when the amount of adsorbed H_2O in the Linde 5A sieve material exceeds 7% by weight, the capacity of the sieve material for N_2 is almost completely suppressed.[29] This corresponds to sorption of ~70 Torr-$\mathcal{L}$/g of sieve material. From this we conclude that bakeout temperatures in excess of 250° C are required to eliminate sufficient water vapor to regenerate the sieve material. In practice, bakeout at a temperature of ~300° C proves adequate.[80] Also, we conclude that some form of bulk roughing might be helpful in the regeneration process. However, this does not prove to be essential.

Metal bakeout jackets are provided by vendors who sell these pumps. They take the form of split sleeves which, as shown in Fig. 5.2.5, fit over the cylindrical pumps. The resistance of the heater is usually matched to available domestic voltages. For example, one need only plug in the power cable to a 110 VAC, 1ϕ outlet and come back in an hour or so to a regenerated pump. These sleeved heaters may be immersed in LN_2 without damage. Therefore, they may be permanently installed on the sorption pumps.

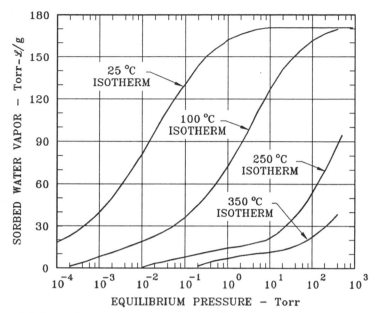

Figure 5.2.11. Adsorption isotherms of water vapor on Linde 5A molecular sieve as reported by Bannock,[80] and Turner.[74]

Two types of dewars are commonly used. One is constructed of an inexpensive plastic foam, or polystyrene material. The second comprises a stn. stl. vacuum *thermos bottle*. The plastic dewars will be damaged if not removed during bakeout. However, in low profile vacuum systems, the bottoms of the sorption pumps are often located so close to the floor that it is impossible to remove the dewars. In this case, the more expensive stn. stl. dewars are used and remain in place during pump regeneration.

5.2.3 Safety Considerations

Of course, there are certain inherent hazards associated with the use of liquid cryogens such as LN_2. Many establishments wisely require the use of protective goggles, *etc.*, when handling and using these cryogens. Above all, pressure relief provisions must exist when using these pumps. Pressure relief provisions which are an integral part of a purchased pump should not be altered or tampered with in any manner. As shown in Figures 5.2.3 and 5.2.4, pressure relief valves on commercially sold pumps are always an integral part of the pump vacuum envelope. This is by design so as to make it impossible for the pump to be installed on a system without the pressure relief provision.

The importance of these pressure relief valves is related in the following potential *horror story* which fortunately had a happy ending. In days past, these pressure relief *valves* comprised rubber corks which were inserted into small tubes, in turn welded to the flange or neck of the sorption pump (*e.g.*, see Figures 5.2.3 and 5.2.4). When the pressure built up in the pump as the result of regeneration, the cork would pop out of the small tube and the gas therein, *presumably not toxic*, harmlessly vent into the laboratory. The corks were loosely attached to the neck of the pump with some sort of light chain or cable. Of course, in time the chain would break or be somehow lost and the user would find himself on his hand and knees, in a dark corner of the lab, looking for the missing cork which had been launched across the room during the last regeneration. Also, in the quiet hours of the night, as you *puttered* with some other work, a cork popping out of the vent tube could be very startling. Because of this, a Viton® sleeve vent valve was invented as a substitute for the cork and the cork was eliminated (see Fig. 5.2.3).

Many production facilities making use of these sorption pumps have provisions for automatic regeneration and LN_2 dewar filling. One such company used three, triple-length sorption pumps (*i.e.*, each charged with ~4.2 kg of sieve material), arrayed on a manifold similar in configuration to that shown in Fig. 5.2.5. In scheduled preventive maintenance programs, they typically changed the sieve material in their pumps each year, while at the same time replacing all of the elastomer o-rings in the pump valves. On one occasion, having done this maintenance, they activated the automatic bakeout and post LN_2 filling provisions. Some time later, but prior to the staged roughing of the main chamber, the base pressure of each sorption pump was determined by opening the valve over that pump and measuring the pressure with a thermocouple gauge located in the main roughing manifold. Inexplicably, the base pressure of one of the pumps was ≥ 1 Torr, where the other two evidenced the expected low base pressures. It was therefore concluded that perhaps the *troublesome* pump had not been given sufficient time to cool to LN_2 temperature. The two well-behaved pumps were used for the staged roughing of the main vacuum chamber and then valved out of the system. The operator then left the scene, with the automatic LN_2 dewar filling cycle still activated so as to further cool the *troublesome* pump.

Several hours later the operator returned and found that the base pressure problem still existed in the third pump. From this he concluded that the pump bakeout cycle had not adequately regenerated the new sieve material in this third pump. He activated the bakeout regeneration cycle. Moments later, the bottom of the *troublesome* pump exploded. Fortunately, the explosion was contained by the stn. stl. dewar, but a *fountain* of sieve material and cryogen spewed out from the top of the dewar. Most fortunately, the operator wore safety glasses, and because of this was not seriously injured.

The event was later duplicated under laboratory conditions. It was determined that: 1) in the process of the preventive maintenance work, the technician had mistakenly assembled the valve over the *troublesome* pump without a bonnet seal; 2) this pump had pumped on a leak at near atmospheric pressure for ~8 hours, while the dewar was automatically topped off with LN₂; 3) due to this, the pump had filled with a mixture of LN₂ and LO₂; 4) on turning on the bakeout heater, cold gases and condensed liquid from within the pump caused a contraction of the Viton® vent sleeve when escaping between the sleeve and vent tube; 5) the flow of cryogen was so restricted by the ever-contracting Viton® sleeve that the pump exploded at a pressure of ~700 Nt/cm². The solution to the problem was simple. It was determined, that when using a Viton® cork, rather than the sleeve, under identical conditions, the cork would pop out at a pump pressure ≲ 50 Nt/cm². The cork was restored, while keeping the sleeve venting provision.

5.3 Liquid Helium Cryopumps

The creation and handling of liquid helium is a mix of both art and science. Excellent references exist on these two subjects, including a book by Scott,[8 1] and a second by Barron.[8 2] These two books are also excellent references on properties of materials for use in cryogenics applications. The only method of practically achieving temperatures ≲ 10° K is through the use of LHe. Some of the early methods used to liquefy this gas will be briefly discussed in another section. For now, it is sufficient to note that all cryo-pumping experiments reported in the literature at temperatures ≲ 10° K were done with LHe. A quick glance at Tables 5.1.3 and 5.1.5 indicates that a good portion of the data therein was obtained with LHe cryopumps. However, many of these pumps were of the nature of experimental apparatus rather than practical cryopumps.

5.3.1 Classification of LHe Cryopumps

Before first classifying these pumps, I will describe some of their salient features. Figure 5.3.1 shows a cross section of a *cryogen-fed* cryopump for which Claude made patent application in 1968.[8 3] *Boiling pool* LHe cryopumps, defined below, had been used much earlier. However, this particular cryopump has some important general features.

At first glance, it has a striking resemblance to features of many of the CLGHe cryopumps reported on in the literature (*e.g.*, see Fig. 5.4.1). First, the pump has a flanged, RT body (1). A *first stage array* (2), cooled with LN₂, is used to shield the *second stage array* (3) which is maintained at LHe temperatures. The first stage array is a chevron-type structure which serves to prevent RT radiation, coming from the system to which the pump is appended, from shining directly on the LHe cooled surfaces. However, this chevron is somewhat transparent to the passage of gas. (It is the convention in cryopumping vernacular to refer to the warmest stage as the *first stage* and thereafter, in multi-stage cryopumps, to number successive stages in the order of their decreasing temperatures. This precedence stems from multi-stage gas cooling and refrigeration processes.)

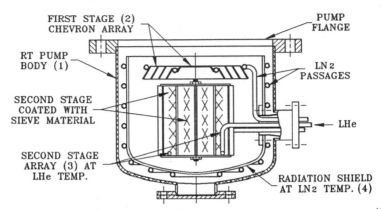

Figure 5.3.1. Early cryogen–fed cryopump patented by Claude.[83]

A third metal, bowl-like member (4) closely follows the contour of the pump body, but is thermally isolated from the body. It too is cooled with LN2 and serves as a radiation shield between the RT pump body and the second stage. Being at the same temperature as the first stage chevron, it is also classified as part of the first stage of the cryopump. The second stage, suspended by wire cables, is completely surrounded by surfaces which are cooled to LN2 temperature. Using a parallel plane approximation and unity spectral emissivities for all surfaces, the equation is given in problem 8 of Chap. 4, we calculate that thermal radiation on the second stage as a consequence of the interposition of the first stage is reduced by ~×210.

The second stage of Claude's pump was coated with a sintered Ni powder sieve material to facilitate cryosorption pumping of certain gases. The cryogens were *force fed* through tubes attached to the arrays. In Claude's case, they were LHe and LN2. They could just as well have been cold gases, or one cryogen a cold gas and the second a liquid, *etc.*[67,68]

Had the arrays themselves been the containers of pools of liquid cryogen, this pump would have been classified as a LHe *boiling pool* cryopump. An example of a combined *boiling pool* and *cryogen-fed* cryopump is shown in Fig. 5.3.2. Variations of this cryopump, once manufactured and sold by the now defunct Excalibur Corporation, were reported on in numerous cryo-pumping articles.[25,26,38,43,62,63,65,85,86]

The pump shown in Fig. 5.3.2 is actually a *three-stage* cryopump. Two stages are *boiling pool* stages (*i.e.*, the first (1) and the third (3) stages) and the middle stage (2) a *cryogen-fed* stage. In this example, an intermediate 20° K stage (2) is interposed between the first LN2 stage and third LHe stage. The gaseous He, which boils off the pool in the third stage, passes through tubing attached to the chevron comprising the second stage, thereby cooling this stage to ~20° K. The third stage was coated with sieve material as described in a previous section.

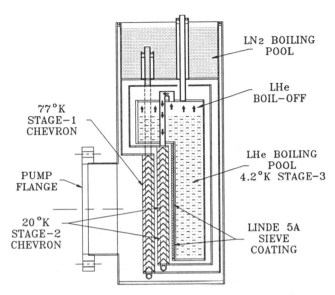

Figure 5.3.2. Three—stage Excalibur CVR—100 boiling pool and cryogen—fed cryopump.

This three-stage cryopump evidenced thermal instabilities under certain operating conditions. When operating at very high pressures, gaseous He, boiling off the LHe pool in the third stage, super-cooled the second stage to near that of LHe. Therefore, gases which would normally not condense on this stage at 20° K (*e.g.*, H_2) did so at temperatures near 4.2° K. However, on reducing the influx of gas being pumped, the boil-off from the third stage would decrease and the second stage increase in temperature. This caused gases condensed thereon to be liberated, which raised the pressure, *etc.* This phenomenon, reported on by Fisher,[63] was at times evidenced as cyclical pressure instabilities, pressure run-away,[65] and long-term pumpdown subsequent to high pressure operation. You will note in a following section that anomalous long-term pumpdowns under some circumstances are also evidenced by CLGHe cryopumps after high pressure operations. This is sometimes referred to as a *pressure clamping effect*.

Improved variations of three-stage LHe cryopumps have been reported on in the literature. For example, Hseuh and Worwetz reported on a three-stage cryopump which had three boiling pool stages.[61] The second and third stages were LHe stages and the first stage was a LN_2 stage. A schematic representation of this pump is shown in Fig. 5.3.3.

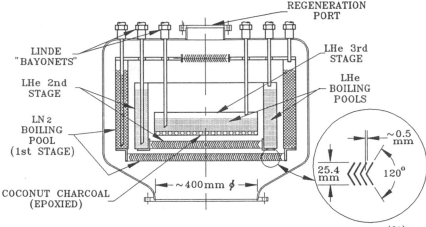

Figure 5.3.3. Three−stage cryopump of Hseuh and Worwetz.[61]

The intermediate boiling pool stage served three purposes. First, it avoided the pressure instability problems associated with the pump shown in Fig. 5.3.2. Secondly, the third stage was coated with sieve material. This material was needed in order to pump substantive quantities of He. By placing an optically dense second stage array, cooled to LHe temperatures, in front of the third stage, H_2 and its isotopes would be cryocondensed on the second stage. This served to prevent plugging of the sieve material by H_2 and reserved the sieve material for the pumping of He. Thirdly, Chubb, *et al.*, had reported that stray IR would cause desorption of H_2 on LHe surfaces.[87] The LHe array interposed in front of the final stage precluded all possibility of this occurring on the third stage. The H_2 possibly desorbed from the second stage would be pumped by the third stage. As we will see, Benvenuti, *et al.*, showed that under certain conditions such intermediate IR shielding stages were not required in LHe cryocondensation cryopumps (*i.e.*, those not using sieve materials).[89-92]

A third type of cryopump is the *linear-arrayed* pump. Pumps of this variety are used primarily in large space simulation chambers and fusion-related apparatus. Hood provides an excellent review of the design and applications of these pumps.[93] An example of the configuration of these pumps is shown in Fig. 5.3.4. They are usually two-stage, cryogen-fed pumps. Many arrays such as shown in Fig. 5.3.4 comprise extruded Aℓ panels, with integral cryogen piping. Again, the cryogen-fed LHe cryopanels are optically shielded from the RT walls of the chamber and equipment within the chamber by cryogen-fed LN_2 panels. Such pumps are used in linear module configurations in the pumping of neutral beam lines in fusion applications.[94] Coupland describes the design and performance of some of these pumps.[70,95] (Also see Duffy and Oddon,[96] and Thibault, *et al.*[51])

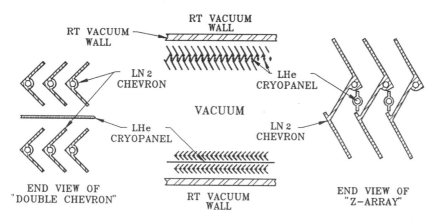

Figure 5.3.4.   Two linear cryopanels described by Hood.[93]

In fusion-related applications it is necessary that these pumps be able to pump He and the isotopes of hydrogen.[51,64,97] It was noted in section 5.1.7 that sieve materials at LHe temperatures can become plugged by the isotopes of hydrogen and become ineffective in the pumping of He. Because of this plugging problem some have resorted to the use of the concurrent cryotrapping of He, on 4.2° K surfaces, with Ar. Batzer, *et al.,*[46] describe such a system. The cryotrapping scheme, as further described by Scdgley, is shown in Fig. 5.3.5.[64]

As noted, the Ar cryotrapping of He and H_2 on 4.2° K surfaces was reported on much earlier.[45] In the above case, jet streams of Ar were directed onto 4.2° K cryopanels. Helium pumping speeds of the order of ~1.0 $\mathcal{L}$/sec-cm^2 were achieved in so doing. The speed of a similar panel coated with coconut charcoal produced speed of the order of ~5.0 $\mathcal{L}$/sec-cm^2.[64]

From the above we are now able to lend some organization to our classification of LHe cryopumps and place them in categories as follows:

1. *cryocondensation,*
2. *cryosorption,*
3. *cryogen-fed,*
4. *boiling pool,*
5. *arrayed* and *lumped* (*i.e.,* large appendage pumps),
6. *one-stage, two-stage* or *three-stage,* and
7. combinations of the above.

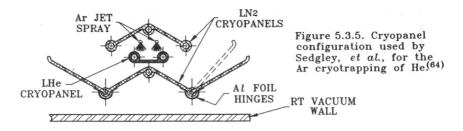

Figure 5.3.5. Cryopanel configuration used by Sedgley, *et al.*, for the Ar cryotrapping of He.[64]

5.3.2 Chevron Design

It has been noted that in multi-staged cryopumps some form of higher temperature cryopanel is often used to shield the lower temperature stages or cryopanels. This serves in some cases to protect sieve materials from plugging effects. In other cases, the intermediate cryopanel is used primarily to shield the lower temperature panel from thermal radiation. In this latter instance it is merely an issue of economics. Though the gas boil-off from LHe cryopumps is always captured and recycled, the very cost of liquefying this gas (*i.e.*, the energy required) is $\times 25$ to $\times 30$ greater than that of producing, for example, LN_2. Therefore, LHe cryopumps make use of LN_2 cooled louvers or chevrons as radiation shields for LHe cryopanels. Similar arguments exist for CLGHe cryopumps; that is, it is much more difficult to produce 50 W of refrigeration at 10° K than 50 W at 77° K. Therefore, louvers or chevrons attached to the first stages of these pumps are used to shield the second stages of CLGHe cryopumps.

A perfect shielding louver or chevron would be one where no thermal radiation is able to pass through and impinge on the colder stage of the cryopump. However, excluding what are called the *highly condensible* gases (*e.g.*, H_2O), most of the gases must be pumped on the colder stages of cryopumps. If *no* radiation can possibly pass by the shielding louver or chevron, molecules will also be impeded from being pumped on the colder stages. Therefore, design of the shielding louvers or chevrons is a compromise between acceptable radiation loads on the colder stage and the conductance of the shielding stage for the passage of gas. The gas conductance of shielding cryopanels is often called its *molecular transmissivity*. Similarly, the measure of thermal radiation or *light* which can pass the shielding cryopanel is called the *thermal* or *optical transmissivity*. The molecular or thermal transmissivity of a cryopanel is merely the ratio of the passage of particles or light through the chevron relative to that passing through a comparably sized aperture.

There are three basic types of optically dense cryopanels. These are shown in Fig. 5.3.6. The first is the *louvered* cryopanel, the second the *chevroned* cryopanel and the third the *z-chevroned* cryopanel. The *louvered* cryopanel, nestled behind the louvered cryoshield (*i.e.*, in the upper left corner of the figure) was first proposed by Santler.[110] Because of this, it is sometimes referred to as the *Santler chevron*. I am not sure of the origin of the remaining configurations. A comprehensive study of the transmissivities of louvers, chevrons and other labyrinth configurations was reported on by Levenson, Milleron and Davis.[111] They used Monte Carlo type calculations to predict the theoretical passage of gas through the optically dense structures, and then empirically verified their findings with vacuum measurements. *Monte Carlo* calculations involve the use of a computer to calculate, on average, the random flow of individual molecules from one place to another in a vacuum system. It is assumed in these calculations that the molecules, when momentarily sticking to some surface along the way, come back off the surface according to the cosine law (*i.e.*, see Chap. 3).

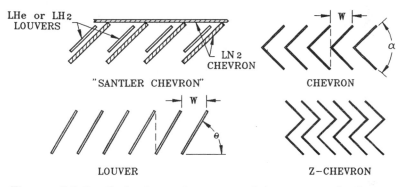

Figure 5.3.6. End view of cryopanel louvers and chevrons.

Define the spacing between the chevrons or louvers shown in Fig. 5.3.6 as W, and their lengths as L. Levenson, *et al.*, determined that the molecular transmissivity of the louver, with zero overlap, $L \gg W$, and an angle $\theta = 60°$, converges to ~0.5. This finding is directly applicable to cryopumps such as shown in Figs. 5.3.1 and 5.4.1. Also, for the conventional chevrons, with $L \gg W$, and for angles $\alpha = 60°$, $90°$ and $120°$, the molecular transmissivities converged to ~0.19, ~0.25 and ~0.28, respectively.

Hydrogen desorption by IR (*i.e.*, infrared) radiation, first reported by Chubb, *et al.*,[87] was particularly troublesome when attempting to cryocondense H_2 on LHe cooled surfaces. There was an apparent departure of the hydrogen vapor pressure from the Clausius-Clapeyron equation. This was noted by Lee,[12] and Benvenuti and Calder.[90] Evidently, this departure stemmed from IR radiation impinging onto the cryopanel on which the H_2 was pumped. Benvenuti, *et al.*, determined that there was a photon (*i.e.*, light) energy threshold of ~45 μm below which the stray IR passing through a shielding chevron would not desorb cryocondensed H_2.[92] They postulated that the IR photons created phonons in the metal substrate which in turn *bumped* off cryocondensed H_2. They arrived at this conclusion by noting that if the H_2 was first cryocondensed on a cryofrost of a more readily cryocondensed gas, the H_2 pressure, on subsequently cryocondensing on the former gas, more closely fit the predicted Clausius-Clapeyron equation. From this they concluded that the predeposited cryofrost established the equivalent of a *phonon cushion* between the metal surface and the cryocondensed H_2. More importantly, they determined just how sensitive the cryocondensed H_2 was to possible desorption by low energy photons.

As an aside comment, this effect may have very important implications in both proton and heavy ion colliders. That is, where synchrotron radiation may not be of sufficient energy to desorb photoelectrons, which in turn desorb gas, phonons or plasmons created by very weak synchrotron radiation may cause desorption of surface gases and pose beam scattering problems.

Because of their findings, Benvenuti, *et al.*, made refined calculations and measurements of the molecular and light transmissivities of chevrons of various geometries. Their initial conclusion was that some form of *z-chevron* would be needed to properly shield the surface on which H_2 was cryocondensed.[91] These *z-chevrons* have molecular transmissivities of 50-75% of the conventional chevrons, and because of this are less desirable for high pumping speeds. However, in subsequent studies, they determined that by *blackening* the first stage conventional chevron with a low reflectivity material, and silver plating the H_2 cryocondensing surface, they were able to achieve high molecular transmissivities with very low light transmissivities. Results of their findings of molecular transmissivities are summarized in Fig. 5.2.7.[89] Depending on the surface photon reflectivity, they were able to achieve photon transmissivities of the order of ~1-5 × 10^{-4} for chevrons with α = 90° and 120°.

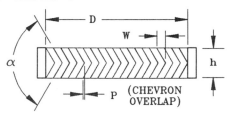

CHEVRON TRANSMISSIVITIES vs. D/h, THE ANGLE α AND WITH P = ZERO.

D/h →	2.5	5.0	10.0	20.0	40.0
α = 90°	0.174	0.217	0.231	0.235	0.238
α = 120°	0.222	0.258	0.269	0.274	0.275

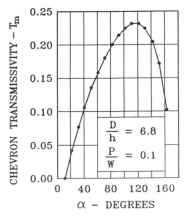

Figure 5.3.7.   Chevron molecular transmissivities vs. geometry as reported by Benvenuti, Blechschmidt and Passardi.[89]

5.4 Closed-Loop, Gaseous Helium Cryopumps (CLGHe)

I will first discuss the design and operation of these cryopumps, then some of the history behind the development of the refrigerators, and then discuss some of the design and operational features of these CLGHe refrigerators. Closed-loop, gaseous helium cryopumps (cryopumps hereafter) are very simple devices. The key elements are the refrigerator and, of secondary importance, the cryopanels and pump body. This order of importance stems from the relative cost and complexity of the refrigerators, compared to that of the cryopanels and pump body. The cryopanels comprise simple, fabricated metal components. Their design merely requires that a little thought be given to considerations discussed in the previous sections. The cross section of a typical cryopump is shown in Fig. 5.4.1.

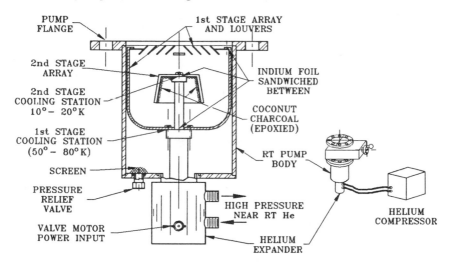

Figure 5.4.1. Cross section of a typical two–stage, closed-loop, gaseous helium cryopump with a Gifford–McMahon refrigerator.

The refrigerator comprises an expander, to which the arrays are attached, and a high pressure He compressor. For the present, assume that the refrigerator is merely a device which provides refrigeration capacity at two stages. The term *refrigeration capacity* means that for a given, *simultaneous* input power to the two stages, the refrigerator will sustain a given temperature at the respective stages. In the absence of cryopanels, the refrigeration capacity might be measured by inserting the expander into a vacuum dewar. Inputting power to the two stages could be accomplished by passing current through electrical resistors mounted to the two cooling stations. This thermal loading with resistors is often used as a final refrigeration inspection test by manufacturers.

From convention, the capacity of the first (*i.e.*, warmest) stage is measured at 77° K, and the second stage at 20° K. CLGHe refrigerators are capable of achieving temperatures as low as ~9 to 10° K at the second stage and ~30° K at the first stage. As will be shown, achieving too low a first stage temperature can sometimes cause problems, and is usually not desirable. These refrigerators have a finite capacity. For example, the second stage refrigeration capacity of the typical ~20 cm ϕ cryopump ranges from 0.7 to 4 W, with the simultaneous loading of 2 to 16 W at the first stage. Similarly, the capacities of ~30 cm ϕ cryopumps range from 3 to 5 W at the second stage and 40 to 50 W at the first stage.

Chevron Design

The first and second stage arrays are attached to the respective cooling stations with threaded bolts. A thin sheet of indium foil (~0.25 mm) is sandwiched between the cooling stations and arrays to provide good thermal intimacy between the two surfaces. This indium foil retains its ductility at very low temperatures, thus preventing a thermal *jump condition* from existing between the arrays and cooling stations.

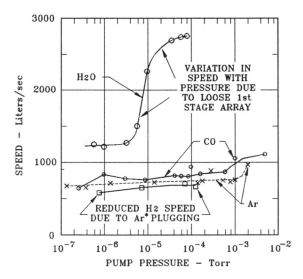

Figure 5.4.2. Pump H2O and H2 speed anomalies stemming from a loose first stage chevron and Ar* (3000 Torr-£) sieve plugging.

If the arrays are not securely fastened to their respective cooling stations, anomalous pumping effects will occur. For example, Fig. 5.4.2 shows pumping speed data of a 20 cm ϕ cryopump taken on a modified CERN dome (see section 1.16.4 of Chap. 1). The speed for H_2O was observed to be inexplicably low. Though measuring the speed of a pump for H_2O is a very difficult undertaking, we knew that results were steady-state and the effect was real. We subsequently discovered that the first stage louver was loosely attached to the first stage array. As a consequence, at pressures $\leq 4 \times 10^{-6}$ Torr, most of the water vapor was pumped on the second stage of the cryopump. Above this pressure, water vapor cryocondensed on the first stage louver.

Recalling the vapor pressure data of Figures 1.7.3 and 1.7.4, we note that, with the exception of He, H_2 and Ne, all of the gases would be readily cryocondensed on the second stage of a cryopump, operating at temperatures of $\leq 20°$ K. The gases He, H_2 and Ne can only be effectively cryosorption pumped at these temperatures. Because of this, activated coconut charcoal is epoxied to the underneath surface of the second stage array. The purpose of *hiding* the charcoal in this location is to protect it from plugging effects of those gases which pass through the first stage chevron and are cryocondensed on the second stage array. The perfect configuration for hiding the charcoal from potentially plugging gases is one in which no gas has access to the sieve. Of course, this would also preclude the pumping of He, H_2 and Ne. Therefore, protecting the sieve material from the plugging gases, and yet making it accessible to gases which must be cryosorption pumped, is a compromise.

A good *rule of thumb* in the design of a cryopump is that on removing the first stage array you should not be able to see the coconut charcoal from any viewing angle. This rule is often broken, and leads to measurement data which can be misleading, to say nothing about limiting the capacity of the cryopumps for He, H_2 and Ne. For example, I once observed a high partial pressure of H_2 in the pumpdown of a system. This high H_2 partial pressure was not, as some thought, due to outgassing of H_2 from the walls of the system. Rather, it stemmed from the dissociation of H_2O, into $O + H_2$, in a quadrupole residual gas analyzer. The high H_2 pressures existed because the cryopump could not pump the H_2. This, in turn, stemmed from an improper second stage design resulting in excessive exposure of the sieve material, and its consequential plugging during the pumpdown. Of course, if one plans to exclusively use a cryopump for the cryosorption pumping of He, H_2 and Ne, then the above *rule of thumb* would not apply.

The best measure of the design effectiveness in shielding the sieve is the measure of the change in speed of a cryosorption-pumped gas as a function of the amount of a second gas which has been cryocondensation pumped. For example, Ar is cryocondensed on the second stage of a cryopump, where H_2 must be cryosorption pumped. If we intermittently pump one gas and then the other, we should have a good measure of the effectiveness with which the sieve material is being protected for the sorption pumping of H_2.

I conducted such a test on a cryopump in the late 1970s. The second stage array is shown in Fig. 5.4.3. It had the advantage of being very directional to the pumping of cryocondensables, but the sieve material was fairly open to the pumping of He and H_2.[116] The sieve material was bonded to the underneath side of 36 metal *leaves*, which were in turn individually attached to a fined brazement.

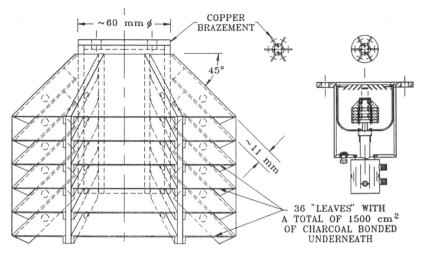

Figure 5.4.3. A second stage array with properly shielded charcoal.[116]

One idea was that if the sieve material was inadvertently plugged by oil, the leaves could be discarded and replaced. Secondly, the layered configuration provided inherent protection from cryocondensing gases. This array afforded an extended surface area (*i.e.*, ~1500 cm²) of bonded activated coconut charcoal. We first quantified the speed and capacity of this array for He and H₂. These results are shown in Fig. 5.4.4.

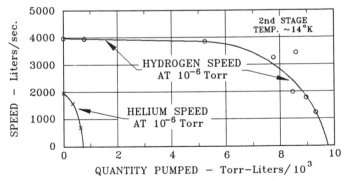

Figure 5.4.4. Relative speed and capacity of a closed loop, gaseous helium cryopump for the gases He and H2, with ~1500 cm² epoxy-bonded coconut charcoal on second stage.

Speed data and capacity data shown in Fig. 5.4.4 were taken on a modified CERN dome. The He and H₂ pressures were held constant at 10⁻⁶ Torr throughout the tests. We arbitrarily defined the capacity of the pump as that point at which the speed for the given gas had decreased to one *e-fold* of the initial speed. This was totally arbitrary. That is, we might have held the pressure constant at 10^{-8} Torr or 10^{-5} Torr, *etc.*, and similarly defined capacity. For very low throughputs, the H_2 speed of this pump was ~4700 $\mathcal{L}$/sec. However, as has been noted, the cryosorption speed will vary markedly with throughput (*e.g.*, see Halama, *et al.*,[25,43] Hseuh, *et al.*[38]). We note that the *capacity* of the cryopump for He, as herein defined, is ~8% of that of H_2. Again, I emphasize the importance of specifying the pressure at which the speed and capacity measurements are taken when making claims about these values.

We then measured the speed of the same cryopump for H_2 as a function of the amount of Ar which had been pumped. These data are shown in Fig. 5.4.5. The H_2 speed was measured at 10^{-6} Torr. A total of ~8 × 10^5 Torr-$\mathcal{L}$ of Ar was pumped before the speed of the cryopump for H_2 had decreased to ~50% of its initial value. I believe that the initial, very pronounced, drop in H_2 speed, with Ar pumped, stemmed from Ar plugging of the lower-most leaves in the second stage array. These lower leaves account for ~17% of the H_2 capacity, but because they are so much more accessible to both Ar and H_2, they accounted for ~25% of the initial H_2 speed. These lower leaves form the equivalent of an inverted cup, the inside of which is coated with activated coconut charcoal. The almost linear nature of the subsequent decrease H_2 speed, with the pumping of Ar, suggested that other mechanisms were at play. For example, it is more probable that for the greater part, the decrease in H_2 speed stemmed from the physical blockage of access of H_2 to the sieve material by thick layers of cryocondensed Ar *ice*.

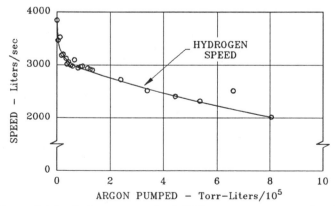

Figure 5.4.5. Hydrogen speed of CLGHe cryopump as a function of the amount of argon pumped; the sieve "plugging" effect.

Sticking Coefficients

We now have the basic *tools* to calculate the performance of a cryopump based on our knowledge of the type of first stage chevron or louver used, the surface area of the second stage and the sticking coefficient of gases on this stage. The sticking coefficients for several gases, as reported by Dawson and Haygood,[18] are given in Table 5.4.1.

Table 5.4.1. The sticking coefficients of some of the common gases as a function of gas and surface temperature.[18]

CRYOSURFACE TEMP. °K	GAS AND GAS TEMPERATURE									
	N2		CO		O2		Ar		CO2	
	77°K	300°K	77°K	300°K	77°K	300°K	77°K	300°K	195°K	300°K
10	1.0	0.65	1.0	0.90			1.0	0.68	1.0	0.75
12.5	0.99	0.63	1.0	0.85			1.0	0.68	0.98	0.70
15	0.96	0.62	1.0	0.85			0.90	0.67	0.96	0.67
17.5	0.90	0.61	1.0	0.85	1.0	0.86	0.81	0.66	0.92	0.65
20	0.84	0.60	1.0	0.85			0.80	0.66	0.90	0.63
22.5	0.80	0.60	1.0	0.85			0.79	0.66	0.87	0.63
25	0.79	0.60	1.0	0.85			0.79	0.66	0.85	0.63
77									0.85	0.63

With the exception of CO, they noted that the sticking coefficients of the above gases, at gas temperatures of 400° K, were ~0.5 for the above surface temperatures. For CO at 400° K, α ~0.73. Also, they reported a sticking coefficient of 300° K H_2O, on a 77° K surface, to be 0.92. From these data, our knowledge of molecular transmissivities through louvers, and with the use of (5.1.18) we are able to make fairly reasonable estimates of the speed of a two-stage, CLGHe cryopump for condensable gases. From the data of Fig. 5.4.4, we deduce that the speed per unit area of the activated coconut charcoal, to first order, is ~3 $\mathcal{L}$/sec-cm^2 for H_2 and ~0.5 $\mathcal{L}$/sec-cm^2 for He. This is the initial speed with zero loading of the sieve, a sieve temperature of 14° K, an H_2 and He temperature of ~60° K, and at a pressure of 10^{-6} Torr.

Thermal Loading of Cryopumps

There are three possible sources of refrigeration loading or heat input to the two stages: 1) *thermal radiation*; 2) *gaseous convection*; and 3) *heat of sorption* losses. Regarding radiation loading, the sizing of the cryopanels must be properly matched to the refrigeration capacity. Also, the mass of the first and second stage array will impact on the cool-down time of a cryopump. This is because of the specific heat of the materials used to construct the arrays. In the early days of cryopump development, many of us overlooked this aspect. On the one hand, massive arrays provide a measure of *thermal inertia* when pumping impulsive gas loads. On the other hand, this very inertia will cause protracted machine down-time (*e.g.*, coaters and implanters) during cool-down subsequent to a regeneration cycle. Another compromise is made.

First stage chevrons or louvers of many commercial cryopumps are often either electro-polished or brightly nickel plated. This gives the customer a good feeling, looks good, but is of little functional value. This is because the total spectral emissivity of the chevron approaches unity after pumping a few μm of H$_2$O.[1 1 2 -1 1 4] Of course, if your system has no water vapor ...

On the other hand, brightly polished surfaces are of benefit on the outer surface of the first stage cryopanel (i.e., the bowl-like structure attached to the first stage cooling station shown in Fig. 5.4.1). Gases which would readily condense on the chevron of the first stage are quickly condensed as they pass between the pump wall and the first stage cryopanel. Because of this, the highly condensibles do not reach the surfaces deep within the gap between the pump body and first stage cryopanel. Brightly polished surfaces will serve to reduce radiation losses in this region.

I learned this lesson *the hard way* when inventing a unique first stage cryopump array. The design featured a three-dimensional, first stage array that filled a solid angle of ~3π steradians.[1 1 5] The idea was to make the projected surface area of the first stage chevron exceed that of a circle having an area equivalent to the 20 cm ϕ pump. This array, within which the second stage array resided, was housed in a bulged pump body. The first stage chevron was blackened. This configuration had decided molecular transmissivity advantages, but the first stage thermal radiation load was excessive, and required refrigeration capacity which could be better otherwise used.

A little judgment must be used in the proper application of a cryopump. For example, while at Varian Associates, a customer once called me on the phone to discuss a problem he was having with one of our cryopumps. He exclaimed, "It won't pump." I was unable to solve the problem over the phone, so I suggested that he send the pump back to the factory. On disassembling the cryopump, we discovered that the indium foil, sandwiched between the second stage station and associated array, had melted. Indium melts at ~156° C. From this we learned that the customer had mounted the pump so that it had a direct view of the hot zone of a high temperature furnace. This was an obvious misapplication of cryopumping. Also, quartz lamps are sometimes used to preheat and outgas substrates prior to coating. Light from these lamps should not shine directly on the first stage chevrons.

The thermal load due to operation at high pressures is doubly troublesome to cryopumps. First, there are the losses associated with the heat of sorption of the gases being pumped. Secondly, there is the thermal load stemming from gas convection losses. Another *rule of thumb* is that the thermal load stemming from cryocondensation pumping on the second stage amounts to ~0.7 W per Torr-$\mathcal{L}$/sec. You can construct more precise values using the data of Table 5.1.2.

Thermal loading on the first and second stages due to operation at high pressures causes the temperature of these stages to increase. These temperature-refrigerator loading effects are quantified in a following section. Though many of the gases are readily cryocondensed at 10 - 20° K, slight increases in the temperature of the second stage will cause gases which are cryosorption pumped thereon to be liberated from the sieve material. Therefore, the base pressure of the system for a cryosorbed species may have a low value under conditions of moderate throughput, but increase markedly with only modest increases in the temperature of the second stage.

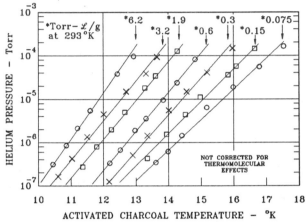

Figure 5.4.6. Adsorption isosteres of helium on coconut charcoal.[102]

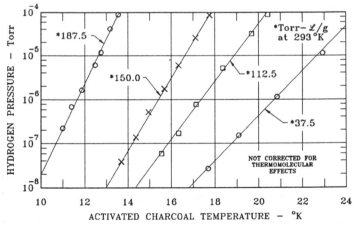

Figure 5.4.7. Adsorption isosteres of hydrogen on coconut charcoal.[102]

For example, the adsorption isosteres for He and H_2, as reported by Lessard, are given in Figs. 5.4.6 and 5.4.7, respectively.[102] These data have not been adjusted for thermomolecular effects (see problem 14). Assume that you have accumulated ~150 Torr-$\mathcal{L}$ of H_2, per gram of charcoal, on the second stage, and the operating temperature of this stage is ~14° K. From

Fig. 5.4.7, we note that the H_2 equilibrium pressure would be $\sim 10^{-7}$ Torr. Note that the H_2 speed at this pressure is *zero*. Now, assume that you introduce a steady flow of Ar at a rate of 5 Torr-$\mathcal{L}$/sec. With this throughput of Ar, we calculate, using the above *rule of thumb*, that the increase in thermal load on the second stage will be ~ 3.5 W. It is reasonable to expect that the temperature of the second stage will increase by several degrees Kelvin as a consequence of the increase in thermal load, due to heat for sorption of Ar, to the second stage. Assume that it increases to $\sim 18°$ K. If this happens, the H_2 pressure will equilibrate at $> 10^{-4}$ Torr. Therefore, as a consequence of pumping high throughputs of Ar, we have rendered the pump useless in the pumping of H_2 at pressures $\leq 10^{-4}$ Torr. For purposes of illustration, I have used an extreme case.

It is possible that a condition of thermal run-away might occur as a consequence of the release of H_2 in the above example. That is, the thermal conductivities of gases are proportional to $m^{-\frac{1}{2}}$, where m is the molecular weight of the gas. If we are able to stably operate at a given pressure of Ar, while *hovering on the fringes* of swamping the pump, thermal convective losses due to operation at the same pressure of H_2 will be $\sim \times 4.5$ greater. Many times, this causes problems in the starting of cryopumps subsequent to regeneration. More will be said about this.

Sputtering Applications

Because of the problems associated with heat of sorption loading of the second stage at high operating pressures (*e.g.*, sputtering pressures), some form of throttling device is often used to interpose a high impedance between the pump and gas source. Because of the pressure drop across this impedance, the required high sputtering pressure is thereby achieved in the chamber, while the pump operates at a much lower pressure. This is also done when using diffusion pumps at excessively high sputtering pressures. The common practice is to put an aperture plate directly over the diffusion pump flange. However, the LN_2 trap is positioned above the aperture so as to afford high pumping speed for H_2O. It is important to maintain high water vapor speed during sputtering applications. Similar schemes have been adapted to cryopump sputtering applications, both with and without the use of LN_2 traps.[117,118]

One such configuration is shown in Fig. 5.4.8. The *variable aperture* is mounted directly to the first stage of the cryopump. During the initial phases of pumpdown, the aperture is adjusted to be full-open, thus affording high speeds for all gases during the pumpdown and component preheating phase. During the sputtering phase, the aperture is throttled back to limit the Ar speed delivered to the coating chamber. However, because of the cold surfaces of the variable aperture (*i.e.*, $\sim 70°$ K), the speed for H_2O remains very high. The speed for H_2, a byproduct of all sputtering operations, will be

~×4.5 that of the Ar speed in the sputtering chamber. This variable aperture has been extensively used on hundreds of sputtering machines.

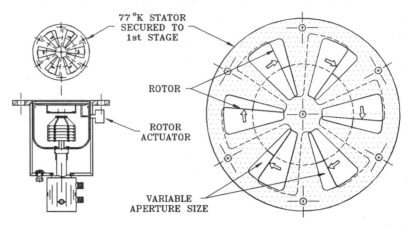

77 °K STATOR
SECURED TO
1st STAGE

ROTOR

ROTOR
ACTUATOR

VARIABLE
APERTURE SIZE

Figure 5.4.8. Variable aperture used in sputtering applications.[117]

Operation of cryopumps at very high pressures can lead to some very interesting effects. One such effect I have coined the *pressure clamping effect*. In the early days of cryopumping, I designed a cryopump specifically for sputtering applications. It featured the previously noted 3π steradian first stage array, housed in the center of a cylindrical LN₂ reservoir. This was a three-stage cryopump. The first stage of the cryopump was an LN₂-cooled cylinder, similar to that of Hseuh's pump shown in Fig. 5.3.3. The second stage was the 3π-array which comprised the first stage of the CLGHe cryopump, and the third stage was the 20° K stage of the CLGHe cryopump.

A variable aperture was mounted on the LN₂ stage.[117] When Ar sputtering, the pressure in the region of the second stage was of the order of 0.01 Torr. Because the 3π-array was so effectively shielded by the LN₂ stage, it operated at a temperature of ~40° K. At this temperature and sputtering pressure, Ar cryocondensed onto the 3π-array. On turning off the Ar source, the temperature of the 3π-array equilibrated at ~35° K. Therefore, the Ar *ice,* formed on the 3π-array during the sputtering operation, had an equilibrium vapor pressure of ~10^{-4} Torr. The pressure *clamped* near this value until all of the Ar *ice* had sublimed away from the 3π-array, and was pumped onto the 20° K stage. This took several hours to occur. The lesson learned here was that the *first stage* of the CLGHe cryopump was operating at *too low a temperature.* This resulted in an Ar *pressure clamping effect.* The fix was simple. We merely put a thermal shunt of the first stage of the CLGHe cryopump so that it always remained at temperatures ≳60° K. However, only you and I know that I made an error in designing this pump. Actually, there were two problems. At the time, the 3π-array was mounted on a refrigerator of moderate first stage capacity. Because of this, cool-down times after regeneration were excessive (*i.e.*, ~2½ hours).

Similar effects are noted when operating a cryopump at very high pressures. In some instances, the speed of the pump seems to mysteriously increase at higher pressures. This effect is sometimes erroneously attributed to increases in first stage louver transmissivities as a function of pressure.[119] In fact, these effects usually stem from the start of cryocondensation on the first stage of the cryopump. This effect is shown in Fig. 5.4.9, where Ar speed of a 20 cm ϕ cryopump is given as a function of pressure. The base pressure of the system was ~10^{-10} Torr prior to the measurement. The measurements were taken on a modified CERN dome. The pump was appended with a 20 cm ϕ ConFlat® flange. The dome was baked to 250° C and the cryopump warmed to ~85° C. The cryopump was turned off and the system pumped on with an auxiliary pump during this bakeout.

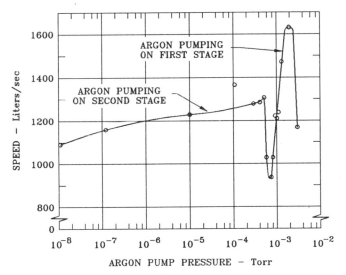

Figure 5.4.9. Increase in argon speed due to first stage pumping.

The refrigeration capacity of the first stage of the cryopump used in tests shown in Fig. 5.4.8 was somewhat marginal. The clue that it was the *pressure clamping effect* was that even though the second stage cooled to <20° K, and therefore had high speeds for Ar, there was a protracted pumpdown after the tests. This was due to the gradual sublimation of Ar off the first stage.

The reverse of this effect was noted by Lessard when pumping Ne on the second stage of a CLGHe cryopump.[102] That is, he initially observed an inexplicably high pressure when pumping Ne. However, with the gradual heating of the second stage to higher temperatures, there was a precipitous drop in the Ne pressure. This was due to the fact that at the initial lower temperatures, Ne cryocondensed on the second stage. However, when heating the second stage to slightly higher temperatures, the cryocondensed Ne came off the outer surface of the second stage and was cryosorption pumped by the sieve material.

Cryopump Applications

Cryopumps are presently manufactured and sold by a number of companies. In the early days of CLGHe cryopumping, some companies offered cryopumps with reverse Stirling-cycle refrigerators.[106,120-122,139] These refrigerators had vibration problems, operated at high rpm, and expander seal wear (*i.e.*, reliability) was impacted. Today, refrigerators of commercial CLGHe cryopumps are machines exclusively using some variation of the Gifford-McMahon refrigeration cycle.[123,124] The literature abounds in publications dealing with various applications of these latter cryopumps.[125-142]

The driving impetus for commercialization of these cryopumps has been the semiconductor industry. It was recognized very early that hydrocarbon contamination caused problems with the properties of thin films.[139] Some have argued with me that the primary reason for the use of cryopumps in this application was the expense associated with LN_2, required for the trapping of diffusion pumped systems. However, I can recall in the early 1950s, as a microwave tube technician at the General Electric Laboratory in Palo Alto, CA, servicing freon refrigerators used to cool diffusion pump traps. This technique was widely used throughout the General Electric facilities, having first been reported on by Leever in 1954.[143] Cryopumps afford a clean, *dry,* and reliable means of pumping.

Cryopumps of the above varieties are used on most functional (*i.e.*, optical and magnetic media), decorative and semiconductor coaters presently offered throughout the world. They are also used on the beam lines and end stations of most ion implanters. Some of the above references (*i.e.*, 125-142) discuss the advantages and disadvantages of the use of cryopumps over other methods of pumping. Some also discuss the use of cryopumping in combination with pumps including sputter-ion pumps. For reasons to be discussed below, this practice is to be discouraged.

The primary limitation to the base pressure achieved by these pumps is the outgassing stemming from the use of elastomer seals (*e.g.*, the pump flange seal). As noted above, one can readily achieve pressures of $\leq 10^{-10}$ Torr without the presence of these elastomers. This was also demonstrated by deRijke.[131] Kikuchi, *et al.*, used all-metal, Helicoflex Delta seals in place of elastomers and achieved pressures $< 10^{-12}$ Torr with a cryopump.[144] These all-metal seals require very low sealing force, and therefore, may often be used as a direct replacement for elastomer seals.[145,148]

At times, metal burst diaphragms are substituted for the cryopump pressure relief valves. This is done to avoid the outgassing from the elastomer seal used in these relief valves. If you decide to use a metal burst diaphragm, rather than building your own, I suggest you purchase a commercially available assembly. Also, have a fellow technician check your calculations on the possible consequences of the use of this diaphragm. You should assume that the entire system to which the cryopump is attached will at some time be pressurized to the limit of the burst diaphragm.

System Configuration

Several aspects should be considered when designing a cryopumped system. These include functional as well as safety considerations. There are some main guidelines which should be followed when designing or specifying a cryopumped system. The configuration of a typical sputtering system is represented in Fig. 5.4.10. The configuration of your system need not be as complex as this system. It all depends on your application. The valves V_1 - V_4 are the conventional pneumatic or manual roughing valves. Valves V_5 - V_8 are the smaller, combination vacuum and gas handling valves.

Some form of rough pumping is always required when using cryopumps. First, it is required to pump the work chamber down prior to opening the main system valve. Cryopumps are tolerant of large *gulps* or impulsive gas loads. For example, a 20 cm ϕ pump is typically able to accommodate an impulsive gas load of 10-50 Torr-$\mathcal{L}$ with no difficulty. However, should this practice be carried to an extreme, some gases may be viscously swept into the bowels of the pump and conceivably cause partial plugging of the sieve material. Secondly, a rough pump is required during the regeneration of the cryopump (see Regeneration, below).

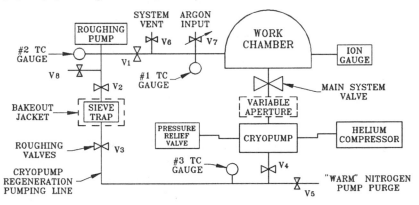

Figure 5.4.10. Schematic representation of a cryopumped sputtering system with pump regeneration provisions.

It is always good vacuum practice to minimize the number of penetrations into the high vacuum system. For this reason, valve V_4 is the only penetration into the cryopump. Also, an additional valve might be placed between the manifold at the point where valve V_7 is attached, and the work chamber. One can then make possible compromises on the quality of these vacuum/gas valves.

Use of a Main Valve Over the Cryopump: A cryopump temporarily stores the gas which it has pumped. In the event of a power failure, these gases will be liberated from the first and second stages of the cryopump. This isn't a catastrophic or instantaneous event. Actually, the cryocondensed gases slowly sublime off the surfaces. The cryosorbed gases are more quickly liberated. The rate at which the gases are liberated depends on the thermal inertia of the arrays, and the heat input to the arrays by convection and radiation. The key is to contain these gases in a safe and ecologically responsible fashion. For example, the gases might comprise an explosive mixture (*i.e.*, see problems 16-18). There are possible sources of ignition in most vacuum systems. Therefore, you don't want this mixture disgorged back into the system. This would eventually happen if there was a power failure and no main isolation valve existed between the work chamber and the cryopump.

Lastly, the main valve should be a pneumatically actuated valve. You have no assurances that you will be present during a power failure. Also, the electronics of the valve actuator should be *latching*. If the power fails and the valve closes, on return of power, the valve should not automatically open.

No Possible Sources of Ignition Below the Main Valve: I have received some criticism in certain areas as being an alarmist regarding this issue. I reason with these criticisms in this manner. Anyone who has worked in vacuum related industries longer than a few years has walked up to a vacuum system, which he thought was under vacuum, and turned on the ionization gauge. More than once I have mistakenly carried this practice to the extreme of attempting to degas a gauge when a system was at atmospheric pressure. The gauge filled up with a white, smokey powder and was ruined. I've done this several times over the years. Perhaps I am more fallible than most. But, had there been an explosive mixture in the chamber to which the gauge was affixed, an explosion would have followed.

While lecturing on cryopumping at a very large corporation in the late 1970s, I stressed the importance of not having possible sources of ignition below the main valve seal. I noted that one person in the audience was nodding his head in a manner which indicated he knew what I was talking about, to the point of having a personal experience in this area. Seeking a testimonial to this effect, I questioned him. He was dressed in a white smock, and was obviously a vacuum technician. He immediately became guarded in his response. His supervisor, dressed in a dark suit and tie, was seated several rows behind him. On pressing the issue, the technician noted that though they hadn't had an explosion of a cryopump, "... the pump came apart very fast as a consequence of gas combustion". For this reason, I stress that one should never locate an ionization gauge or other possible sources of ignition (including sputter-ion pumps) below the main valve seal of a cryopump.

While at Varian Associates, in the late 1970s we conducted an experiment to determine the effect on a cryopump in which such an explosion had occurred. The pressure relief valve in the pump was adjusted to relieve at a pressure of ~900 Torr. Using remotely actuated valves, we concocted a stoichiometric mixture of 900 Torr of O_2 and H_2 within a cryopump. The pump was housed within a sandbagged bunker constructed in the Varian parking lot. We remotely ignited the gas mixture. Indeed, the pump *came apart very fast*. We called this an explosion. The results of this test demonstrated to us that there was no practical method of constructing a cryopump which would be immune to the released energy of such an explosion. Secondly, it made us zealots regarding the avoidance of the use of possible sources of ignition below the main valve of any cryopump.

Toxic Gases: In many instances, cryopumps are used to pump toxic gases, such as carrier gases containing dopants for ion implantation. For this reason, consideration should be given to the manner of disposing of these gases on regeneration of the cryopump. A closed, off-gas venting system is needed in such cases.

Sustained Untrapped Pumping on a Cryopump: When using a mechanical pump to pump on a cryopump during some phase of the regeneration cycle, care should be taken that oil backstreaming into the cryopump is not occurring. Such oil backstreaming will most certainly cause sieve plugging.

Regeneration of Cryopumps

Because of their finite capacity for certain gases, cryopumps must be regenerated from time to time. Regeneration simply involves the warming up of a cryopump so that cryopumped gases may be pumped away by some auxiliary pump. Studies of methods of accomplishing this have recently appeared in the literature.[146,147] Regeneration is usually the first thought that comes to mind when cryopump pressures seem inordinately high. Except in high-throughput sputtering applications, regeneration is required on a very infrequent basis (see problem 20). More *rules of thumb* are applicable.

First, there is no such thing as a quick, partial regeneration of a cryopump. That is, if you start the regeneration process, you must complete the process. If you warm the pump sufficiently to liberate some of the cryosorbed gases, and then cool the pump down, some of the cryocondensed gases will be liberated in the process. These gases will find their way to the sieve material and cause plugging problems. Theoretically, partial regeneration is possible. However, from a practical standpoint, you do not normally have sufficient control of the temperatures of the various surfaces of the cryopump to achieve partial regeneration.

On warming a cryopump for regeneration, the charcoal on the second stage is the last surface in the cryopump to reach room temperature. Because of this, no matter what technique you use, water vapor originating from other surfaces within the cryopump will ultimately end up on the charcoal. Because of this, during the regeneration process, you must allow sufficient time for this water vapor to be pumped away. Slightly warming the pump will accelerate the process.

Flushing the cryopump with a warm purge-gas accelerates the regeneration process. Dry argon or nitrogen are suitable for this purpose. The same valve through which the cryopump is rough pumped during regeneration is often used as the inlet for the purge gas. Some form of screen or shield is needed to make sure that activated coconut charcoal, which is known to fall off second stage arrays, does not cause sealing problems with this valve or the pressure relief valve. Use of a stand-pipe arrangement will eliminate such problems with the roughing valve. Some form of screen is usually used to protect the pressure relief valve seat. Without a similar stand-pipe arrangement for the pressure relief valve, it is not advised that the purge gas be of sufficient pressure to cause release of this valve. This is because the flushing gas may carry debris to the valve seat and cause subsequent sealing problems.

The warm purge-gas serves another purpose. This is to dilute the partial pressures of He or H_2 which might be resident in the cryopump during regeneration. A partial pressure of 10^{-3} Torr of these gases may be sufficient to make it difficult to start the cryopump. Therefore, the purge-gas is helpful in assuring the dilution and viscous conveyance of these gases to the rough pump during regeneration. One uses the purge-gas to dilute He and H_2 partial pressures, through the sequential venting and roughing of the cryopump over several cycles.

Sales people often extol the fact that *their* cryopumps can be started at N_2 or air pressures as high as $2 - 3 \times 10^{-2}$ Torr. Indeed, this may be the case. However, when starting a cryopump at this pressure, speculate on where you think this gas is pumped. Yes, it is all pumped on the charcoal, and will partially plug the charcoal for the subsequent pumping of He and H_2. For this reason, one should achieve the best possible base pressures prior to starting the cryopump.

The roughing pump used during regeneration must be trapped or you will have problems with oil plugging of the sieve material. Sieve traps are often used for this purpose. Note in Fig. 5.4.10 that an isolation valve is located on both sides of the sieve trap. This is done for a very specific reason. Mechanical roughing pumps back-stream oil at operating pressures $\lesssim 0.1$ Torr. Assume that the sieve trap shown in Fig. 5.4.10 is regenerated by a high temperature bakeout. Of course, during this process valve V_2 is kept open and valve V_3, closed. If after this regeneration process, the roughing pump is allowed to pump on the sieve trap for an extended period, and at pressures near its blank-off pressure, the sieve trap will become loaded with oil. It will then be more of a source than remedy to the back-streaming of oil into the cryopump on its the subsequent regeneration.

Because of the above considerations, cryopump component vendors and system manufacturers often provide automatic regeneration control provisions with their equipment. These controls are usually micro-processor based systems that take the cryopump through the regeneration sequence, including the regeneration and subsequent isolation of the sieve traps.

I've noted that cryopumps need very infrequent regeneration in many applications (*e.g.*, see problem 20); however, one can easily be misled into thinking one does if after regenerating the cryopump, one observes pressure performance substantiating the perception. I call this the *placebo effect*. A cross section of the expander of a typical cryopump is given in Fig. 5.5.1. The first stage displacer operates at a temperature gradient from RT to ~77° K (*i.e.*, the temperature of the first stage). The second stage displacer operates at temperature gradient of from 77° K to say 10 - 20° K (*i.e.*, the temperature of the second stage).

Any gas other than He, and perhaps a little H_2, which is in the high pressure He stream, will cryocondense on surfaces within the second stage displacer-regenerator. This may take hours or even days to happen, and may be a gradual accumulation process. It all depends on the amount and type of contaminant gas. When the contaminant gas is cryocondenced on these beds, it obstructs heat exchange processes which occur in the regenerator-displacer beds. This results in a gradual increase in the second stage temperature. The effect of the increase in second stage temperature is a gradual increase in system pressure. You observe the pressure increase and deduce that it is causing the temperature increase. Therefore, you *regenerate* the pump. Sure enough, after the regeneration of the cryopump, the subsequent system pressure is observed to return to its usual low value. You conclude that the pump needed regeneration. What in fact has happened is that the contaminant gas, which collected in the regenerator-displacer bed, was liberated back into the high pressure He on pump regeneration. The process of contaminant collection on the regenerator-displacer bed will start all over again.

I have also noted cyclical pressure variations, with a periodicity of an hour or so, in cryopumps as a consequence of gas being collected in the regenerator beds, then being liberated as a consequence of second stage temperature increases, then again collecting, etc. Neon contaminant in the He will cause this effect. If a slight second stage temperature increase causes a slight system pressure increase, the system gas is probably either He or H_2. Assuming that the sieve on the second stage of the cryopump has not been plugged by oil, you should be able to make rough calculations of the probable H_2 loading which would results in this pressure increase. Is it consistent with the cryopump capacity and operating pressures? *Count the molecules* before assuming that regeneration is required.

5.5 Closed-Loop, Gaseous Helium Refrigerators

Some of the History[149]

Most of us have seen the diagram which shows the chronology of the development of the semiconductor industry. The commercial genesis of this industry was the Fairchild Corporation. On a much smaller scale, there are counterpart origins of many of the technologies in the vacuum industry. We saw in Chap. 4 that SAES Getters, in Italy, was the origin of the significant contributions made to NEG technology. Varian Associates played the major role in the development of sputter-ion pumps. The late Harrold Farrow started the Excalibur Corporation which took liquid helium cryopumping from a laboratory curiosity into commercialization. The work of Linde, a Division of the Union Carbide Corporation, in liquid helium cryopumps and artificial zeolites was similarly notable.

Arthur D. Little (ADL), working in concert with the Massachusetts Institute of Technology (MIT), was the corporate origin of the closed-loop gaseous helium refrigerators, variations of which are used in today's cryopumps. MIT owned ADL from 1935 to 1952, as A.D. Little bequeathed controlling interest of the corporation to MIT on his passing. Therefore, there was a close relationship between ADL and MIT. They worked jointly during WWII on many technical projects.

The impetus for the refrigerator work at ADL had nothing to do with vacuum. Rather, it was directed at finding some more convenient and safe way of producing liquid helium. Prior to the start of WWII there were two places in the United States where one could conduct experiments at 4.2° K. One was at the University of Chicago, the second was at the Naval Research Laboratory. Researchers had to take their experiments to these facilities to conduct 4.2° K work. The liquid helium at these facilities was produced by high pressure, cascading Joule-Thomson expansion of N_2, H_2 and He. This process was potentially hazardous because of the necessity of working with high pressure H_2 in the second stage of expansion. Joule-Thomson (JT) expansion is merely the cooling effect stemming from the expansions of high pressure gas to lower pressures and below the inversion curve.[150]

In 1946, the late Professor Samuel C. Collins and his colleagues at MIT and ADL, successfully developed what became known as the *Collins Helium Cryostat*. This was a simple, less hazardous machine for the production liquid helium.[151] This technology made use of a gaseous helium, mechanical refrigerator to produce refrigeration in two stages, and equivalent to that achieved by the JT-expansion of N_2 and H_2 of the former process. The third stage was the JT-expansion of the He resulting in its liquefaction.

Industry and laboratories came to MIT thereafter and asked them to produce more of these helium liquefiers. MIT was not in the business of building and selling equipment. Therefore, ADL took a licence under the

MIT patent to build and sell these machines. By the mid-1960s the were 250 of these machines in use at laboratories throughout the world. Of course, not all of the machines were produced by ADL.

In 1955 Dudley Buck, at MIT, demonstrated a superconducting *switch* for computer applications.[155] He called it the cryotron memory. Insurance companies and other large institutions needed reliable and fast methods of making calculations. The cryotron memory device presented this potential. It required the use of LHe, and had good market potential for the sale of additional Collins Helium Cryostats.

The Collins Helium Cryostat needed a full-time operator to keep it up and running. It was belt-driven, required considerable care and attention, and would operate about 1000 hours before it had to be rebuilt. Therefore, ADL undertook the development of a continuous duty, unattended, closed-cycle machine which would achieve 4.2° K, and which would operate continuously for one year between maintenance intervals. The third stage of this machine was the JT-expansion of He. The late Professor W.E. Gifford was a mechanical engineer at ADL at the time. Professor Gifford and Dr. Howard O. McMahon made major refinements in these helium mechanical refrigerators.[123,124] They developed a closed cycle machine which is presently called the *GM-cycle* (Gifford-McMahon-cycle) refrigerator.

Applications for these refrigerators blossomed. For example, they found use in defense electronics and satellite communications. These applications inherently brought issues of reliability and low maintenance to the forefront. In 1952 ADL purchased back the stock of the company, owned by MIT, as part of the investment portfolio of a retirement income plan for its employees. In 1967, ADL created an 80% ADL subsidiary of its cryogenic activities. The name of this subsidiary was *500, Incorporated*. Two years later, in 1969, the 80% ADL ownership was transferred to the ADL employee retirement portfolio, and the name was changed *Cryogenic Technology, Inc.* (CTI).

CTI worked for a number of years with vacuum technologists in an to gain acceptance of these machines in cryopumping applications. They built ten of a variety, on speculation, and specifically for cryopumping applications. The refrigerator was referred to as called the *Creacher*. The name came from a combination of the words cryogenics and teacher. Presumably, CTI sought with this machine to teach vacuum technologists about the merits of cryopumping with mechanical refrigerators. The machine, first reported on in 1963, was a modified Taconis-cycle refrigerator, capable of achieving 10° K at the second stage, and making use of LN_2 at the first stage.[153] Three years later, Hogan and Turner reported on use of this refrigerator in a vacuum application.[71] However, because it lacked the use of sieve material on the second stage, it could not pump He, H_2 or Ne. The idea of *gluing* coconut charcoal to a second stage array was literally incomprehensible to a UHV vacuum technologists at that time.

In 1967, Boeing, at Kent, Washington, had the task of testing the NASA Ranger and Surveyor, unmanned satellites that went to the moon. These satellites had TV and IR cameras. They were trying to test them in a diffusion pumped, 12.2 m ϕ × 24.4 m tall space chamber (*i.e.*, a 12.2 m ϕ right angle cylinder subtended by two 12.2 m ϕ hemispheres). The chamber was pumped by two diffusion pumps, each of ~1.2 m ϕ. The diffusion pump backstreaming problem was causing havoc with the optics of the systems. CTI, working with Douglas McKinney of Boeing, built the world's first GM-cycle cryopumped space chamber. The cryopumps were augmented with two, 2400 $\mathcal{L}$/sec sputter-ion pumps and TSP pumps, as vacuum technologists had still not rediscovered activated coconut charcoal. The chamber had an LN2 shrouded wall.

According to John Harvell, of CTI, it was James A. O'Neil, of CTI, who made the *breakthrough* of gluing activated coconut charcoal to the second stage of a GM-cycle refrigerator. He was experimenting with the cryotrapping of He with Ar. He constructed the equivalent of a diffusion pump, where the pumping medium was Ar, rather than oil. He had little success with this cryotrapping apparatus with a second stage temperature of 20° K. Therefore, he resorted to gluing activated charcoal to the second stage.

In 1961, W. E. Gifford left CTI and took a chair of Professor of Mechanical Engineering at Syracuse University. While there, he started a small company called Cryomech. Professor Gifford was advisor to Ralph C. Longsworth who received his Ph.D. at Syracuse University. Longsworth managed Cryomech thereafter. Gifford and Longsworth developed a GM-cycle, pneumatically driven, rotary valve machine while at Syracuse. An agreement was reached between CTI and Gifford to cross-licence this design. Ralph Longsworth eventually left Cryomech and headed up a cryogenics development group at Air Products and Chemicals, Inc. (APCI), in Allentown, PA. While there, he developed a refined version of the pneumatically driven, rotary valve, GM-machine which featured a modified Solvay-cycle. Samuel Collins also consulted for the Advanced Products Division of APCI in the late 1950s and early 1960s.

The above scientists and technologists are the people who developed and refined the refrigerators which are presently used in all closed-loop gaseous helium cryopumps presently manufactured and sold throughout the world. It all started with development of the Collins Helium Cryostat. In the years that followed, hundreds of articles appeared in cryogenics journals describing various versions of the basic GM-cycle machines. Also, in excess of 200 United States patents were subsequently issued on various improvements to these machines.

While at Varian Associates, I was for a time in charge of R&D for the Pumps and Instruments Group. In 1975 my marketing counterpart, John McLaughlin and I visited Cryomech, APCI and CTI to learn what these new cryopumps were all about. We had the pleasure of meeting for the first time Professor Gifford at Cryomech, and Dr. Ralph Longsworth at APCI. We quickly became *converts* to this new technology. The semiconductor industry was blossoming, and we knew that cryopumping would play a major role in this area.[1 2 7] In July of 1975 we put together a five-year business plan for Varian's participation in this market. The key to success was the GM-cycle refrigerator. CTI's pricing strategy was such that it was impossible for us to purchase their machines and integrate them into cryopumps, while at the same time compete with them in the open market. For the next few years, Varian purchased cryopump refrigerators from APCI, and developed a line of cryopumps and cryopumped systems. The first system was a batch coater, but the most successful was the *Varian-3180*, the first cassette-to-cassette coater of the semiconductor industry. Varian eventually developed their own line of cryopump refrigerators as did many other vacuum companies. But, the origin of all these machines was the GM-cycle refrigerator developed at CTI (ADL), for the Collins Helium Cryostat.

The GM-Cycle Refrigerator

The key to success in the development of the refrigerator resided in two areas. First, one had to develop a method of separating out contaminants from the He gas stream. As noted, any gas or contaminant other than He will cause problems in the refrigerator expander. This includes oil vapors, water, and all of the air gases. The second challenge was to develop a reliable expander seal which would operate at 40 - 80° K for several thousand hours.

An example of a typical expander, with regenerator-displacer bed, is shown in Fig. 5.5.1. There are two regenerator-displacer pistons coupled in series. These pistons reciprocate within the thin-wall cylinder heads. In case of the CTI design, they are driven by mechanical coupling to an electric motor, through a scotch-yoke assembly. In the Gifford-Longsworth (APD hereafter) approach, the pistons are driven pneumatically by exerting a He pressure difference on a third, slack-piston fixed to the RT end of the first stage expander. From a reliability and functional standpoint, I perceive of no disadvantage to either approach used to drive the regenerator-displacer pistons. Though the reciprocating frequency of the APD machines is $\sim \times 2$ that of the CTI approach, the stroke of the latter is $\sim \times 2$ that of the former. Therefore, seal wear will be the same in either case.

The regenerator bed comprises either a lead-alloy shot (the second stage) or a combination of lead shot and bronze screen material. Materials in these regenerator beds have high surface-to-volume ratios. They alternately serve as a means of storing heat calories, on one phase of the refrigeration cycle and giving up heat calories in a different phase of the cycle. The specific heat of the regenerator bed material is important in this function. The magnitude of the specific heat of the regenerator bed materials, at reduced temperatures, is the primary limitation to the minimum temperatures which can be achieved by these machines. Also, any contaminant which deposits on the surface of the regenerator material, so as to create a thermal barrier between it and the gaseous He, will impact on the refrigeration capability (capacity) of the refrigerator. For this reason, high purity He is required.

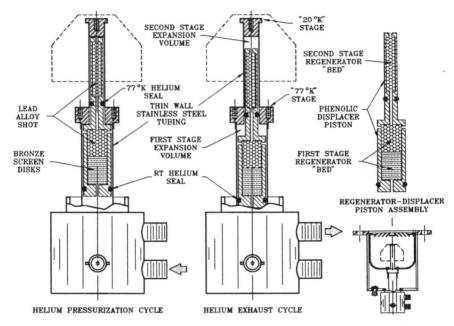

Figure 5.5.1. Closed-loop, gaseous helium expander assembly.

High and low pressure He valves, within the expander assembly, open and close in a manner synchronous with the reciprocating displacer pistons. When the pistons are near the top of the stroke, high pressure He fills the expander. The high pressure valve is then closed, and the low pressure valve opened. At the same time, the regenerator-displacer piston assembly is withdrawn, and the He is allowed to expand and exhaust through the low pressure He valve. This expansion process causes cooling of the He in the volume created above the respective pistons, and that He filling the void volume in the regererator-displacer beds. Some of the refrigeration created by the expansion is *used* at the first and second stage cooling stations to cool the arrays attached thereon. Some of the remaining refrigeration produced by the expansion process precools the regenerator beds, in anticipation of the next portion of the cycle. That is, the exhausting He picks up calories from the regenerator bed on passing over the surfaces of the bed material. During the next pressurization cycle, the He is precooled by the regenerator beds. In this case, the regenerator beds *pick up* calories from the high pressure, RT

He. The role of the regenerator bed is now clear. It functions as a thermal ballast in the refrigeration cycle.

The amount of refrigeration produced at the two stages depends on the amount and effectiveness of the regenerator beds, the volume before and after expansion, and the pressure of He at the two stages of the cycle. This, in very simple terms which even I can understand, is how these expanders work. If the He seals leak, He will bypass the regenerator beds, and cause a reduction in refrigeration capacity. Also, the shunting He will result in thermal losses. The reliability of these seals is very high in todays GM-cycle machines. Usually the seals require replacement every 15,000 - 20,000 hours, though Harvell notes in one particular instance the seals lasted 65,000 hours, and still had not failed.[154]

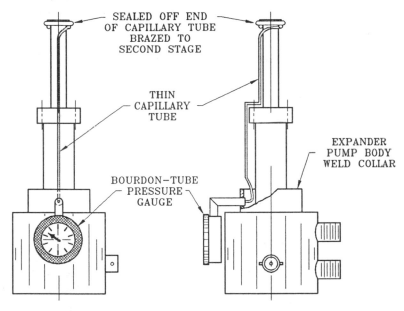

Figure 5.5.2. Hydrogen vapor pressure bulb used to measure the temperature of the second stage of the expander.

Two methods are used to measure the temperature of the second stage of the expander. One method involves the use of what is called a *hydrogen vapor bulb*. As shown in Fig 5.5.2, a small tubular pipe, sealed of on the end, is wrapped around and brazed to the second stage cooling station. The long, thin capillary tube, posing negligible thermal shutting of the second stage, is then brought out through the bottom of the expander pump body weld collar. It is then attached to a conventional high pressure gauge. By measuring the hydrogen (vapor) pressure one is thereby able to deduce the temperature of the second stage (see problem 22). The second method of measuring the temperature of the second stage is through use of silicon diodes. The forward voltage drop in these diodes changes significantly as a function of temperature. It is nearly linear, and $-dV/dT$ is ~50 mV/° K in the 1 to 35 ° K range. Also, the current source need not be highly regulated.[158]

The Compressor

The compressor is needed to supply high purity, high pressure He to the expander. A schematic representation of a typical compressor assembly is shown in Fig. 5.5.3. For the last few decades, the He compressor modules of most cryopump refrigerators have been modified air-conditioning compressors. For example, in the late 1970s, the CTI Model-21, APCI (APD) Model-202 and Model-204 refrigerators used piston-type, Tecumseh air-conditioning compressors. The CTI Model-350SC refrigerator used an RCA Whirlpool, rotary vane compressor. Longsworth, who had worked for two years at Carrier Air Conditioning, in Syracuse, proposed the use of these compressor modules in this application, to Gifford. About the same time, the same compressor modules came into use at ADL. Prior to that, oil lubricated He compressors had been used in this application. The technology for removing the oil vapors from the He stream was developed at ADL.

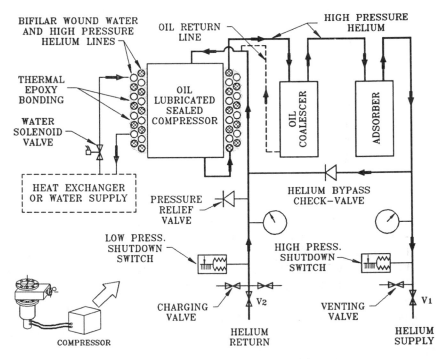

Figure 5.5.3. Closed loop, gaseous helium cryopump compressor.

The total He mass flow required for a typical 20 cm ϕ cryopump is ~4.7 Atm.-$\mathcal{L}$/sec (10 SCFM) and that of a 30 cm ϕ cryopump, twice this mass flow. The He compressor system is pressurized to a static pressure of ~140 Nt/cm^2 (~200 psig). When operating, the pressure difference between the low pressure and high pressure ports of the expander is ~140 Nt/cm^2. The above static and dynamic pressures are only approximate values and will differ with the size and type of refrigerator.

These sealed compressor modules were designed to operate with freon®. The compressor motor is an integral part of the compressor and is therefore, also housed in the compressor case. The heat of compression of He is much greater than that of freon®. Therefore, the compressor module is usually modified to include provisions for additional cooling of both the compressor and the high pressure He prior to its conveyance to the expander. This is sometimes accomplished by bifilar-wrapping the high pressure He and water cooling lines and bonding them to the compressor casing with thermal epoxy.

Infant mortality compressor problems in cryopump refrigerators were rampant in the mid and late 1980s. Problems were encountered with these compressors because new powdered metallurgy techniques used by a particular compressor manufacturer to fabricate many of the compressor components.[1 5 6] Cryopump compressor failure problems nearly put some cryopump companies out of business. Also, this venture in powdered metallurgy cost the compressor manufacturer many hundreds of millions of dollars because of home refrigerator problems.

After removing the freon® from the sealed units, the units are flushed clean and charged with a low vapor pressure oil. The *coalescer*, immediately downstream of the compressor, serves to remove oil droplets or *fog* from the He stream. This oil returns to the compressor through a bypass, oil return line. This takes the form of a capillary tube or some form of aperture from the high, to low pressure He circuit. It is important that this oil return circuit not become plugged or, in time, oil will carry over to the expander.

The *adsorber*, located downstream of the coalescer, serves to filter out the remaining gas impurities in the He stream. As you might guess, this adsorber is filler with either, and sometimes both activated charcoal and artificial zeolite. It has a finite capacity. Because of this, it must be changed every 5,000 - 10,000 hours. Some manufacturers claim adsorber life of as much as 18,000 hours. This is not the place to economize. Therefore, I always change the adsorber every 5,000 hours. Most compressor packages have some form of elapsed time clock to serve as warning that preventative maintenance is required.

Because the motor is housed within the compressor casing, problems of electrical breakdown or discharge will occur if the He pressure becomes too low. For this reason, a low pressure shutdown switch is located on the He return line. Also, if the compressor is charged with too high a He pressure, the compressor will overheat. For this reason, a high pressure shutdown switch is installed on the He supply line.

A He bypass valve is used to shunt the supply and return He lines. This serves a very important function. Oil which might have been carried over into the He lines, as a consequence of shipping or handling, poses a threat to the expander. Therefore, immediately after receipt of the compressor, the compressor unit should be operated for a couple of hours prior to installation of the expander. This age-in of the compressor will afford assurance that oil that might have gotten in the lines, will be flushed back into the compressor by the He stream.

Valves V_1 and V_2 may be special sealed-off fitting which retain the high pressure He, rather than valves manipulated by the user. The *charging* and *venting* valves are self-explanatory. These are also used to top off the He supply in the event leaks in the compressor package. Also, in the event of He contamination, the compressor package is pressurized and then partially vented through the respective valves. For reasons by now obvious, high purity He should be used.

As noted, the He expander will not work properly if the regenerator beds become contaminated. A *contamination* is any gas or vapor other than He. The expander is similar to a *dialysis machine*. All the system impurities, if not filtered out by the compressor adsorber, will eventually collect in the regenerator beds.

While at Varian, many of the cryopumps we manufactured were integrated into Varian ion implanters. We once had a contamination problem with a batch of compressor adsorbers, and didn't know it at the time. Some were contaminated by air gases, and some were not. In refrigerators with contaminated adsorbers, after 10 - 20 hours of cryopump operation, sufficient gas contamination accumulated in the expanders to seriously degrade their refrigeration performance. We had measured the refrigeration capacity of each refrigerator by thermal loading them with resistors. But this, and the final test pump procedure then involved only 3-4 hours of cold operation. The problem became evident when dozens of the cryopumps were installed and operating on ion implanters in various stages of manufacturing. If the manufacturing staff observed some anomalous cryopump behavior on an implanter, they did one of two things. They first exchanged compressors. If the problem persisted, they then regenerated the cryopump. Because of the previously described *placebo effect* and the above *dialysis effect*, in a matter of a few days every refrigerator system in the ion implanter factory became contaminated.

The solution to the problem involved exchanging all the adsorbers, and the multiple flushing of the expanders, hoses and compressors with high purity He. Thereafter, every cryopump refrigerator was operated for 50 hours as part of the cryopump final test. The lesson in this for you is that if you are observing some anomalous behavior in your cryopump, never swap compressors to determine if the problem is due to this unit. If the cryopump expander has cryosorbed the contaminant on the regenerator beds, and you then swap compressors, you now have two contaminated compressors as well as one contaminated expander.

Refrigerator Capacity

The first and second stages of a Gifford-McMahon refrigerator are in series. Helium passes through the first stage regenerator bed prior to reaching the second stage regenerator bed. Therefore, we intuitively know that the thermal loading of the first stage will impact on the thermal capacity and temperature of the second stage.

The size and aspect ratios (*i.e.*, length to diameter) of the regenerator beds and the amount of bed material (*e.g.*, lead alloy shot) in the regererator-displacer pistons impacts on the refrigeration capacity of the respective stages. Also, the amount of He flow and its expansion in the two stages impacts on refrigeration at the respective stages. An example of the interdependence of the simultaneous thermal loading of the stages of a Gifford-McMahon refrigerator is shown in Fig. 5.5.4.[157]

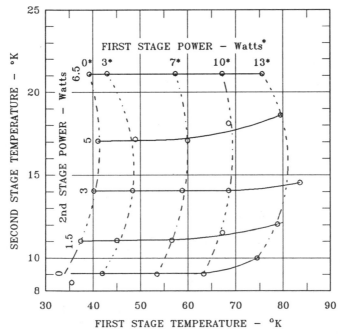

Figure 5.5.4. Temperature variations in the first and second stages of a Gifford-McMahon refrigerator with thermal loading of both stages.[157]

(Data reproduced with the pemission of Leybold-Heraeus: Refrigerator Mod. # RDG-510.)

These data were provided by Dr. Dieter Müller of Leybold. Following the convention, the simultaneous capacity is ~6 W at 20° K with ~13 W at 77° K. With this capacity, one would have more then ample refrigeration for the first stage of a 20 cm ϕ pump, with a very robust second stage capacity.

Problem Set

1) Why do the curves for Ar, Xe and Kr, in Hobson's data, shown in Fig. 5.1.7, abruptly cease their gradual trends with surface coverage, and become vertical lines with continuing coverage?

2) Use (1.12.7) and (1.12.3) to obtain the results of (5.1.3). Assume that the gas on entering the cold chamber is instantaneously thermalized to temperature T_2.

3) Using (5.1.4) and (5.1.5), and given the values of T_1, A_a and A_s, derive S_{eff}, the effective speed produced in the warm chamber due to the pumping of the cold chamber, in terms of α, C_1, A_s and A_a. Assume that the gas on entering the cold chamber is instantaneously thermalized to the unknown temperature T_2.

4) Using (5.1.4) and (5.1.5) and given the values of T_1, A_a and A_s, derive the sticking coefficient α, assuming you have measured the values of we know Q and P_1. Assume that the gas on entering the cold chamber is instantaneously thermalized to the unknown temperature T_2.

5) Using (5.1.4) and (5.1.5) and given the values of T_1, T_2, A_a and A_s, derive the sticking coefficient α, assuming you have measured the values of P_1 and P_2.

6) Calculate the change in kinetic energy of 1 Torr-$\mathcal{L}$ of Ar when taken from RT to zero velocity. If this is done once every second, what is the rate at which work must be done bring this RT gas to rest (i.e., watts of power)? What is the ratio of the above power to that of the heat of sorption for Ar?

7) It was noted in Section 5.2.5 that when roughing a system with two sorption pumps, one should valve out the first pump at ~0.1 Torr. What is so special about this pressure level?

8) Calculate the base pressure which could be theoretically achieved by the simultaneous use of 100 sorption pumps, each containing ~1350 g of Linde 5A sieve materials and cooled to LN_2 temperature, and used to pump a one liter volume of Ne at pressure of 10 Torr.

9) Repeat the calculation in problem 8. However, this time use only two sorption pumps, and stage the first at its equilibrium pressure.

10) Referring to Fig. 5.1.8, calculate the effective speed delivered to the RT chamber. Assume that gases are immediately thermalized on entering the respective chambers, and that no pumping occurs on the walls of the chamber at temperature T_2.

Problem Set

11) The chamber to the left of Fig. 5.1.8 is at RT and the one to the right is very cold. A test gas introduced in the RT chamber is therein thermalized and subsequently pumped by the cold chamber. The RT gas on being pumped by the cold chamber has a sticking coefficient of α_1. Assume now that a chevron is placed in the aperture separating the two chambers, and that this chevron is at an intermediate temperature so as to prechill the gas prior to it reaching the cold chamber. Assume that α_2 is the new sticking coefficient of the prechilled gas in the cold chamber. What is the conductance of the chevron, assuming the ratio of measured speeds, in the RT chamber and with and with out the chevron, is ~0.5.

12) Assume the two chambers in Fig. 5.1.3 are initially at RT $= T_0$. The volumes of the two chambers are given as V_1 and V_2. Assume that there is an initial He pressure in both chambers is given as P_0. If the temperature of the chamber of volume V_2 is cooled to temperature T_2, and there is no surface pumping in either chamber, express the final pressure in both chambers in terms of the givens.

13) What was the probable temperature of the first stage louver of the cryopump for which speeds are given in Fig. 5.4.2?

14) Assume that during measurement of the isosteres of Figures 5.4.6 and 5.4.7, the gases were at a temperature of ~77° K. However, assume that the pressure gauge used to measure the pressures was located in an adjacent RT chamber. How do you compensate for this temperature difference?

15) Construct adsorption isotherms for He and H_2 using the data of Figures 5.4.6. and 5.4.7. Compensate for thermomolecular effects by assuming conditions given in problem 14.

16) Assume that the vessel in Fig. 5.4.10 has a volume of 200 liters. Assume that this is a batch coater and that you are using an e-gun source to deposit $A\ell$ films on some substrate. After loading the bell-jar with the substrates which are to be coated, you rough the bell-jar to 0.1 Torr, isolate the roughing pump, and then valve in the cryopump. You limit the roughing pressure of the bell-jar to this value to avoid back-streaming of oil from the roughing pump. Assume that the coating process takes ~20 minutes, and that the partial pressures of H_2O and H_2 during this process are 2×10^{-7} Torr and 5×10^{-6} Torr, respectively. Assume the H_2 pumping speed in the vessel is 10^3 ℓ/sec. Assume that you complete 100 batch cycles prior to deciding to regenerate the cryopump. On warming the cryopump to room temperature, what is the partial pressure of O_2 and H_2 in the cryopump if the isolated pump has a volume of 50 liters.

Problem Set

17) Hydrogen and air comprise an explosive mixture in proportions ranging from 4.0 to 75 percent air, by volume. Conceive of a scenario in the example given in problem 16, wherein the mixture of O_2 and H_2 is not potentially explosive. (Hint: in real evaporative coating applications, this is difficult.)

18) Assume that no pressure relief valve was used on the cryopump in problem 16. What would the total pressure be in the cryopump after the 100 coating cycles. If ~290 Torr-$\mathcal{L}$ of H_2 and half this much O_2, when burned, corresponds to ~1.0 g of TNT in explosive energy,[94] how much potential equivalent of TNT explosive energy is stored in the cryopump.

19) Derive the He and H_2 pumping speeds per unit area of activated coconut charcoal at 14° K by using the zero loading data of Fig. 5.4.4 and through use of (5.1.18). Assume the second stage is configured as shown in Fig. 5.4.3, but that the slots between the leaves are on average rectangular. Assume the cryopump is a 30 cm ϕ pump and that the first stage is shielded by 60° louvers. (Hint: start by setting up an equation similar to (2.10.8).)

20) Assume that a second stage array similar to that described in Fig. 5.4.3 is used in the cryopump in problem 17, and that the partial pressure of H_2 is ~10^{-6} Torr during the coating operation. Assume that we delay regeneration of the cryopump until the speed of the cryopump has decreased to 2000 $\mathcal{L}$/sec. How many batch coatings will we be able to make prior to the requirement of cryopump regeneration?

21) Work problems 16-18 for the case described in problem 20.

22) What is the pressure indicated by a hydrogen vapor bulb of a cryopump if the temperature of the second stage was 20° K?.

References

1. Redhead, P.A., Hobson, J.P., and Kornelsen, E.V., The Physical Basis of Ultrahigh Vacuum (Chapman and Hall, Ltd., London, 1968), p. 61.
2. Chubb, J.N. and Pollard, I.E., "Experimental Studies of Hydrogen Condensation on to Liquid Helium Cooled Surfaces", Vacuum 15, 491 (1965).
3. Levenson, L.L., "Condensation Coefficients of Gases Measured with a Quartz Crystal Microbalance, II, Ar, Kr, Xe, and CO_2", Proc. 14th Nat. AVS Symp., 1967 (Herbrick and Held Printing Company, Pittsburgh, 1968), p. 95.
4. Brown, R.F., Trayer, D.M., and Busby, M.R., "Condensation of 300-2500 K Gases on Surfaces at Cryogenic Temperatures", J. Vac. Sci. Technol. 7(1), 241 (1970).
5. Bentley, P.D. and Hands, B.A., "The Condensation of Atmospheric Gases on Cold Surfaces", Proc. Roy. Soc. (London), A359, 319 (1978).
6. Dushman, S., Scientific Foundations of Vacuum Technique (John Wiley and Sons, Inc., New York, 1949), p. 389.
7. Honig, R.E. and Hook, H.O., "Vapor Pressure Data for Some Common Gases", RCA Review, 360 (1960).
8. Barron, R., Cryogenic Systems (McGraw-Hill Book Company, New York, 1966), p. 271.
9. Dushman, S., op. cit. (1949), p. 742.
10. Heafer, R.A., Cryopumping - Theory and Practice (Clarendon Press, Oxford, 1989), p. 61.
11. Eisenstadt, M.M., "Condensation of Gases During Cryopumping - The Effect of Surface Temperature on the Critical Energy for Trapping", J. Vac. Sci. Technol. 7(4), 479 (1970).
12. Lee, T.J., "The Condensation of H_2 and D_2: Astrophysics and Vacuum Technology", J. Vac. Sci. Technol. 9(1), 257 (1972).
13. Honig, R.E., "Vapor Pressure Data for the Solid and Liquid Elements", RCA Review, 567 (1962).
14. Redhead, P.A., et al., op. cit., p. 25.
15. Bennett, M.J. and Tompkins, F.C., "Thermal Transpiration: Application of Liang's Equation", Trans. Faraday Soc. 53, 185 (1957).
16. Podgurski, H.H. and Davis, F.N., "Thermal Transpiration at Low Pressure. The Vapor Pressure of Xenon Below 90° K", J. Phys. Chem. 65(2), 1343 (1961).
17. Edmonds, T. and Hobson, J.P., "A Study of Thermal Transpiration Using Ultrahigh-Vacuum Techniques", J. Vac. Sci. Technol. 2(4), 182 (1965).
18. Dawson, J.P. and Haygood, J.D., "Cryopumping", Cryogenics 5(2), 57 (1965).

References

19. Bentley, P.D. and Hands, B.A., "The Initiation of Deposition of Gases on Cryopumps", Proc. 7th Int. Vac. Cong. and 3rd Int. Conf. on Solid Surfaces, 1977 (R. Dobrozemsky, F. Rüdenaur, F.P. Viehböck, A. Breth, Postfach 300, A-1082 Vienna, Austria, 1977), p. 73.
20. Hobson, J.P., "Physical Adsorption Isotherms Extending from Ultrahigh Vacuum to Vapor Pressure", J. Phy. Chem. 73(8), 2720 (1969).
21. Büttner, P., "Measuring Pumping Speed, Condensate Thickness and Condensate Stability in a Large Cryopump", Proc. 8th Int. Vac. Cong., 1980 (Supplément à la Revue, "Le Vide, les Couches Minces, No. 201", 1981), p. 316.
22. Guthrie, A. and Wakerling, R.K., Vacuum Equipment and Techniques (McGraw-Hill Book Company, Inc., New York, 1949), p. 249.
23. Redhead, P.A., et al., op. cit., p. 40.
24. Longsworth, R.C. and Webber, R.J., "Cryopump Vacuum Recovery After Pumping Ar and H_2", Submitted to J. Vac. Sci. Technol., February, 1991.
25. Halama, H.J. and Aggus, J.R., "Measurement of Adsorption Isotherms and Pumping Speed of Helium on Molecular Sieve in the 10^{-11} - 10^{-7} Torr Range at 4.2° K", J. Vac. Sci. Technol. 11(1), 333 (1974).
26. Halama, H.J. and Aggus, J.R., "Cryosorption Pumping for Intersecting Storage Rings", J. Vac. Sci. Technol. 12(1), 532 (1975).
27. Robens, E., "Physical Adsorption Studies with the Vacuum Mirobalance", J. Vac. Sci. Technol. 17(1), 92 (1980).
28. Dushman, S., Scientific Foundations of Vacuum Technique (John Wiley and Sons, New York, 1962), p. 376.
29. Stern, S.A. and DiPaolo, F.S., "The Adsorption of Atmospheric Gases on Molecular Sieves at Low Pressures and Temperatures. The Effect of Preadsorbed Water", J. Vac. Sci. Technol. 4(6), 347 (1967).
30. Hobson, J.P., "Theoretical Isotherms for Physical Adsorption at Pressures Below 10^{-10} Torr", J. Vac. Sci. Technol. 3, 281 (1966).
31. Dushman, S., op. cit. (1962), p. 388.
32. Dushman, S., op. cit. (1962), p. 384.
33. Brunauer, S., Emmett, P.H., and Teller, E., "Adsorption of Gases in Multimolecular Layers", J. Amer. Chem. Soc. 60(1), 309 (1938).
34. Stern, S.A., Mullhaupt, J.T., Hemstsreet, R.A., DiPaolo, F.S., "Cryosorption Pumping of Hydrogen and Helium at 20° K", J. Vac. Sci. Technol. 2, 165 (1965).
35. Sedgley, D.W., Tobin, A.G., Batzer, T.H., and Call, W.R., "Characterization of Charcoals for Helium Cryopumping in Fusion Devices", J. Vac. Sci. Technol. A5(4), 2572 (1987).
36. Dubinin, M.M., Radushkevich, Proc. Acad. Sci. USSR 55, 331 (1947).[25]

References

37. Hobson, J.P. and Williams, B.R., "Cryosorption Pumping of Helium on Porous Silver at 4.2° K", J. Vac. Sci. Technol. 6(6), 965 (1969).
38. Hseuh, H.C., Chou, T.S., Worwetz, H.A., and Halama, H.J., "Cryosorption Pumping of He by Charcoal and a Compound Cryopump Design for TSTA", Proc. 8th Symp. on Engineering Problems of Fusion Research, 1979 (IEEE, New York, 1980), p. 1568.
39. Danilova, N.P. and Shal'nikova, A.I., "Adsorption of Helium on a Copper Surface at 4.2° K", Instrum. Experim. Tech. No. 6, 1464 (1967).
40. Hobson, J.P., "Cryopumping", J. Vac. Sci. Technol. 10(1), 73 (1973).
41. Redhead, P.A., et al., op. cit., p. 27.
42. Hilleret, N. and Calder, R., "Ion Desorption of Condensed Gases", Proc. 7th Int. Vac. Cong. and 3rd Int. Conf. on Solid Surfaces, 1977 (R. Dobrozemsky, F. Rüdenauer, F.P. Viehböck, A. Breth, Postfach 300, A-1082 Vienna, Austria, 1977), p. 227.
43. Halama, H.J., Hseuh, H.C., and Chou, T.S., "Cryopumping of Hydrogen and Helium", Advances in Cryogenic Engineering, 27 (Plenum Press, Inc., New York, 1981), p. 1125.
44. Chou, T.S. and Halama, H.J., "Cryopumping of Deuterium and Helium Mixtures on Smooth 4.2° K Surfaces", Proc. 7th Int. Vac. Cong. and 3rd Int. Conf. on Solid Surfaces, 1977 (R. Dobrozemsky, F. Rüdenauer, F.P. Vichböck, A. Breth, Postfach 300, A-1082 Vienna, Austria, 1977), p. 65.
45. Hengevoss, J. and Trendelenburg, E.A., "Continuous Cryotrapping of Hydrogen and Helium by Argon at 4.2° K", Proc. 10th Nat. AVS Symp., 1963 (The Macmillian Company, New York, 1964), p. 101.
46. Batzer, T.H., Patrick, R.E., Call, W.R., "A TSTA Compound Cryopump", J. Vac. Sci. Technol. 18(3), 1125 (1981).
47. Tempelmeyer, K.E., Dawbarn, R., and Young, R.L., "Sorption Pumping of Hydrogen by Carbon Dioxide Cryodeposits", J. Vac. Sci. Technol. 8(4), 575 (1971).
48. Arakawa, I., Kobayashi, M., and Tuzi, Y., "Effects of Thermal Spikes on the Characteristics of Cryosorption Pumps with Condensed Carbon Dioxide Layers", J. Vac. Sci. Technol. 16(2), 738 (1979).
49. Hengevoss, J., "Influence of the Temperature History of Condensed Argon on its Hydrogen Adsorptivity at Low Temperatures", J. Vac. Sci. Technol. 6(1), 58 (1969).
50. Arakawa, I. and Tuzi, Y., "Temperature Dependence of Pumping Speed for H2 of a Cryosorption Pump with a Condensed Gas Layer Sorbent", J. Vac. Sci. Technol. A4(3), 293 (1986).
51. Thibault, J.J., Disdier, J.M., Arnaud, G., and Boissin, J.C., "Design and Preliminary Tests of the Liquid Helium Cooled MFTF Cryopumping System", J. Vac. Sci. Technol. 18(3), 1140 (1981).

References

52. Heafer, R.A., Cryopumping - Theory and Practice (Clarendon Press, Oxford, 1989), p. 410.
53. Dewar, Sir James, "Charcoal Vacua", Nature 12(7), 217 (1875).
54. Dushman, S., op. cit. (1949), p. 462.
55. Dushman, S., op. cit. (1949), p. 466.
56. Dushman, S., op. cit. (1949), p. 387.
57. Anderson, R.B. and Dawson, P.T., Editors, Experimental Methods in Catalytic Research (Academic Press, Inc., New York, 1976).
58. Hemstreet, R.A. and Kenmore, N.Y., U.S. Patent No. 3,296,773, "Adsorbent-Coated Thermal Panels", awarded January 10, 1967.
59. Redhead, P.A., et al., op. cit., p. 51.
60. Steele, W.A. and Halsey, G.D., "The Interaction of Gas Molecules with Capillary and Crystal Lattice Surfaces", J. Phy. Chem. 59, 57 (1955).
61. Hseuh, H.C. and Worwetz, H.A., "Performance of BNL-TSTA Compound Cryopump", J. Vac. Sci. Technol. 18(3), 1131 (1981).
62. Dillow, C.F. and Palacios, J., "Cryogenic Pumping of Helium, Hydrogen, and a 90% Hydrogen-10% Helium Mixture", J. Vac. Sci. Technol. 16(2), 731 (1979).
63. Fisher, P.W. and Watson, J.S., "Cryosorption Vacuum Pumping of D_2, He, and H_2 at 4.2° K for CTR Applications", Proc. 24th Conf. on Remote Systems Technol. (ANS, Washington, D.C., 1976), p. 12.
64. Sedgley, D.W., Batzer, T.H., and Call, N.R., "Helium Cryopumping for Fusion Applications", J. Vac. Sci. Technol. A6(3), 1209 (1988).
65. Watson, J.S. and Fisher, P.W., "Cryosorption Vacuum Pumping Under Fusion Reactor Conditions", Proc. 7th Int. Vac. Cong. and 3rd Int. Conf. on Solid Surfaces, 1977 (R. Dobrozemsky, F. Rüdenauer, F.P. Viehböck, A. Breth, Postfach 300, A-1082 Vienna, Austria, 1977), p. 363.
66. Dushman, S., op. cit. (1949), p. 484.
67. Stern, S.A., Hemstreet, R.A., and Ruttenbur, D.M., "Cryosorption Pumping of Hydrogen at 20° K II. Development and Performance of Cryosorption Panels", J. Vac. Sci. Technol. 3(3), 99 (1966).
68. Grenier, G.E. and Stern, S.A., "Cryosorption Pumping of Helium at 4.2° K", J. Vac. Sci. Technol. 3(6), 334 (1966).
69. Tobin, A.G., Douglas, W.S., Batzer, T.H., and Sedgley, D.W., "Evaluation of Charcoal Sorbents for Helium Cryopumping in Fusion Reactors", J. Vac. Sci. Technol. A5(1), 101 (1987).
70. Coupland, J.R., Hammond, D.P., Bächler, W., and Klein, H.H., "Experimental Performance of a Large-Scale Cryosorption Pump", J. Vac. Sci. Technol. A5(4), 2563 (1987).
71. Turner, F.T. and Hogan, W.H., "Small Cryopump with Internal Refrigerator", J. Vac. Sci. Technol. 3(5), 252 (1966).

References

72. Forth, H.J. and Frank, R., "Installation of Cryopumps in the Estec-Space Simulation Chambers", Proc. 7th Int. Vac. Cong. and 3rd Int. Conf. on Solid Surfaces, 1977 (R. Dobrozemsky, F. Rüdenauer, F.P. Vienböck, A. Breth, Postfach 300, A-1082 Vienna, Austria, 1977), p. 61.
73. Neal, R.B., "The Stanford Two-Mile Linear Electron Accelerator", J. Vac. Sci. Technol. $\underline{2}$(3), 149 (1965).
74. Turner, F., "Cryosorption Pumping", Varian Tech. Pub., VR-76, 1972, Varian Associates, Inc., Palo Alto, CA.
75. Baker, M.A. and Laurenson, L., "The Application of Foreline Sorption Traps", Vacuum $\underline{19}$(2), (1969).
76. Jepsen, R.L., U.S. Patent No. 3,116,764, "Sorption Pumping Apparatus", filed March 1959, awarded January 1964.
77. Gareis, P.J. and Hagenbach, G.F., "Cryosorption", Ind. and Engng. Chem. $\underline{57}$(5), 27 (1965).
78. Welch, K.M., "A New Dry Vacuum Roughing Pump", Research/Development, February, 1972, p. 42.
79. Longsworth, R.C., "Performance of a Cryopump Cooled by a Small Closed-Cycle 10-K Refrigerator", Adv. Cryo. Engng. $\underline{23}$, 658 (1978).
80. Bannock, R.R., "The Role of Molecular Sieve Pumping", Vacuum $\underline{12}$, 101 (1962).
81. Scott, R.B., Cryogenic Engineering (D. Van Nostrand Company, Inc., Princeton, New Jersey, 1959).
82. Barron, R., Cryogenic Systems (McGraw-Hill Book Company, New York, 1966).
83. Claude, G., British Patent No. 1,225,608, "Improvement in or Relating to Cryogenic Pumping Devices", filed August 1968, awarded March 1971.
84. Longsworth, R.C. and Lahav, Y., "A Cryopumped Leak Detector", J. Vac. Sci. Technol. $\underline{A5}$(4), 2646 (1987).
85. Fisher, P.W. and Watson, J.S., "Cryosorption Pumping of Deuterium by MS-5A at Temperatures above 4.2 K for Fusion Applications", J. Vac. Sci. Technol. $\underline{15}$(2), 741 (1978).
86. Powers, R.J. and Chambers, R.M., "A Clean Cryo-Vacuum System with High Pumping Speed for All Gas Species", J. Vac. Sci. Technol. $\underline{8}$(1), 319 (1971).
87. Chubb, J.N., Gowland, L., and Pollard, I.E., "Condensation Pumping of Hydrogen and Deuterium on to Liquid-Helium-Cooled Surfaces", Brit. J. Appl. Phys. $\underline{1}$(2), 361 (1968).
88. Tobin, A.G., private communication, March 29, 1991.
89. Benvenuti, C., Blechschmidt, D., and Passardi, G., "Molecular and Radiation Transmissivities of Chevron Type Baffles for Cryopumping", J. Vac. Sci. Technol. $\underline{19}$(1), 100 (1981).

References

90. Benvenuti, C. and Calder, R.S., "The Desorption of Condensed Hydrogen from Various Substrates by Infrared Thermal Radiation", Phys. Lett. 35A(4), 291 (1971).

91. Benvenuti, C., "Characteristics, Advantages, and Possible Applications of Condensation Cryopumping", J. Vac. Sci. Technol. 11(3), 591 (1974).

92. Benvenuti, C., Calder, R.S., and Passardi, G., "Influence of Thermal Radiation on the Vapor Pressure of Condensed Hydrogen (and Isotopes) Between 2 and 4.5 K", J. Vac. Sci. Technol. 13(6), 1172 (1976).

93. Hood, C.B., "The Development of Large Cryopumps from Space Chambers to the Fusion Program", J. Vac. Sci. Technol. A3(3), 1684 (1985).

94. Graham, W.G. and Ruby, L., "Cryopumping Measurements Relating to Safety, Pumping Speed, and Radiation Outgassing", J. Vac. Sci. Technol. 16(3), 927 (1979).

95. Coupland, J.R., Hammond, D.P., Obert, W., "Experimental Performance of an Open Structure Cryopump", Vacuum 32 (10/11), 613 (1982).

96. Duffy, T.J. and Oddon, L.D., "Beam-Line Cryopump", UCRL Preprint No. UCRL-77236, 1975.

97. Langhorn, A.R., Kim, J., Tupper, M.L., Williams, J.P, and Fasolo, J., "Performance of Doublet III Neutral Beam Injector Cryopump System", J. Vac. Sci. Technol. A2(2), 1193 (1984).

98. Redhead, P.A., et al., op. cit., p. 46.

99. Stern, S.A. and DiPaolo, F.S., "Cryosorption Pumping of Air on Molecular Sieves at 77 K - The Ultimate Achievable Vacuum", J. Vac. Sci. Technol. 6(6), 941 (1969).

100. Dubinin, M.M., "The Potential Theory of Adsorption of Gases and Vapors for Adsorbents with Energetically Nonuniform Surfaces", Chem. Rev. 60(1), 235 (1960).

101. Gareis, P.J. and Stern, S.A., Liquid Helium Technology, "Pumping of Helium and Hydrogen in the High Vacuum Range by Means of Cryosorption Arrays", Proc. Intl. Inst. of Refrigeration Commission, 1966 (Pergamon Press, Oxford, 1966), Vol. 6, p. 429.

102. Lessard, P.A., "Cryogenic Adsorption on Noncondensibles in the High-Vacuum Regime", J. Vac. Sci. Technol. A7(3), 2373 (1989).

103. Halama, H.J., Lam, C.K., Bamberger, J.A., "Large-Capacity Cryogenic Pumping of D_2 and H_2 for Fusion", J. Vac. Sci. Technol. 14(5), 1201 (1977).

104. Kuluva, N.M. and Knuth, E.L., "Sorption Pumping at Pressures Less Than 10^{-5} Torr", Proc. 9th Nat. AVS Vac. Symp., 1962 (The Macmillian Company, New York, 1963), p. 237.

105. Liu, B-K., Ren, J-S., Cui, X-H., "On Cryosorption Pumping of Hydrogen with the ZDB-150 Type Cryopump Cooled by a Two-Stage Closed-Cycle Refrigerator", J. Vac. Sci. Technol. 20(4), 1000 (1982).

References

106. Visser, J., Symersky, B., Geraerts, A.J.M., "A Versatile Cryopump for Industrial Vacuum Systems", Vacuum 27(3), 175 (1977).

107. Endow, N. and Pasternak, R.A., "Physisorption of Xenon and Krypton on Glass and on Molybdenum Films", J. Vac. Sci. Technol. 3(4), 196 (1966).

108. Troy, M. and Wightman, J.P., "Physisorption of Ar, Kr, CH_4, and N_2 on Stainless Steel at Very Low Pressures", J. Vac. Sci. Technol. 8(6), 743 (1971).

109. Be, S.H., "New Cryopumps for a Resonator of the RIKEN Ring Cyclotron", Vacuum 38(7), 543 (1988).

110. Santeler, D.J., "Pressure Simulation of Outer Space", Proc. Sixth National Symposium on Vacuum Technology Transactions, 1959 (Pergamon Press, New York, 1960), p. 129.

111. Levenson, L.L., Milleron, N., and Davis, D.H., "Optimization of Molecular Flow Conductance", Proc. Seventh National Symposium on Vacuum Technology Transactions, 1960 (Pergamon Press, New York, 1961), p. 372.

112. Caren, R.P., Gilcrest, A.S., Zierman, C.A., "Thermal Absorptances of Cryodeposits for Solar and 290° K Blackbody Sources", Proc. 1963 Cryogenic Engineering Conference, 1963 (Plenum Press, New York, 1964), Advances in Cryogenic Engineering, Vol. 9, p. 457.

113. Cunningham, T.M. and Young, R.L., "The Radiative Properties of Cryodeposits at 77° K", Proc. 1962 Cryogenic Engineering Conference, 1962 (Plenum Press, New York, 1963), Advances in Cryogenic Engineering 8, p. 85.

114. Merriam, R.L. and Viskanta, R., "Radiative Characteristics of Cryodeposits for Room Temperature Black Body Radiation", Proc. 1968 Cryogenic Engineering Conference, 1968 (Plenum Press, New York, 1969), Advances in Cryogenic Engineering, Vol. 14, p. 240.

115. Welch, K.M., U.S. Patent No. 4,311,018, "Cryogenic Pump", filed 9/22/80, awarded 1/19/82.

116. Welch, K.M., U.S. Patent No. 4,295,338, "Cryogenic Pumping Apparatus with Replaceable Pumping Surface Elements", filed 10/18/79, awarded 10/20/81.

117. Welch, K.M., U.S. Patent No. 4,285,710, "Cryogenic Device for Restricting the Pumping Speed of Selected Gases", filed 9/18/78, awarded 8/25/81.

118. Klein, H.-H., Heisig, R., and Augustine, C.M., "Use of Refrigerator-Cooled Cryopumps in Sputtering Plants", J. Vac. Sci. Technol. A2(2), 187 (1984).

119. Bentley, P.D., "The Modern Cryopump", Vacuum 30(4/5), 145 (1980).

References

120. Haarhuis, G.J., "Stirling Cryogenerators and Cryopumping", LeVide No. 171-172, 351 (1974).
121. Visser, J. and Scheer, J.J., "Twenty-Kelvin Cryopuming in Magnetron Sputtering Systems", J. Vac. Sci. Technol. 16(2), 734 (1979).
122. Visser, J. and Scheer, J.J., "A Cryosorption Pumped Rapid Cycle UHV System", J. Vac. Sci. Technol. 19(1), 122 (1981).
123. McMahon, H.O. and Gifford, W.E., "A New Low-Temperature Gas Expansion Cycle", Advances in Cryogenic Engineering 5 (Plenum Press, Inc., New York, 1960), p. 354.
124. Gifford, W.E., "The Gifford-McMahon Cycle", Advances in Cryogenic Engineering 11 (Plenum Press, Inc., New York, 1966), p. 152.
125. Becker, G.E., "Operation of a Cryopumped UHV System", J. Vac. Sci. Technol. 14(1), 640 (1977).
126. Benini, M. and Morbidi, P., "Performance of Closed-Cycle Refrigerator Pump", Proc. 8th Int. Vac. Cong., 1980 (Supplé ment à la Revue, "LeVide, les Couches Minces, No. 201", 1981), p. 275.
127. Bridwell, M.C. and Rodes, J.G., "History of the Modern Cryo-pump", A3(3), 472 (1985).
128. Dennison, R.W. and Gray, G.R., "Cryogenic Versus Turbomolecular Pumping in Sputtering Applications", J. Vac. Sci. Technol. 16(2), 728 (1979).
129. Denision, D.R., "Characteristics of Cryo- and Cryo/Ion Pumped Vacuum Systems", Proc. 7th Int. Vac. Cong. and 3rd Int. Conf. on Solid Surfaces, 1977 (R. Dobrozemsky, F. Rüdenauer, F.P. Viehböck, A. Breth, Postfach 300, A-1082 Vienna, Austria, 1977), p. 69.
130. de Rijke, J.E., "Performance of a Cryopump-Ion Pump System", J. Vac. Sci. Technol. 15(2), 765 (1978).
131. de Rijke, J.E., "Factors Affecting Cryopump Base Pressure", J. Vac. Sci. Technol. A8(3), 2778 (1990).
132. Hands, B.A., "Recent Developments in Cryopumping" Vacuum 32(10/11), 603 (1982).
133. Longsworth, R.C., "Characteristics of a 550-mm ID Low Profile Refrigerated Cryopump", Advances in Cryogenic Engineering 27, 1980 (Plenum Press, Inc., New York, 1981), p. 1135.
134. Longsworth, R.C.,. "Refrigerators for Cryopumps" for AVS Course "Fundamentals of Cryopumping", 1978.
135. Longsworth, R.C., "Characteristics of Cryopumps Cooled by Small Closed-Cycle 10-K Refrigerators", Advances in Cryogenic Engineering 23 (K.D. Timmerhaus, Editor, Plenum Press, New York, 1978), p. 658.
136. Minata, M. and Stolz, J., "System Level Vibration of a Low-Vibration Cryopump for Semiconductor Processing Equipment", J. Vac. Sci. Technol. A7(3), 2361 (1989).

References

137. Muhlenhaupt, R.C., "A Comparison of Cryopump and Diffusion Pump Performance on MMA 29x45 ft. Vacuum Chamber", J. Vac. Sci. Technol. 20(4), 1005 (1982).
138. Ikegami, K., Nakajima, S., Be, S.H., Ohsako, N., Morimoto, K., Kikuchi, T., and Morishita, S., "Design and Performance Characteristics of Refrigerator-Cooled Cryopumps for the RIKEN Ring Cyclotron", Vacuum 38(2), 99 (1988).
139. Ruthlein, H. and Forth, H.J., "The Application of Cryopumps with Integrated K 20 Refrigerators in Vacuum Deposition Technology", Proc. 7th Int. Vac. Cong. and 3rd Int. Conf. on Solid Surfaces, 1977 (R. Dobrozemsky, F. Rüdenauer, F.P. Viehböck, A. Breth, Postfach 300, A-1082 Vienna, Austria, 1977), p. 77.
140. Schäfer, G. and Häfner, H.-U., "Cryopumps for Evaluating Space Simulation Chambers", J. Vac. Sci. Technol. A5(4), 2359 (1987).
141. Welch, K.M. and Flegal, C., "Elements of Cryopumping", Industrial Research/Development, March, 1978.
142. Welch, K.M. and Longsworth, R.C., An Introduction to the Elements of Cryopumping (K.M. Welch, Editor, Varian Associates, Palo Alto, CA, 1980).
143. Leever, R.C., "A Low Temperature Mechanically Refrigerated Cold Trap", Proc. 1st Nat. AVS Symp., 1954 (W.M. Welch Manufacturing Company, 1955), p. 19.
144. Kikuchi, T., Ohsako, N. and Hayashi, Y., "Capability of Obtaining Extreme High Vacuum by Commercial G-M Refrigerator-Cooled Cryopump", Vacuum 41(7/9), 1941 (1990).
145. Welch, K.M., McIntyre, G.T., Tuozzolo, J.E., Skelton, R. Pate, D.J., and Gill, S.M., "Metal and Elastomer Seal Tests for Accelerator Applications", Vacuum 41(7/9), 1924 (1990).
146. Longsworth, R.C., and Booney, G.E., "Cryopumping Regeneration Studies", J. Vac. Sci. Technol. 21(4), 1022 (1982).
147. Häfner, H.-U., Klein, H.-H., and Timm, U., "New Methods and Investigations for Regenerating Refrigerator Cryopumps", Vacuum 41(7/9), 1840 (1990).
148. Lefrancois, M., Montuclard, J., and Rouaud, C., "Low-Load Metal Seals to Replace Elastomer O-rings: the Helicoflex-delta (Δ) Seals", Vacuum 41(7/9), 1879 (1990).
149. The author is indebted to Mr. John T. Harvell and Mr. John Peterson, of CTI, who were interviewed in April, 1990, to obtain some of this historical data. Also, I am indebted to Dr. Ralph Longsworth and Mr. Richard L. Rerig for a similar interview in May, 1990. All have been my friends for many years, even with recorder in hand.
150. Barron, R., op. cit. p. 81.

References

151. Smith, J.L. and Robinson, G.Y., "A Tribute to Samuel C. Collins: September 28, 1898-June 19, 1984", Advances in Cryogenic Engineering 31 (R.W. Fast, Editor, Plenum Press, New York, 1986), p. 1.

152. Barron, R., op. cit., p. 122.

153. Chellis, F.F. and Hogan, W.H., "A Liquid-Nitrogen-Operated Refrigerator for Temperatures Below 77° K", Advances in Cryogenic Engineering 9, (K.D. Timmerhause, Editor, Plenum Press, New York, 1964), p. 545.

154. Harvell, J.T., private communications, April, 1990.

155. Buck, D.A., "The Cryotron: A Superconductive Computer Element", Proc. Inst. Radio Engrs. 44(4), 482 (1956).

156. O'Boyle, T.F., "Chilling Tale, GE Refrigerator Woes Illustrate the Hazards in Changing a Product'", The Wall Street Journal, Eastern Edition, May 7, 1990, p. 1.

157. Müller, D., private communications, June, 1990.

158. Sondericker, J., "Production and Use of High Grade Silicon Diode Temperature Sensors", Advances in Cryogenic Engineering 27, 1980 (Plenum Press, Inc., New York, 1981), p. 1163.

SUBJECT INDEX

AUTHOR INDEX